Pharmacokinetics
in
Drug Discovery and
Development

Pharmacokinetics
in
Drug Discovery and
Development

Edited by
Ronald D. Schoenwald

CRC PRESS

Boca Raton   London   New York   Washington, D.C.

FIRST INDIAN REPRINT, 2009

Library of Congress Cataloging-in-Publication Data

Pharmacokinetics in drug discovery and development/edited by Ronald D. Schoenwald.
   p. cm.
Includes bibliographical references and index.
ISBN 1-56676-973-6 (alk. paper)
1. Pharmacokinetics. 2. Drug development. I. Schoenwald, Ronald D.

RM301.5 .P437 2002
615'.7--dc21   2001052887

Dedication

In gratitude to those who have gone before us:
Bennett, Dettli, Gibaldi, Kruger-Theimer, Levy,
Nelson, Riegelman, Wagner, and others

Preface

In the last six or more decades pharmacokinetics has evolved from its origins into a complex discipline with numerous subspecialties and applications. According to many scientists, pharmacokinetics began with the publication of two research articles by Torsten Teorell in the *International Archives of Pharmacodynamics* in 1937. These articles established the basis for structural or classical pharmacokinetics with the conception of compartments and first-order kinetics to explain the movement of drugs into, within, and out of the body. With the help of many scientists, such as Bennett, Dost, Dettli, Gibaldi, Kruger-Theimer, Levy, Nelson, Riegelman, Wagner, and others, pharmacokinetics evolved from its inception to an established discipline, which by the 1970s saw the introduction of new models and approaches to explain the kinetics of drugs in the body. These were pharmacodynamic, noncompartmental, and physiological modeling, as well as population, linear systems, and convolution/deconvolution approaches. Also, at about this time, clinical pharmacokinetics evolved into an established discipline applied to the development of new drugs by the pharmaceutical industry and in various clinical settings for the adjustment of dosing regimens. Once established, the Food and Drug Administration incorporated pharmacokinetics as a tool to aid in certifying the safety and efficacy of new drugs from detailed studies of the absorption, distribution, metabolism, and excretion behavior of a drug. In 1977, the bioavailability regulations were passed into law, enabling generic manufacturers to establish therapeutic equivalency based on knowledge learned from pharmacokinetics, particularly absorption phenomena.

With the expansion of pharmacokinetics, it has become exceedingly difficult for any one individual to become a full-fledged expert in all areas. Consequently, this book was devised to summarize pertinent areas in the field and to provide a reference for further study. The book is divided into four sections: I, Basic Principles; II, Industrial and Regulatory Applications; III, Clinical Applications; and IV, Research Applications. In each chapter, an expert provides extensive details about the subspecialty, with an emphasis on the theme or focus of that section.

Terminology, concepts, and approaches used in a particular subspecialty or application should be useful as a source to students, health-care specialists, newly trained pharmacokineticists, and experts in one particular pharmacokinetic subspecialty who want to become proficient in another.

Ronald D. Schoenwald

The Editor

Ronald D. Schoenwald, R.Ph., Ph.D., is currently Professor Emeritus, College of Pharmacy, The University of Iowa. He was Professor of Pharmaceutics from 1978 to 2001, serving as head of the Pharmaceutics Division from 1994 through 2000. Dr. Schoenwald received his master's and Ph.D. from Purdue University and became a Distinguished Alumnus. He is a current recipient of the Dale E. Wurster Research Fellow Award from The University of Iowa College of Pharmacy.

Dr. Schoenwald gained industrial experience at Alcon Laboratories from 1973 to 1978. He holds 35 patents, including 3 that have resulted in clinical trials. He has written two textbooks, chapters in various textbooks, and numerous articles in refereed journals. He now resides in Arizona.

Contributors

Gail D. Anderson, Ph.D.
Professor
Department of Pharmacy
School of Pharmacy
University of Washington
Seattle, Washington

Jeffrey S. Barrett, Ph.D.
Director, Global Head
  of Biopharmaceutics
Aventis Pharmaceuticals
Bridgewater, New Jersey

David W. A. Bourne, Ph.D.
Professor
College of Pharmacy
University of Oklahoma
Oklahoma City, Oklahoma

John M. Dopp, Pharm.D.
Cardiovascular Research Fellow
Division of Clinical
  and Administrative Pharmacy
College of Pharmacy
The University of Iowa
Iowa City, Iowa

Mark G. Eller, Ph.D.
Senior Director
  of Clinical Pharmacology
Quintiles, Inc.
Kansas City, Missouri

Erika J. Ernst, Pharm.D.
Assistant Professor
Division of Clinical
  and Administrative Pharmacy
College of Pharmacy
The University of Iowa
Iowa City, Iowa

James M. Gallo, Ph.D.
Associate Professor
Department of Pharmacology
Fox Chase Cancer Center
Philadelphia, Pennsylvania

Ronald A. Herman, Ph.D.
Assistant Professor
Division of Clinical
  and Administrative Pharmacy
College of Pharmacy
The University of Iowa
Iowa City, Iowa

Michael E. Klepser, Pharm.D.
Associate Professor
Division of Clinical
  and Administrative Pharmacy
College of Pharmacy
The University of Iowa
Iowa City, Iowa

Bernd Meibohm, Ph.D.
Assistant Professor
Department of Pharmaceutical Sciences
University of Tennessee
Memphis, Tennessee

Nishit B. Modi, Ph.D.
Director/Research Fellow
Department of Clinical Pharmacology
ALZA Corporation
Mountain View, California

Paul J. Perry, Ph.D.
Professor of Psychiatry and Pharmacy
Division of Clinical
  and Administrative Pharmacy
Colleges of Pharmacy
Department of Psychiatry
College of Medicine
The University of Iowa
Iowa City, Iowa

Bradley G. Phillips, Pharm.D.
Associate Professor
Division of Clinical
  and Administrative Pharmacy
College of Pharmacy
The University of Iowa
Iowa City, Iowa

Ronald D. Schoenwald, R.Ph., Ph.D.
Professor Emeritus
Division of Pharmaceutics
College of Pharmacy
The University of Iowa
Iowa City, Iowa

Veneeta Tandon, Ph.D.
Senior Pharmacokinetic Reviewer
Food and Drug Administration
Rockville, Maryland

Peter Veng-Pedersen, Ph.D.
Professor
Division of Pharmaceutics
College of Pharmacy
The University of Iowa
Iowa City, Iowa

**Honghui Zhou, Ph.D., F.C.P.,
  A.B.C.P.**
Manager, Clinical Pharmacokinetics
Global Clinical Pharmacokinetics
Janssen Research Foundation
Johnson & Johnson Pharma R&D
Titusville, New Jersey

Contents

SECTION IV   Research Applications

Section I

Basic Principles

1 Basic Principles

Ronald D. Schoenwald

CONTENTS

INTRODUCTION

Pharmacokinetic processes are broadly defined as absorption, distribution, metabolism, and excretion, giving rise to the frequently used acronym — ADME. Each of these processes has a kinetic component and an extent component. The former refers to the rate of movement or how fast the process occurs over time, whereas the extent

TABLE 1.1
Definitions of Pharmacokinetic Processes (ADME)

Process	Definition
Absorption	Absorption is defined as the net transfer of drug from the site of absorption into the circulating fluids of the body. For oral absorption, two steps are required: (1) crossing the epithelium of the gastrointestinal membrane either by transcellular or paracellular pathways and entering the bloodstream via capillaries, and (2) passing through the hepatoportal system intact into systemic circulation. If the drug is metabolized prior to reaching systemic circulation, it is said to have undergone presystemic or first-pass metabolism.
Distribution	Distribution is defined as the net transfer of drug from the circulating fluids of the body to various tissues and organs.
Metabolism	Metabolism is the bioconversion of drug to another chemical form or metabolite, mostly by endogenous enzyme systems involving phase 1 reactions, such as oxidation (often by the cytochrome P-450 system), reduction, hydrolysis, or dealkylation, or by phase 2 reactions, such as acetylation, sulfation, or glucuronidation.
Excretion	Excretion is the removal of drug from the body primarily via urine and occasionally via feces, bile, sweat, or exhaled air.

component refers to the amount of drug or fraction of the dose that is absorbed, distributed, metabolized, or excreted. In general, the processes (ADME) describe the movement into, within, and out of the body. Metabolism and excretion are collectively referred to as elimination, whereas distribution, metabolism, and elimination are referred to as disposition. These processes, along with the pharmacological effect of the drug, the condition of the patient's organ system, and release of drug from the dosage form, determine an adequate dose and dosing interval for a patient. Specific definitions are provided in Table 1.1.

Drug development relies significantly on acquiring a knowledge of pharmacokinetics for a new drug entity. For example, if a new drug is sufficiently potent, but does not reach the target site in the correct concentration for a specified time period, as determined by the dosing regimen, it will be of little value to the patient. During the drug discovery phase, pharmacokinetic information in animals provides feedback to research scientists so that a structure can be designed that optimizes both structure–activity relationships (SAR) and transport to the target site. Once the drug is considered a potential candidate for human use, a multitude of research and development efforts are undertaken to ensure the safety and efficacy of the new drug and an Investigational New Drug (IND) application is filed with the Food and Drug Administration (FDA). If approved, studies in human subjects begin. The pharmacokinetic data obtained from the animal studies filed under the IND provides a source of information for initiating dosing regimens in humans. As Phase I–IV studies of a New Drug Application (NDA) are completed, a therapeutic range, as well as a dosing regimen, is established for various patient conditions.

LINEAR PHARMACOKINETICS

The term *linear pharmacokinetics* is defined from the differential equations that express the change in the amount or concentration of drug over time:

$$\frac{dC}{dt} = -k_{el}C \tag{1.1}$$

In Equation 1.1, dC/dt represents the rate of change of concentration of drug in blood or plasma over time and k_{el} is the first-order rate constant for elimination out of the body. In this equation, linear refers to the fact that the rate is directly proportional to concentration and if the rate is plotted on rectilinear coordinates vs. time, a straight-line relationship would result. Rate equations are more often written with respect to the amounts of drug in a particular compartment. Nonlinear, in mathematical terms, applies to rate equations in which the rate is no longer linearly related to concentration. In pharmacokinetics this often applies to drugs for which metabolic pathways or plasma protein binding become saturated at concentrations usually within the therapeutic range.

In the case of saturation of a metabolic pathway, the area under the plasma concentration–time curve increases nonlinearly with dose and often presents a sharp increase in plasma concentrations potentially leading to toxicity. In the case of high affinity to plasma proteins, total drug concentration in plasma also increases, and the fraction of drug that is free and capable of distribution and elimination may increase because the plasma binding site is saturated. Above the binding saturation level, nonlinear distribution occurs, which may lead to toxicity. Unrelated to non-linearity, the case exists in which a second drug may be administered with affinity for the same binding site. It can displace the bound drug and increase the concentration of free drug markedly and also lead to toxicity. Warfarin and other antico-agulant drugs are often cited as examples. Consequently, other drugs given along with warfarin should be done so with caution.

CLASSICAL MODELING

Teorell[1,2] is often credited with introducing the concepts that allow for classical pharmacokinetic modeling to be useful in describing drug kinetics in the body. Teorell theorized that (1) drug kinetics in the body occur by consecutive first-order processes in which drug is first absorbed, then distributed and eliminated, and (2) tissues and organs can be compartmentalized and separated by first-order rate constants representing transport across rate-determining barriers. Figure 1.1 shows three pharmacokinetic models, which are common to nearly all drugs. The original papers defining classical pharmacokintic modeling assigned tissues and organs to compartments based upon blood flow and blood volume characteristics. Drugs in circulating fluids and rapidly perfused tissues were assigned to the central compartment (e.g., heart, liver, kidney, lungs), whereas drugs in fluids of distribution (i.e., lymph and extracellular water) and poorly perfused tissues (e.g., large muscle mass, skin, and the central nervous system) were assigned to the peripheral compartment. However,

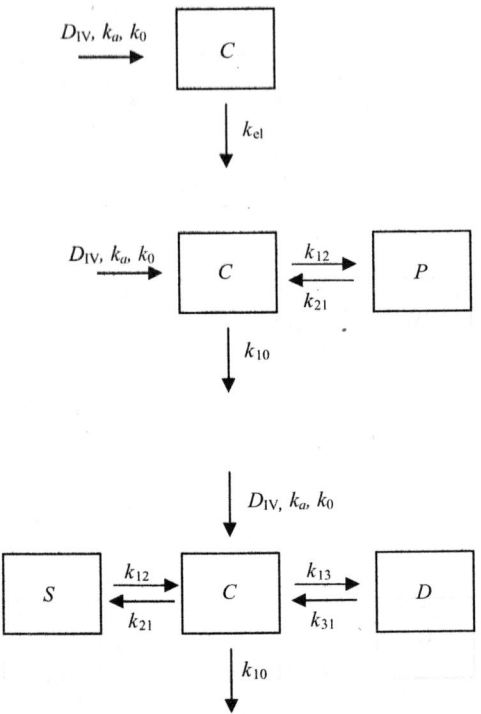

FIGURE 1.1 The three most commonly used pharmacokinetic models in explaining the pharmacokinetic behavior of drugs. The symbols C, P, S, and D represent central, peripheral, shallow, and deep compartments, whereas the first-order rate constants, symbolized by k_{ij}, represent drug transport from compartment i to compartment j. D_{IV}, k_a, and k_0 represent a bolus IV dose, the absorption rate constant, and constant rate infusion, respectively.

it became apparent from experiments with various drugs that a pharmacological response from intravenous injection sometimes produces an immediate effect, whereas other drugs when injected require hours to reach a peak response. For many of the latter drugs, a delay in effect is due to the time required for drug to enter and accumulate within the peripheral compartment. This observation indicates that the biophase (i.e., the anatomical location of the receptors) in located within the peripheral compartment. On occasion, the kinetics of a drug may follow a three-compartment model for which the two peripheral compartments represent "shallow" and "deep" compartments. The deep compartment is considered less accessible than the shallow compartment so that in assigning relative values to the rate constants, $k_{21} > k_{31}$. A less accessible deep compartment might represent teeth, bone, hair, nails, or cartilage (e.g., accumulation of tetracycline derivatives in bone).

    In addition to blood flow and blood volume, partitioning and binding are also determinants for drug distribution. Partitioning, a rapid phenomenon, is responsible for drug reaching a rapid equilibrium with all tissues in a compartment. The concentration at equilibrium is, in part, due to the hydrophilic/lipophilic properties of the structure of the drug. Regardless of blood flow or blood volume, lipophilic drugs

TABLE 1.2
Properties of a Classical Pharmacokinetic Compartment

1. Kinetic homogeneity. A compartment contains tissues that can be grouped according to similar kinetic properties to the drug allowing for rapid distribution between tissues.
2. Although tissues within a compartment are kinetically homogeneous, drug concentrations within a compartment may have different concentrations of drug depending on the partitioning and binding properties of the drug.
3. Within each compartment, distribution is immediate and rapidly reversible.
4. The barriers between compartments, depicted as arrows in Figure 1.1, are considered diffusion rate limiting and can represent transport across an anatomical barrier (i.e., epithelium of the gastrointestinal tract or the glomerulus in the kidney), a metabolic conversion, or, if unknown, transport into and out of groups of tissues with different kinetic properties.
5. Distribution to any tissue within a compartment depends upon blood flow, blood volume, and partitioning, as well as tissue and plasma protein binding.
6. Compartments are interconnected by first-order rate constants. Input and exit rate constants may be zero order.

may accumulate in lipophilic tissues in high concentrations. Conversely, drugs that are hydrophilic may remain in circulating fluids with minimal distribution to lipophilic tissues. The term *kinetically homogeneous* is often used to define a compartment and refers to drug and tissue properties that allow for rapid distribution of drug between tissues so that various tissues and organs can be grouped kinetically constituting a compartment. Other properties of a compartment are summarized in Table 1.2. Drugs are also capable of binding to plasma proteins, which can reduce or slow distribution into tissues. Conversely, drugs may bind significantly to tissues allowing for a greater accumulation than explained by partitioning. Not all rapidly perfused tissues can be grouped together in the central compartment; similarly, not all peripheral tissues can be grouped within the peripheral compartment. Partitioning, tissue, and plasma protein binding depend not only on tissues, but also on drug properties. Therefore, depending on the particular drug, some tissues may reside in either the central or the peripheral compartments. Detailed experiments may be conducted to determine which tissue should be assigned to a compartment. Blood and extravascular fluids are nearly always assigned to the central compartment along with organs responsible for elimination, e.g., kidneys, liver, and the lungs.[3-7]

USE OF A SIMPLIFIED ONE-COMPARTMENT MODEL

If a bolus dose is administered and time intervals are adequate and well spaced, the drug will most often follow the pharmacokinetics of a two- and sometimes a three-compartment model. However, the one-compartment model, with its greatly simplified equations, is often applied in the clinic for use as an estimate of plasma concentrations of drug. Although the biophase often resides in tissues (which may or may not reside in the central compartment), plasma concentrations are chosen for measurement because of ease of sampling and because they are a reference to proper dosing, via the establishment of a therapeutic range.[6] Theoretically, the

TABLE 1.3
Justification for Using a One-Compartment Model in the Clinic Instead of a Theoretically Correct Multicompartment Model

1. The drug distributes only to rapidly perfused tissues and circulating fluids located in the central compartment (e.g., hydrophilic drugs, such as the penicillins and cephalosporins).
2. The drug distributes to tissues in both the central and peripheral compartments; however, drug concentrations in the peripheral compartment are below the therapeutic range (e.g., aminoglycosides).
3. The drug distributes to the peripheral compartment in significant concentrations. By ignoring the distribution to peripheral compartment(s) and using the simpler one-compartment model equations, the error is within acceptable limits (e.g., theophilline).

justification for using the simpler one-compartment model set of equations instead of more complex equations associated with a more precise multicompartment model is explained in Table 1.3.

PARAMETERS THAT DEFINE DRUG DISPOSITION

Drug disposition refers to how the drug distributes to tissues, in the central compartment as well as peripheral compartments, and how it is eliminated from the body. The disposition of drug over time can be determined from a bolus intravenous dose providing the time intervals are well spaced and characterize 3–5 half-lives of each kinetic process. Although full characterization of a drug's pharmacokinetic disposition behavior includes knowledge of both equation and model parameter values, three parameters, at a minimum, characterize disposition. These are the drug's half-life, volume of distribution, and clearance.

Half-Life

Half-life is sometimes referred to as the biological half-life and represents the time required for half the drug to be eliminated. It can be estimated quickly from data by visually determining the time that the concentration falls to one half of any initial concentration within the elimination phase, as shown in Figure 1.2. Half-life is a dependent variable that is measured from the terminal elimination rate constant, $t_{1/2\beta}$.

$$t_{1/2\beta} = \frac{0.693}{\beta} \qquad (1.2)$$

Half-life depends on clearance and the volume of distribution. Equation 1.3 defines the relationship between the three variables:

$$t_{1/2\beta} = \frac{0.693\,V_d}{Cl} \qquad (1.3)$$

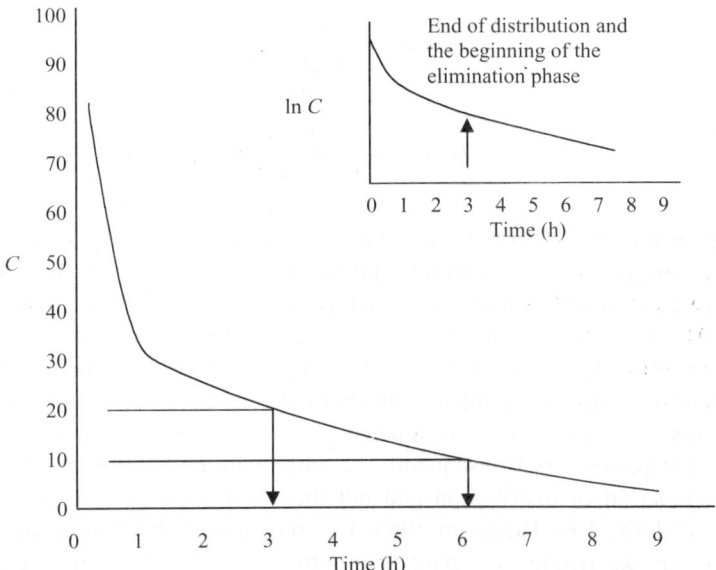

FIGURE 1.2 Graphical illustration of determining a half-life. Since 10 is one half of 20, the time required for drug to decline in that time frame represents a half-life (3 h). The insert emphasizes the requirement that drug measurement of half-life must be determined within the elimination phase.

Changes in half-life can change with clearance or with the volume of distribution. The volume of distribution measured at steady state, V_{dss}, is an accurate measure of physiological space, i.e., the volume in which the drug is contained. This is because the latter volume term is a function of distribution only. Other volume terms, V_d(area) and V_d(ext), which are defined below, can change as a function of elimination.

Volume Terms

The volume of distribution is a measure of physiological volume in which the drug is contained. It is called a volume term because it relates the amount of drug in the body to the concentration in plasma, C,

$$\text{amount} = V_d C \tag{1.4}$$

In more general terms, it could be referred to as a proportionality constant between the two variables, amount and concentration. However, from its inception, researchers and clinicians have endeavored to define V_d in physiological terms and not mathematical terms.

Early attempts at mathematically relating V_d to model parameters have resulted in equations that included elimination terms and not just distribution or physiological space terms. Consider the following mathematical definitions for V_d, which are valid for a drug following a two-compartment model.

$$V_d(\text{area}) = \frac{D_{IV}}{\beta(\text{AUC})} \tag{1.5}$$

$$V_d(\text{ext}) = \frac{V_c(\alpha - \beta)}{(k_{21} - \beta)} \tag{1.6}$$

In Equations 1.5 and 1.6, D_{IV} is a bolus intravenous dose; β is the terminal elimination rate constant representing elimination out of the body when the drug follows a two-compartment model; α, which is larger than β, is the distribution rate constant; AUC is the area under the concentration–time curve for drug in plasma following the intravenous dose; and k_{21} is a model parameter representing distribution of drug from the peripheral compartment into the central compartment. In both of the above equations, the V_d term is equated to an elimination term, β in Equation 1.5 and α in Equation 1.6. Consequently, an important assumption in defining V_d, that it is a function of distribution and not the elimination of drug, is violated. Equation 1.7, derived by Riggs[8] in 1963, is a function of distribution parameters only and is thus independent of changes in elimination, changing with drug distribution only.

$$V_d(\text{SS}) = V_c\left[\frac{k_{21} - k_{12}}{k_{21}}\right] \tag{1.7}$$

Equations 1.5 through 1.7 are model dependent; therefore, a different set of equations is available for one- or three-compartment models.

There are also volume terms defined for each compartment. However, only the volume of the central compartment, V_c, holds much relevance. It is sometimes referred to as the "initial volume of distribution" because theoretically, following a bolus intravenous dose (i.e., disposition considered only), drug immediately distributes throughout the central compartment but not to the peripheral compartment. Equation 1.8 gives the equation for V_c:

$$V_c = \frac{D_{IV}}{C_0} = \frac{D_{IV}}{A + B} \tag{1.8}$$

In Equation 1.8, C_0 is the concentration of drug in plasma at time zero, and A and B are equation parameters that result from curve fitting a biexponential curve (i.e., a two-compartment model).[7,8]

Clearance

Clearance has become an important parameter over the last decade because of its use in clinical pharmacokinetics. In clinical terms, it is defined as the milliliters of blood cleared of drug per minute. In basic pharmacokinetic terms, clearance can be defined as the milliliters of V cleared of drug per unit of time (see Equation

1.13); however, this definition is difficult to conceptualize and is not used much today. A major advantage of clearance is that working equations are model independent and expressed in parameter or variable terms that are not difficult to obtain experimentally.

$$Cl = \frac{F \, \text{Dose}}{\text{AUC}} \tag{1.9}$$

$$Cl = \frac{K_0}{C_{ss}} \tag{1.10}$$

$$Cl_R = \frac{\frac{\Delta X_u}{\Delta t}}{C_{t_{mid}}} \tag{1.11}$$

$$Cl = Cl_R + Cl_M \tag{1.12}$$

$$Cl = k_{el} V_d \tag{1.13}$$

In Equations 1.9 through 1.13, F is the fraction of drug absorbed; K_0 is constant rate infusion; C_{ss} is the level of drug in plasma at a steady-state concentration; $\Delta X_u/\Delta t$ is the change concentration of drug in urine over a specified time interval, and $C_{t_{mid}}$ is the concentration of drug in plasma over the same specified time interval; the subscripts used to specify clearance, R and M, represent renal and all nonsaturated metabolism in the body. Equation 1.11 is sometimes used with Equation 1.12 by subtracting Cl_R from Cl to estimate Cl_M.[8–10]

ABSORPTION ANALYSIS

At its simplest level, absorption is not difficult to interpret. The fraction absorbed, F, and the time for absorption to be completed are relatively simple to define. The fraction absorbed is determined from experiments in which the same subject is given the same dose of a bolus IV dose (a solution may suffice) and an oral dose. The ratios of the AUC for the oral and IV routes of administration, AUC_O and AUC_{IV}, respectively, yield a correct value.

$$F = \frac{AUC_O D_{IV}}{AUC_{IV} D_O} \tag{1.14}$$

Corrections to F are shown in Equation 1.14 for differences in dose, which come from the ratio D_{IV}/D_O.

The oral absorption process requires two steps for completion. The drug must first cross the epithelial layer of the gastrointestinal tract. Because of the increased surface area, drugs are more readily absorbed across the duodenum and the ilium than other regions of the gastrointestinal tract. Most drugs appear to cross the epithelium by a passive or transcellular mechanism, which depends on the lipophilicity and rapid partitioning of the drug into epithelial cells. If a drug is primarily ionized at the absorption site, either because of its pK_a or the pH of the absorbing environment, it may cross the epithelium by a paracellular mechanism. Although not well defined, carrier-mediated transport may be responsible for the absorption of drugs, such as peptides. Once drug crosses the epithelium and enters the bloodstream, it must cross the hepatoportal system, which after passing through the liver intact, enters the inferior vena cava. Once the drug enters systemic circulation and is available for distribution to various tissues and organs, it is considered completely absorbed. If a drug is metabolized prior to reaching the circulating fluids of the body (i.e., inferior vena cava), the term *presystemic metabolism* defines the process. This phenomenon may also be referred to as a first-pass effect, which is a measure of the fraction of drug that passes between the portal and hepatic vessels after drug is absorbed on its first pass through systemic circulation during which time drug is completely subjected to liver metabolism. After drug is absorbed (i.e., reaches systemic circulation) only about 25 to 30% of the blood volume flows through the liver; therefore, biotransformation is reduced on each passage through the liver.

The time for absorption to be completed and an estimation of k_a is shown in Figure 1.3. Although a graphical estimate of k_a might suffice for an approximation, curve-fitting techniques are more accurate and can assess k_a from a nonlinear assessment of oral data for either a one-compartment (Equation 1.15) or a two-compartment (Equations 1.16 through 1.19) model.

$$C = \frac{k_a FD}{V(K - k_a)}(e^{-Kt} - e^{-k_{aa}t})\tag{1.15}$$

$$C = Le^{-\alpha t} + M^{-\beta t} + Ne^{-k_a t}\tag{1.16}$$

$$L = \frac{k_a FD(k_{21} - \alpha)}{V_c(k_a - \alpha)(\beta - \alpha)}\tag{1.17}$$

$$M = \frac{k_a FD(k_{21} - \beta)}{(V_c(k_a - \alpha)(\beta - \alpha))}\tag{1.18}$$

$$N = \frac{k_a FD(k_{21} - \beta)}{V_c(k_a - \alpha)(\beta - \alpha)}\tag{1.19}$$

Equations 1.15 and 1.16 represent the concentration of drug following an oral dose when the pharmacokinetics follows either a one- or two-compartment model, respectively. Equations 1.17 through 1.19 define the equation parameters, *L*, *M*, and

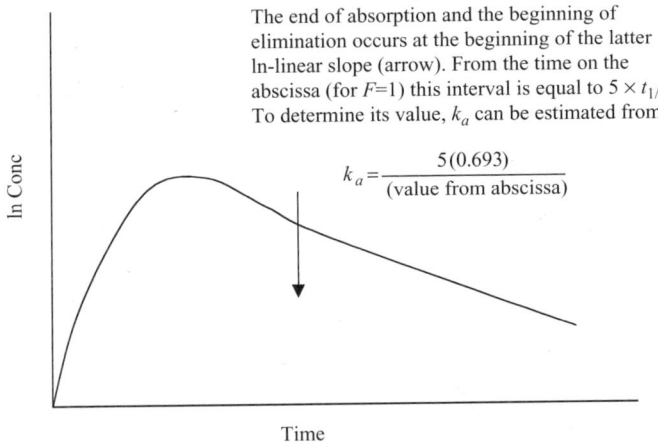

The end of absorption and the beginning of elimination occurs at the beginning of the latter ln-linear slope (arrow). From the time on the abscissa (for $F=1$) this interval is equal to $5 \times t_{1/2a}$. To determine its value, k_a can be estimated from:

$$k_a = \frac{5(0.693)}{\text{(value from abscissa)}}$$

In Conc

Time

FIGURE 1.3 Determination of the absorption rate constant, k_a, can be estimated by an approximation from a graph.

N, which are collections of equation and model parameters for the triexponential relationship (i.e., a two-compartment model).

For an oral dose, a biexponential relationship is characteristic of a drug following one-compartment model pharmacokinetics, whereas a triexponential defines a three-compartment model. For various reasons (i.e., $k_a \rightarrow \alpha$ or $k_a \rightarrow k_{21}$), a triexponential oral curve can become an apparent biexponential curve and lead to incorrect modeling interpretations.[11] Furthermore, both Equations 1.15 and 1.16 assume that the absorption process is a continuous first-order phenomenon, which may not be correct.

To circumvent these difficulties, researchers have developed deconvolution techniques that make no assumption regarding the order of the absorption process. With the use of deconvolution techniques, it is possible to separate the absorption process from disposition, allowing for the expression of absorption as % unabsorbed vs. time or % absorbed vs. time. Analysis of these plots permits an evaluation of the absorption process with fewer assumptions[12–16] than an analysis of the plasma concentration–time curve using the method of residuals. A lag time, first-order or parallel first-order processes (i.e., absorption and metabolism) and a zero-order absorption can sometimes be unambiguously determined from % unabsorbed vs. time plots (Figure 1.4). A minimum of five plasma concentrations measured in the absorption phase is usually necessary to reliably determine k_a by deconvolution methods.

BIOAVAILABILITY/BIOEQUIVALENCE

In the late 1970s, the FDA proposed legislation to require the drug companies to submit data for new generics that met certain bioavailability requirements. As a result of legislation passed during that time, the FDA has since published a yearly list of multisource drugs (i.e., generics), with monthly updates. The publication, titled "Approved Drug Products with Therapeutic Equivalence Evaluations" (USP

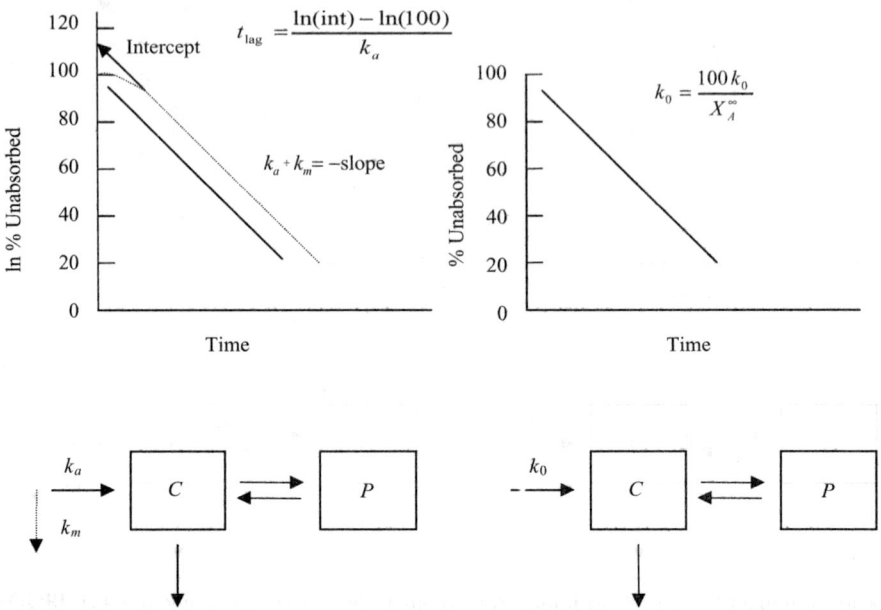

FIGURE 1.4 Graphical representation of the determination of the input function, k_a or k_0, the absorption first-order rate constant and the constant rate infustion, respectively, along with the model representation for each representation.

DI® Volume III), identifies products that are approved on the basis of safety and efficacy and lists drugs that are rated with regard to therapeutic equivalency. It does not include drugs marketed prior to 1938 nor does it include drugs approved based solely on safety. Although measurements of bioavailability and bioequivalence are essential criteria for determining generic substitution, a number of related terms are also frequently used in reference to generic substitution (Table 1.4).

BIOAVAILABILITY MEASUREMENTS

Bioavailability is a measure of the rate and extent of absorption. C_{max}, t_{max}, and AUC are the primary measurements used to determine bioavailability from oral concentration–time curves. The maximum plasma concentration of drug that is observed is defined as C_{max}, occurring at t_{max}. AUC is the area under the plasma concentration–time curve, measured through the last time interval or estimated through infinity. However, these experimental observations only indirectly measure the rate and extent of absorption. Determinations that best describe the rate and extent of absorption are k_a and F, the first-order rate constant for absorption and the fraction of the dose absorbed, respectively.

However, these latter measurements are relatively difficult to obtain and may be too variable to be useful, particularly k_a. The theoretical equations for C_{max}, t_{max}, and AUC are given below to show their relationship to the pharmacokinetic parameters (one-compartment model).

TABLE 1.4
List of Terms Used by the FDA in Defining Therapeutic Evaluations

Terms	Definition
Bioavailability	The term describes the rate and extent that a drug is absorbed (i.e., reaches the inferior vena cava or systemic circulation).
Bioequivalent drug products	The term refers to pharmaceutical equivalents that show similar bioavailability when studied under similar experimental conditions. The FDA describes the conditions under which a test and reference drugs shall be considered bioequivalent (USP DI® Volume III): "The rate and extent of absorption of the test drug do not show a significant difference from the rate and extent of absorption of the reference drug when administered at the same molar dose of the therapeutic ingredient under similar experimental conditions in either a single dose or multiple doses." Blood levels for the generic and standard products should be very similar with the inference that equal blood levels of drug for both products result in an equal therapeutic effect.
Pharmaceutical equivalents	The FDA (USP DI® Volume III) defines the term as: "Drug products are considered pharmaceutical equivalents if they contain the same active ingredient(s), are of the same dosage form, route of administration and are identical in strength or concentration (e.g., chlordiazepoxide hydrochloride, 5 mg capsules)."
Pharmaceutical alternatives	The FDA (USP DI® Volume III) defines the term as: "Drug products are considered pharmaceutical alternatives if they contain the same therapeutic moiety, but are different salts, esters, or complexes of that moiety, or are different dosage forms or strengths (e.g., tetracycline hydrochloride, 250 mg capsules vs. tetracycline phosphate complex, 250 mg capsules; quinidine sulfate, 200 mg tablets vs. quinidine sulfate, 200 mg capsules)."
Therapeutic equivalents	The FDA lists products as therapeutic equivalents if they are pharmaceutical equivalents that are bioequivalent.

$$t_{max} = \frac{\ln(k_a/K)}{(k_a - K)} \qquad (1.20)$$

$$C_{max} = \frac{FDk_a}{V(k_a - K)}(e^{-Kt_{max}} - e^{-k_a t_{max}}) \qquad (1.21)$$

$$AUC = \frac{FD}{VK} \qquad (1.22)$$

The purpose of defining t_{max}, C_{max}, and AUC in terms of rate and extent of absorption, as shown in Table 1.5, can be reasoned from Equations 1.20 through 1.22. Equations 1.20 and 1.21 for t_{max} and C_{max} show their relationship to k_a, which lends credence for their use in representing rate of absorption. Equations 1.21 and 1.22 contain F and D, which justifies the use of AUC and C_{max} as measurements for extent of absorption. Table 1.5 gives the interpretation of the experimental observations, t_{max},

---

TABLE 1.5
Bioavailability Parameters and Their Influence on Rate and Extent of Absorption

Experimental Determination	Interpretation
C_{max} (plasma)	Change in rate and extent
t_{max} (plasma)	Change in rate
AUC (plasma)	Change in extent
A_U^∞ (urine)	Change in extent

C_{max}, and AUC, in defining rate and extent of absorption for use in bioavailability and bioequivalence.

The amount of absorbed drug that is ultimately excreted (A_U^∞) can be related to the amount of the absorbed dose. For example, for the same individual, if more drug is absorbed from product A compared to product B, then more drug should appear in the urine. This principle is used in bioavailability studies to determine bioequivalence as a measure of extent of absorption and expressed as "relative bioavailability." One requirement of the use of A_U^∞ is that enough time is allowed to collect the total amount of the absorbed drug in the urine which is $5 \times t_{1/2}$. The experimental procedure and the volunteers that are used in the study also should be the same for products A and B.

Relative Bioavailability

Relative bioavailability is a term that compares the extent of absorption of a test product (i.e., generic) to a standard or reference product, which is often an oral solution or an immediate-release tablet, but not an IV product. The term is referred to as relative because the absolute rate and extent of absorption may not be known for either product. The FDA selects the reference product as a standard, whereas other generic products, to be classified as bioequivalent, must be equivalent in rate and extent of absorption and particularly in therapeutic equivalency. Two of the parameters in Table 1.5, AUC and A_U^∞, can be used to determine relative availability. Equations 1.23 and 1.24, which assume equal doses, show the relationship for relative bioavailability.

$$\text{Relative Bioavailability} = \frac{\text{AUC}_{test}}{\text{AUC}_{reference}} \qquad (1.23)$$

$$\text{Relative Bioavailability} = \frac{(A_U^\infty)_{test}}{(A_U^\infty)_{reference}} \qquad (1.24)$$

Occasionally, the reference products or the test are given IV and the oral dose is not equivalent and a correction is necessary. When that occurs, Equations 1.23 and 1.24 include a correction for the dose as shown in Equations 1.25 and 1.26.

$$\text{Relative Bioavailability} = \frac{\text{AUC}_{\text{test}}}{\text{AUC}_{\text{reference}}} \frac{\text{Dose}_{\text{reference}}}{\text{Dose}_{\text{test}}} \tag{1.25}$$

$$\text{Relative Bioavailability} = \frac{(A_U^\infty)_{\text{test}}}{(A_U^\infty)_{\text{reference}}} \frac{\text{Dose}_{\text{reference}}}{\text{Dose}_{\text{test}}} \tag{1.26}$$

Absolute Bioavailability

Absolute bioavailability refers to the fraction of the oral dose that is absorbed (i.e., reaches systemic circulation). The fraction absorbed is also defined as F so that Equations 1.14 and 1.27 are equivalent. Absolute bioavailability is shown in Equations 1.27 and 1.28, the latter representing a correction for dose. Each product is tested individually to determine its absolute bioavailability in the subjects. The IV dose becomes a reference for which drug is assumed to be 100% available. On occasion, a solution may be substituted for an IV dosage form providing it is 100% bioavailable (i.e., reaches systemic circulation).

$$\text{Absolute Bioavailability} = \frac{\text{AUC}_{\text{oral}}}{\text{AUC}_{\text{IV}}} \tag{1.27}$$

$$\text{Absolute Bioavailability} = \frac{\text{AUC}_{\text{oral}}}{\text{AUC}_{\text{IV}}} \frac{\text{Dose}_{\text{IV}}}{\text{Dose}_{\text{oral}}} \tag{1.28}$$

It is important to distinguish between absolute and relative bioavailability. For example, two oral products, A (generic) and B (reference), may produce an absolute bioavailability of 0.50 and 0.60 for equal doses administered to a group of individuals in a crossover trial, but when compared for their relative bioavailability show 0.83 (0.50/0.60). Under these circumstances, the products would be considered bioequivalent and acceptable for clinical use but could be mistakenly interpreted as well absorbed (i.e., > 0.75).

Bioequivalence

Bioequivalence is determined by showing that no statistical differences exist among C_{max}, t_{max}, and AUC for the generic and reference products. For many drug products a ±20 difference is considered clinically important. However, a statistically significant difference between parameter values may not necessarily represent a clinically significant difference for some drugs. For many years, since the inception of the Bioavailability Regulations in 1977, the FDA has carefully established percent differences between products as ±20, depending on the particular drug and a knowledge of the clinical relevance of the drug. More recently, rate and extent between two formulations that differ by more than –20%/+25% have been accepted. This rule is not applied to all drugs. Some drugs may dictate a different standard, sometimes more and sometimes less stringent. The design of the study must also be carefully considered, including the required number of subjects, so that a statistical significance between products will result if the difference between products is outside the allowable difference that has been established for the drug. In practice, two one-sided statistical tests are used to determine if a difference exists between a particular parameter for each product. One test determines that the average parameter value for the generic product is no more than 20% below that for the reference product. The other test is used to determine that the average parameter for the generic product is no more than 25% above the reference product. A summary is presented in Table 1.5 of the changes in the experimental parameters, C_{max}, t_{max}, and AUC defining their meaning relative to the rate and extent of absorption.[17,18]

MULTIPLE DOSING KINETICS

In a linear pharmacokinetic model, plasma concentrations are additive, which is a useful principle for predicting multiple dosing profiles and for estimating dosing regimens in chronic conditions. This principle is referred to as the superposition principle, which states that the concentration of drug remaining in the body at any time is added to the concentration remaining from previous doses. As accumulation occurs, the concentration remaining in the body increases for each interval until steady state is reached. From this concept, a multiple dosing factor (MDF) can be derived, as shown in Equation 1.29.[19]

$$MDF = \frac{1 - e^{-nk_i\tau}}{1 - e^{-k_i\tau}} \qquad (1.29)$$

A multiple dosing equation is derived by inserting the MDF into the single-dose equation as a coefficient to each exponential term in the equation. The MDF can be reduced to a simpler form at steady state or when the amount administered equals the amount being eliminated. When a number of doses are administered, n becomes large enough (i.e., steady state) so that the term, $e^{-nk_i\tau}$, becomes negligible and Equation 1.29 simplifies to Equation 1.30.

$$MDF = \frac{1}{1 - e^{-k_i\tau}} \qquad (1.30)$$

In Equations 1.29 and 1.30, n is the number of doses administered, k_i is the rate constant associated with the exponential term that is multiplied by MDF, and τ is the dosing interval. For example, Equation 1.21 represents the concentration of drug for a single oral dose of a drug following the pharmacokinetics of a one-compartment model. Equation 1.31 represents a similar equation, which has been adjusted to accommodate the administration of drug at a dose given repeatedly at a fixed interval, τ.

$$C_n = \frac{FDk_a}{V(k_a - K)}\left[\left(\frac{1 - e^{-nK\tau}}{1 - e^{-K\tau}}\right)e^{-Kt} - \left(\frac{1 - e^{-nk_at}}{1 - e^{-k_at}}\right)e^{-k_at}\right] \qquad (1.31)$$

Whenever either k_a or τ becomes large, the second term on the right-hand side of Equation 1.31 becomes negligible and can be ignored. Equation 1.31 can be altered to other, more useful applications. For example, Equations 1.32 and 1.34 represent the maximum and minimum concentrations of drug at steady state.

$$C_{max(SS)} = \frac{FDk_a}{V(k_a - K)}\left[\left(\frac{1}{1 - e^{-K\tau}}\right)e^{-Ktp_\infty} - \left(\frac{1}{1 - e^{-k_a\tau}}\right)e^{-k_a tp_\infty}\right] \qquad (1.32)$$

The parameter tp_∞ in Equation 1.33 represents the time to peak at steady state, which is slightly less than tp, the time to peak for a single dose. The value for tp_∞ can be determined from Equation 1.33.[20]

$$tp_\infty = \frac{\ln}{(k_a - K)}\left[\frac{k_a(1 - e^{K\tau})}{K(1 - e^{-k_a\tau})}\right] \qquad (1.33)$$

$$C_{min(SS)} = \frac{FDk_a}{V(k_a - K)}\left[\left(\frac{1}{1 - e^{-K\tau}}\right)e^{-K\tau} - \left(\frac{1}{1 - e^{-k_a\tau}}\right)e^{-k_a\tau}\right] \qquad (1.34)$$

ACCUMULATION

The right-hand side of Equation 1.30 also equals the accumulation factor, which is the factor by which accumulation occurs in the body upon multiple dosing, predicting the increase in $C_{min(SS)}$ relative to $C_{min(I)}$. This factor can also be used to establish a loading dose by multiplying the estimated value by the usual daily dose or maintenance dose. The loading dose, taken initially only, produces drug concentrations within the first interval that would be achieved by taking the maintenance dose repeatedly until steady state is reached, which is $5 \times t_{1/2}$. The accumulation factor can also be expressed as a ratio of $C_{min(SS)}$ to $C_{min(I)}$ and then rearranged as shown in Equation 1.35 to predict a value for $C_{min(SS)}$ as a function of $C_{min(I)}$, K, and τ.

$$C_{min(SS)} = \frac{C_{min(I)}}{1 - e^{K\tau}} \qquad (1.35)$$

Equation 1.35 can be used to adjust dosages in patients who do not respond to the normal recommended dosing regimen because of kidney or hepatic dysfunction. For example, if two concentrations of drug are measured in the elimination phase of the first interval with the second concentration measured at $C_{min(I)}$, K can be determined and with $C_{min(I)}$ being measured directly, Equation 1.36 yields a predicted $C_{min(SS)}$. If the predicted $C_{min(SS)}$ is not within the therapeutic range, a ratio of the desired concentration (i.e., middle of the therapeutic range) to the calculated $C_{min(SS)}$ can be multiplied by the dose given in the first interval to yield a maintenance dose that will produce the desired concentration.[21]

When Equation 1.30 is interpreted as an accumulation factor, it predicts that if τ is large enough or if the half-life of the drug is relatively short (e.g., cephalosporins) then very little accumulation occurs. Conversely, if τ is short and the half-life is long (e.g., most benzodiazepines), accumulation will be large compared with $C_{min(1)}$. Regardless of the value for the accumulation factor, if the maintenance dose is given in a fixed interval and the half-life does not change, steady state will be reached in five half-lives.

THERAPEUTIC DRUG MONITORING

Because pharmacokinetic processes (ADME) vary among individuals, particularly among patients whose organ systems are compromised, a therapeutic response is not always consistent with regard to dose. Therefore, for drugs with a narrow therapeutic range, a patient's drug concentration in blood is sometimes monitored. The goal of therapeutic drug monitoring (TDM) is to provide optimal therapy by adjusting either the dose or the dosing interval, τ, as mentioned previously. From two plasma concentrations of drug measured in the elimination phase, providing one is at $C_{min(1)}$, it is possible through the use of Equation 1.35 to predict $C_{min(SS)}$. Subsequently, the maintenance dose is adjusted so that by the time $C_{min(SS)}$ is reached at $5 \times t_{1/2}$, the latter is within the therapeutic range. This method of dosing adjustment is referred to as a retrospective approach. It is time-consuming and expensive because blood must be drawn at precise time intervals and an assay must be available for each drug candidate. In the clinic this expense must be weighed against its therapeutic value.

An alternative method is the prospective approach, which uses patient information, such as serum creatinine, patient's age, weight, height, or body surface area to empirically estimate V, K, and creatinine clearance (Cl_{cr}). Adjustments to a maintenance dose are achieved through various approaches. The most significant application of either the retrospective or prospective approach is with regard to dosing adjustment in renal failure.[22]

DOSAGE ADJUSTMENT IN RENAL FAILURE

When the ability of the kidney to remove a drug from the body is impaired, the rate of elimination of the drug from plasma is reduced. Changes in renal drug clearance or Cl_R can be estimated by changes in renal function, which are routinely monitored in hospitalized patients from the measurement of serum creatinine, SrCr, and then converted to creatinine clearance, Cl_{cr}. The graphical representation of this relationship is shown in Figure 1.5. From either the manufacturer's suggested dosage adjustment in renal failure or from a knowledge of basic concepts in pharmacokinetics, it is possible to estimate a proper dose in patients whose renal function is below normal. Two pieces of information are necessary to estimate an adjusted dose — the patient's serum creatinine level (SrCr in units of mg/dl) and the percent or fraction of drug excreted unchanged (f), the latter of which can be obtained from the monograph of the drug.

For drugs that are primarily eliminated by the renal route, the half-life of the drug is increased or, conversely, K or β, the first-order rate constants for drug

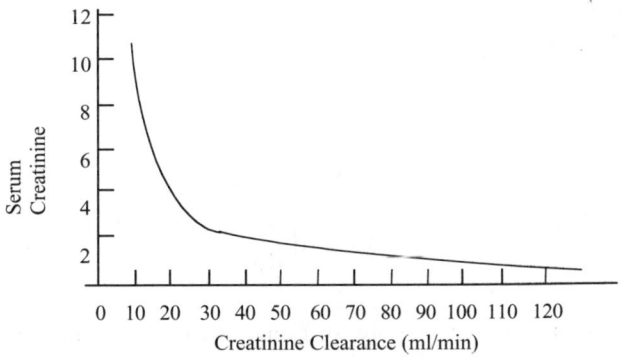

FIGURE 1.5 Graphical relationship between serum creatinine and creatinine clearance.

elimination for a one- or two-compartment model, is reduced. Usually, it can be assumed that k_e, the first-order rate constant for renal excretion, is reduced (and hence renal clearance, $k_e V_d$) and that neither hepatic metabolism nor the volume of distribution is affected. Consequently, Cl_R, and Cl_S become major determinants. Various approaches have been developed to help the clinician adjust doses in patients with renal impairment. These approaches are simple to use, directly related to pharmacokinetic principles, but with certain limitations. One approach requires the use of a ratio of the drug clearances in the patient with renal impairment compared to the normal patient. Another, referred to as the Wagner method,[23] applies a linear relationship between the elimination rate constant, K, and Cl_{cr}.

DOSING ADJUSTMENT BASED ON DRUG CLEARANCE
(RATIO OR PROPORTION METHOD)

To compensate for impaired kidney function, the clinician has the option to either reduce the dose (D), or to lengthen the dosing interval (t). For patients taking chronic medication and whose dose might require an adjustment, a simple ratio of systemic clearances for the renal failure patient (SR) and the normal patient (S) can be used. In this method, the maintenance dose, D, in the normal patient can be adjusted to a corrected dose, D_R, for the renal failure patient.

$$D_R = D\left(\frac{Cl_{sr}}{Cl_S}\right) \qquad (1.36)$$

$$\tau_R = \tau\left(\frac{Cl_S}{Cl_{sr}}\right) \qquad (1.37)$$

In both Equations 1.36 and 1.37, V can be canceled from the numerator and denominator since it is often assumed that V is constant regardless of renal status. Therefore, Equations 1.36 and 1.37 become Equations 1.38 and 1.39, respectively, which can be further manipulated to represent half-life since the elimination rate constant is inversely related to half-life ($K = 0.693/t_{1/2}$). However, the latter equations are not used very often.

$$D_R = D\left(\frac{K_R}{K}\right) \tag{1.38}$$

$$\tau_R = \tau\left(\frac{K}{K_R}\right) \tag{1.39}$$

The assumptions that must apply for Equations 1.36 through 1.39 are that drug concentrations are at steady state, Cl_{cr} is directly proportional to Cl, and the percentage of drug excreted by the renal route is known and significant (e.g., >30%). If a small fraction of the dose is eliminated by the kidneys (and the metabolites are not active), then a reduction in kidney function has little effect on elimination and a dosing adjustment is not necessary (assuming that liver function is normal and does not influence renal function).

WAGNER METHOD

In 1971, Wagner[23] published an approach based on a linear relationship between percent hourly loss ($K_R\%$ in h $\times$ 100) for various drugs and creatinine clearance, normalized to a 70 kg male (ml/min/1.73 m^2). This approach has been modified over the years so that either a nomogram or a table of ranges with accompanying recommendations for dosage adjustment for drugs could be used.[24] These tables are not reproduced here. In Table 1.6, regression parameters (b and k_m) are tabulated for each drug according to Equation 1.40.

$$\%K_R = bCl_{cr} + k_m \tag{1.40}$$

In Table 1.6, K_R represents K for a patient with renal failure, whereas k_m (rate constant for metabolism) is listed as "a" in Table 1.6, which, if relatively large compared to "b," represents a drug that is primarily eliminated by metabolism. Under these conditions (e.g., doxycycline), k_e does not change for the patient with renal failure compared with the patient with normal kidney function. The major disadvantage to the Wagner method is that the list is restrictive and does not include enough current drugs to make it clinically useful. More specifically, the precision and accuracy of Equation 1.40 depends on the patient population that was used to generate the equation. Consequently, the approach requires that the patient closely equal the population values for pharmacokinetic processes other than elimination.

An accurate measurement of kidney function requires that urine be collected for 24 h with both creatinine and urine volume carefully measured. However, 24-h urine collection is labor-intensive and difficult to obtain accurately and therefore is not suitable for clinical practice. (*Note:* It may be possible to use as few as 4 or 6 h to measure Cl_{cr} accurately.)

TABLE 1.6
Adjustment of Dosage in Patients with Impaired Renal Function Using Creatinine Clearance

Drug	Patient $K_R\% = a + b(Cl_{cr})$	
	a	b
Ampicillin	11.0	0.59
Carbenicillin	6.0	0.54
Cephalexin	3.0	0.67
Cephaloridine	3.0	0.37
Cephalothin	6.0	1.34
Chloramphenicol	20.0	0.1
Chlortetracycline	8.0	0.04
Digitoxin	0.3	0.001
Digoxin	0.8	0.009
Doxycycline	3.0	0.0
Erythromycin	13.0	0.37
Gentamycin	2.0	0.28
Isoniazid (fast inactivators)	34.0	0.19
Isoniazid (slow inactivators)	12.0	0.11
Kanamycin	1.0	0.24
Methicillin	17.0	1.23
Oxacillin	35.0	1.05
Penicillin G	3.0	1.37
Polymyxin B	2.0	0.14
Rolitetracycline	2.0	0.04
Streptomycin	1.0	0.26
Sulfadiazine	3.0	0.05
Sulfamethoxazole	7.0	0.0
Tetracycline	0.8	0.072
Trimethoprin	2.0	0.04
Vancomycin	0.3	0.117

Data from Wagner,[3] p. 161.

The amount of creatinine excreted in the urine in a 24-h period is calculated from Equation 1.41.

$$\text{Amount of creatinine excreted per day per kg of bodyweight}$$
$$= \frac{(\text{Urine Volume per 24 h})(\text{Urine Creatinine Conc.})}{(\text{Patient's Weight})} \tag{1.41}$$

A measure of creatinine can be expressed directly as Cl_{cr} by measuring urine output for 24 h and dividing this volume by a measurement of serum creatinine drawn at the midpoint of the 24-h collection period. A simpler determination, but

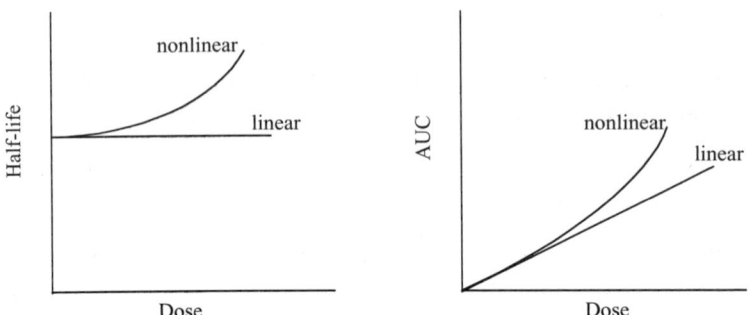

FIGURE 1.6 Nonlinearity of half-life and the AUC with dose.

less precise, is to measure glomerular filtration by measuring serum creatinine levels and converting to creatinine clearance through the use of a nomogram.[24]

The correlation shown in Figure 1.5 allows for a reasonable conversion of $SrCr_{SS}$ to Cl_{cr} from about 5 to about 40 or 50 ml/min. The sensitivity of the relationship is reduced considerably for Cl_{cr} values greater than 40 or 50 ml/min because of the lack of sensitivity of the curve over the higher Cl_{cr} region.

Cockcroft and Gault[25] developed an equation that allows for differences in the patient's weight, age, and gender but which are not included as separate determinants in Figure 1.5.

$$Cl_{cr} = \frac{[140 - \text{age(yr)}][\text{body wt(kg)}]}{72(SrCr_{SS})} \qquad (1.42)$$

The units for Cl_{cr} in Equation 1.42 are expressed in ml/min, whereas $SrCr_{SS}$ is expressed in mg/dl, or milligrams of creatinine contained within a deciliter (100 ml) of serum. To account for the weight of females, Cl_{cr} determined from Equation 1.42 is multiplied by 0.85. Equation 1.42 is an approximate assessment of creatinine output and it is most accurate for patients older than 18 years of age. It is not accurate in special situations such as dosing children, the elderly people (>70 years old), or the very obese. Although other equations to calculate Cl_{cr} have been developed,[26] none has been applied clinically as extensively as Equation 1.42. In the elderly, muscle mass is proportionately less than that of individuals between 18 and 70. Between 18 and 70, corrections for age are accounted for in Equation 1.42. In spite of its limitations, the Cockcroft and Gault equation is widely used clinically.

The two most important points to consider in determining kidney function from Cl_{cr} is that the patient's muscle mass is not appreciably different from normal and that $SrCr_{SS}$ is at steady state (e.g., constant concentration in the body). If kidney function changes, it can take from 1 to 4 days for $SrCr_{SS}$ to reach a steady-state value and stabilize. The time to reach 95% of steady state (e.g., a clinically accepted criterion) depends on the classification of kidney failure. For example, if kidney function is either 25 or 10% of normal, 1.85 and 4.6 days are required to reach steady state, respectively, whereas, if kidney function is 50% of normal, $SrCr_{SS}$ requires only 0.92 days to reach steady state. If an accurate measurement is necessary, two determinations of SrCr can be taken at 12-h intervals, which should be within 0.2 mg/dl of each other to assume a steady state has been reached. Ideal body weight (IBW) can be determined from Equations 1.43 and 1.44 for males or females.[25]

$$IBW \text{ (males)} = 50 + 2.3(\text{height in inches} > 60) \tag{1.43}$$

$$IBW \text{ (females)} = 45 + 2.3(\text{height in inches} > 60) \tag{1.44}$$

For patients below their IBW, actual body weight (ABW) is often used in Equation 1.42.

DOSING ADJUSTMENT FROM THE USE OF CREATININE SERUM LEVELS AND THE FRACTION EXCRETED UNCHANGED

As opposed to using a nomogram, a formal equation using f and Cl_{cr} terms can be expressed for adjusting dose based on Cl_{cr} measurements as shown in Equations 1.45 and 1.46:

$$D_R = D\left[\frac{Cl_{cr}(f)}{100} + (1 - f)\right] \tag{1.45}$$

In Equation 1.36, D_R is the adjusted maintenance dose for the patient with renal failure, D is the maintenance dose for a normal individual, f is the fraction excreted, and Cl_{cr} is the patient's creatinine clearance, which can be determined from the serum creatinine value using the Cockcroft and Gault Equation 1.42. If a new dosing interval for the patient with renal failure (τ_R) is calculated, a reciprocal of the bracketed term in Equation 1.45 is used and the new equation becomes Equation 1.46.[27]

$$\tau_R = r\left[\frac{1}{\dfrac{Cl_{cr}(f)}{100} + (1 - f)}\right] \tag{1.46}$$

The advantage in using Equations 1.45 and 1.46 as opposed to Equations 1.36 and 1.37 is that f can be obtained from the monograph of a drug and Cl_{cr} can be determined for most patients in a clinical setting. A disadvantage to the above two equations is that f is a population value and not specific to the patient and therefore can be in error.

HEPATIC ELIMINATION AND DOSING ADJUSTMENTS

In clinical practice, Cl and Cl_R can be determined directly for patients but not hepatic or metabolic clearance. Therefore, Cl_{nr} is assumed to be the difference between Cl and Cl_R. However, the difference may not represent the liver entirely, since the latter may not be the only metabolizing organ for a drug. Drugs are often categorized according to their extraction ratio (ER) and listed for various drugs in monographs as high, intermediate, or low. Drugs that show high ER values need to be carefully assessed with regard to a dosing regimen. An approximation of ER is determined from the areas under the plasma concentration–time curves from oral and IV dosing.

$$ER = 1 - \frac{AUC_{oral}}{AUC_{IV}} \qquad (1.47)$$

Unlike renal disease, no endogenous estimator of hepatic disease can be relied upon to adjust dosage regimens. This occurs partly because there are many enzyme systems in the liver and not all are equally affected by liver disease. Therefore, depending on the particular enzyme system that acts on a drug, liver disease may not affect all drugs equally in reducing elimination function. A further complication comes from the fact that liver disease may result in a significant change in the volume of distribution because of changes in plasma protein binding. Also, changes in the fraction absorbed can occur for drugs taken orally because of a reduction in hepatic blood flow and its effect on the first-pass effect.

In general, no quantitative approaches to dosing adjustments have yet been developed for patients taking drugs that are highly extracted by the liver. Each drug and patient must be evaluated carefully based upon specific guidelines that are available for these drugs in their monographs.[28]

NONLINEAR PHARMACOKINETICS

Linear pharmacokinetics can be applied to most drugs; however, when nonlinear pharmacokinetics correctly describe the pharmacokinetic behavior of a drug, it is essential that the more-complicated kinetics be applied. Drugs described by nonlinear pharmacokinetic behavior can result in a disproportionate increase in plasma drug concentrations when the dose is increased, which for drugs with narrow therapeutic ranges can be dangerous, particularly in children. Most often nonlinear behavior is a result of saturation of a metabolic carrier-mediated system. The enzyme involved in the metabolic process can no longer accept drug and biotransformation becomes limiting (i.e., saturation), resulting in an increase in plasma concentration of drug and possible toxicity. Other processes can become saturated such as absorption or excretion. The former is not common, whereas the latter may be more common via saturation of an active secretion pathway in the kidney. However, glomerular filtration may not be altered simply because the later predominates in the elimination process. Consequently, minimal change occurs in the overall elimination process of the drug. Whenever an elimination process becomes significantly nonlinear, the elimination rate constant, clearance, and half-life become dose-dependent, as shown in Figure 1.6. Drug elimination by a carrier-mediated process is often referred to as Michaelis–Menten kinetics and is represented by Equation 1.48.

$$\frac{dC}{dt} = -\frac{V_M C}{K_M + C} \tag{1.48}$$

In Equation 1.48, V_M is the maximum elimination rate. K_M is the Michaelis–Menten constant, which is a collection of first- and second-order rate constants. Although an enzymatic process metabolizes many drugs, few reach saturation at therapeutic drug concentrations. At lower concentrations, the drug follows a linear equation of drug reacting with the body's enzyme to yield a metabolite. C is the concentration of drug in plasma. Figure 1.6 shows the relationship of dose with half-life and AUC.

Equation 1.48 predicts whether the kinetics of the drug are linear or nonlinear depending on the relative concentration of C compared to K_M, as shown in Figure 1.7.

$$\frac{dC}{dt} = -\frac{V_M C}{K_M} = -KC \tag{1.49}$$

If the concentration of C is small relative to K_M (i.e., a drug for which K_M is large or the therapeutic concentrations are relatively low), it can be dropped from Equation 1.48 to yield 1.49. In Equation 1.49, $K = V_M/K_M$. Conversely, if K_M is small relative to C, then Equation 1.48 becomes Equation 1.50 and the rate is constant and equal to $-V_M$.

$$\frac{dC}{dt} = \frac{V_M C}{C} = -V_M \tag{1.50}$$

The application of Equation 1.48 is most apparent in dosing anticonvulsant drugs, which for some patients show nonlinear pharmacokinetics over therapeutic concentrations.

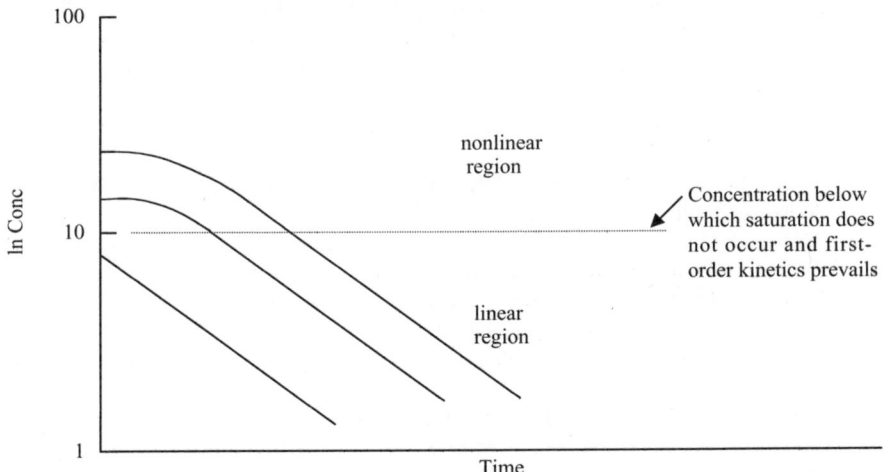

FIGURE 1.7 Graphical representation of three concentrations of drug over time for which drug levels above the dotted line show enzyme saturation and hence nonlinearity.

Various methods have been designed for clinical use to predict doses that remain within the therapeutic range. These are as follows:

1. The two-point method (or Ludden method), which utilizes two steady-state drug levels (each representing a different dose) to predict a subsequent dose that will likely result in therapeutic drug concentrations[29]
2. The one-point method, which is a statistical approach based on previous known blood levels in a particular segment of the patient population and predicts, through the use of a nomogram, a proper dose[30]
3. A Bayesian approach,[31–33] which is also statistically based and depends on a segment of the population, but may use demographic information as well as blood level responses to a particular drug

The latter approach has a wider application and has been useful for linear as well as nonlinear drugs. Numerous articles[34–36] have compared the accuracy and precision of each approach.

REFERENCES

1. Teorell, T., Kinetics of distribution of substances administered to the body I. The extravascular modes of administration, *Arch. Int. Pharmacodyn.*, 57:205–225, 1937.
2. Teorell, T., Kinetics of distribution of substances administered to the body II. The intravascular modes of administration, *Arch. Int. Pharmacodyn.*, 57:226–240, 1937.
3. Wagner, J. G., *Fundamentals of Clinical Pharmacokinetics*, Drug Intelligence Publications, Inc., Hamilton, IL, 1975, 57–120.
4. Shargel, L. and Yu, A. B. C., *Applied Biopharmaceutics and Pharmacokinetics*, 4th ed., Appleton & Lange, Stamford, CT, 1999, 36–40.
5. Rowland, M. and Tozer, T. N., *Clinical Pharmacokinetics Concepts and Applications*, 3rd ed., Williams & Wilkins, Baltimore, MD, 1995, 313–322.
6. Bauer, L. A., Clinical pharmacokinetics and pharmacodynamics, in *Pharmacotherapy a Physiological Approach*, 4th ed., Dipiro, J. T. et al., Eds., Appleton & Lange, Stamford, CT, 1999, 21–43.
7. Gibaldi, M. and Perrier, D., *Pharmacokinetics*, Marcel Dekker, New York, 1975, 45–96.
8. Riggs, D. S., *The Mathematical Approach to Physiological Problems*, Williams & Wilkins, Baltimore, MD, 1963, 193–217.
9. Rowland, M. and Tozer, T. N., *Clinical Pharmacokinetics Concepts and Applications*, 3rd ed., Williams & Wilkins, Baltimore, MD, 1995, 18–19.
10. Rowland, M., Benet, L. Z., and Graham, G. G., Clearance concepts in pharmacokinetics, *J. Pharmacokinet. Biopharm.*, 1:123–136, 1973.
11. Gibaldi, M. and Perrier, D., *Pharmacokinetics*, Marcel Dekker, New York, 1975, 80–84.
12. Wagner, J. G. and Nelson, E., Percent absorbed time plots derived from blood level and/or urinary excretion data, *J. Pharm. Sci.*, 52:610–611, 1963.
13. Wagner, J. G., Pharmacokinetic absorption plots from oral data alone or oral/intravenous data and an exact Loo-Riegelman equation, *J. Pharm. Sci.*, 72:838–842, 1983.

14. Loo, J. and Riegelman, S., New method for calculating the intrinsic absorption rate of drugs, *J. Pharm. Sci.*, 57:918–928, 1968.
15. Wagner, J. G., *Pharmacokinetics for the Pharmaceutical Scientist*, Technomic Publishing Co., Lancaster, PA, 1993, 159–205.
16. Gibaldi, M. and Perrier, D., *Pharmacokinetics*, Marcel Dekker, New York, 1975, 129–145.
17. Shargel, L. and Yu, A. B. C., *Applied Biopharmaceutics and Pharmacokinetics*, 4th ed., Appleton & Lange, Stamford, CT, 1999, 247–279.
18. Rowland, M. and Tozer, T. N., *Clinical Pharmacokinetics Concepts and Applications*, 3rd ed., Williams & Wilkins, Baltimore, MD, 1995, 34–46.
19. Wagner, J. G., *Pharmacokinetics for the Pharmaceutical Scientist*, Technomic Publishing Co., Lancaster, PA, 1993, 103–109.
20. Gibaldi, M. and Perrier, D., *Pharmacokinetics*, Marcel Dekker, New York, 1975, 97–128.
21. Gibaldi, M. and Perrier, D., *Pharmacokinetics*, 2nd ed., Marcel Dekker, New York, 1982, 121–123.
22. Schumacher, G. E., *Therapeutic Drug Monitoring*, Appleton & Lange, Norwalk, CT, 1995.
23. Wagner, J. G., *Biopharmaceutics and Relevant Pharmacokinetics*, Drug Intelligence Publications, Hamilton, IL, 1971, 222–233.
24. Shargel, L. and Yu, A. B. C., *Applied Biopharmaceutics and Pharmacokinetics*, 4th ed., Appleton & Lange, Stamford, CT, 1999, 531–571.
25. Cockcroft, D. W. and Gault, M. H., Prediction of creatinine clearance from serum creatinine, *Nephron*, 16:31–41, 1976.
26. Matzke, G. R. and Millikin, S. P., Influence of renal function and dialysis on drug disposition, in *Applied Pharmacokinetics, Principles of Therapeutic Drug Monitoring*, 3rd ed., Evans, W. E., Schentag. J. J., and Jusko, W. J., Eds., Applied Therapeutics, Vancouver, WA, 1992, 8:18–8:19.
27. Dettli, L., *Elimination Kinetics and Dosage Adjustment of Drugs in Patients with Kidney Disease*, Vol. 1, No. 4., Gustav Fischer Verlag, Stuttgart, Germany, 1977, 25–31.
28. Gibaldi, M., *Biopharmaceutics and Clinical Pharmacokinetics*, 4th ed., Lea & Febiger, Philadelphia, PA, 1991, 284–288.
29. Ludden, T. M. et al., Individualization of phenytoin dosage regimens, *Clin. Pharmacol. Ther.*, 21:287–293, 1977.
30. Tozer, T. N. and Winter, M. E., Phenytoin, in *Applied Pharmacokinetics, Principles of Therapeutic Drug Monitoring*, 3rd ed., Evans, W. E., Schentag, J.J., and Jusko, W.J., Eds., Applied Therapeutics, Vancouver, WA, 1992, 25:1–25:44.
31. Sheiner, L. B. et al., Forecasting individual pharmacokinetics, *Clin. Pharmacol. Ther.*, 26:294–305, 1979.
32. Vozeh, S. et al., Predicting individual phenytoin dosage regimens, *J. Pharmacokinet. Biopharm.*, 9:131–146, 1991.
33. Sheiner, L. B. and Beal, S. L., Bayesian individualization of pharmacokinetics — simple implementation and comparison with non-Bayesian methods, *J. Pharm. Sci.*, 71:1344–1348, 1982.
34. Martin, E. et al., The clinical pharmacokinetics of phenytoin, *J. Pharmacokinet. Biopharm.*, 5:579–596, 1977.
35. Yuen, G. J. et al., Predicting phenytoin dosages using Bayesian feedback: a comparison with other methods, *Ther. Drug Monit.*, 5:437–441, 1983.

36. Sheiner, L. B. and Beal, S. L., Evaluation of methods for estimating population pharmacokinetics parameters, I. Michaelis-Menten model: routine clinical pharmacokinetic pharmacokinetic data, *J. Pharmacokinet. Biopharm.*, 8:553–571, 1980.

Section II

---

*Industrial and Regulatory
Applications*

2 PK/PD Approach

Honghui Zhou

CONTENTS

INTRODUCTION

The classical analyses of dose–response data, in many instances, fail to identify the minimum effective dose of a new drug. To individualize doses further for drugs with high intersubject pharmacokinetic (PK) variability, the physician needs additional information about factors that might contribute to intersubject variability in the dose–response relationship and thus influence clinical response to a drug, such as patients' demographics (age, sex, weight, ethnic origin, renal or hepatic function, genomics, etc.), concurrent diseases, comedication, etc. Pharmacodynamics (PD), the science that relates the time course of drug concentration to the time course of pharmacological effects in humans and animals, has recently gained the attention of increasing numbers of clinical pharmacologists and scientists.

For a long time, PK and PD had been considered separate disciplines of pharmacology. PK described the quantitative relationship of concentration–time profiles in different body fluids — plasma, serum, blood, urine, saliva, and cerebral spinal fluid (CSF), etc. — whereas PD quantified the characterization of the intensity of effects resulting from certain drug concentrations at the hypothetical effect site. PK/PD modeling bridges these two once disassociated branches of pharmacology. An integration of PK/PD models should be able to permit predictions of the effects of dose, desired dosing rate, and time on the temporal pattern of drug effects.

The utilization of PK/PD modeling in all phases of drug development to answer questions is critical to the future of a drug candidate. Thus, the use of PK/PD modeling should be viewed in the context of drug development as a question-driven process.

PK/PD EVOLUTION HISTORY

The earliest attempts to characterize the PK/PD (dose–concentration–effect) relationships were mostly limited to those drugs with direct correlation between observed effect and measured concentrations.[1-4]

When the concentration–time courses and the concentration–effect relationships of these drugs were known, it was often possible to predict the temporal pattern of their pharmacological effects.[5] It had long been believed that there is no relationship between the drug concentration in plasma and time course of action for many drugs, predominantly based on the observation that the pharmacological effect of many drugs lags behind their concentration in plasma. The apparent dissociation between drug concentration and effect was first overcome by Sheiner and co-workers[6,7] based on the concept of Segre[8] who proposed to use a hypothetical effect compartment to account for the lag between concentration and response. This approach led to a collapse of the hysteresis loop for drugs with a temporal delay between effect and plasma concentration in the effect compartment. Since then, the PK/PD modeling has exponentially evolved as a research area with increasing importance and dedication in academia, industry, and regulatory authorities.

Indirect response models are models for drugs whose mechanism of action consists of either inhibition or stimulation of a physiological process involved in the elaboration of the clinical expression of the observed effect. Thus, an indirect response model is also associated with hysteresis between plasma concentration and effect. However, if the mechanism is at least partially known and understood, the link between PK and PD data can be broken down into discrete physiologically based parts, which moves modeling from abstract number crunching to a physiological mechanism-oriented endeavor.[9]

PK/PD CONCEPTS AND PERSPECTIVES

In the field of PD, there are various theories and technical approaches to correlate the time course of pharmacological effects with plasma drug concentrations. This chapter first gives a brief overview of the current mainstay in PK/PD modeling. Nevertheless, since PK/PD modeling has first been developed and is most frequently used for reversibly acting drugs with continuous effect data, this chapter largely focuses on this kind of PK/PD analysis.

DOSE–RESPONSE VS. CONCENTRATION–RESPONSE

One of the primary objectives of clinical dosing-ranging studies is to determine the optimal effective dose based on dose–response relationship. For many drugs, the dose–response relationship can be described by a hyperbolic (sigmoid) curve. Mathematically, the most general PD model describing this kind of curve is the sigmoid E_{max} model, which can be readily derived from the well-known receptor theory in biochemistry. The potency of a drug can be defined as the dose producing one half of the maximal effect (ED_{50}). Ideally, a drug should be effective at doses lower than those that cause serious toxicity. The elucidation of dose–response relationships is feasible for a number of drug classes, including biological response modifiers, anticancer drugs, antihypertensives, metabolic/endocrine drugs, prostaglandin analogues, H_2-receptor antagonists, pulmonary drugs, and immunosuppressors/immunomodulators.

However, there appear to be fewer than the expected number of successful trials really showing an adequate dose–response relationship.[10] A key hindrance of dose-oriented therapy for certain drugs is that it does not ensure consistent drug exposure control among patients. Many factors may contribute to the inharmonious drug exposure among patients, such as intersubject variability in PK or PD, poor compliance, limitations of study design, method of analysis, etc.

POPULATION PK/PD VS. INDIVIDUAL PK/PD

The term *population approach* is often employed to indicate a paradigm that attempts to define important PK and PD differences among population subgroups during drug development. The application of data analysis methods to extract this information from relatively complex data sets is often referred to as a population-style analysis.[11] The ultimate goal of PK/PD analysis is to provide information that can be used to establish guidelines for individualizing drug dosage regimens. In addition to quantifying the means and variances of PK/PD model parameters, such as E_{max} (maximum response) and EC_{50} (concentration producing one half of the maximal effect), another goal of the population PK/PD analysis is to investigate the factors, or covariates, that may lead to differences among individuals and among subgroups of the population with respect to the therapeutic and toxic effects of the drugs. These factors usually include (but are not limited to) age, weight, sex, concomitant medications, disease state, polymorphic phenotypes, etc. The population PK/PD analysis can be applied to each phase of drug development.

The population PK/PD analysis, rather than individual PK/PD analysis, allows for reliable data analysis following sparse sampling, correctly accounts for sources of variability, allows for pooling of data across individuals and studies, and provides for an integrated model for PK/PD data (even across species). However, data analysis should be driven by the question of interest, rather than the method of analysis. It is not always necessary to use complicated methods of analysis. Sometimes a simpler individual PK/PD model approach can be used if the PK/PD data collected are relatively rich in a quite homogeneous population (especially in Phases I and IIa settings).

Mechanism-Based and Empirical PK/PD Models

The effect compartment model has successfully established the link for the temporal dissociation of PK and PD for a number of drugs. However, it is viewed as a "black-box" method compared with the mechanism-based PK/PD methods, such as an indirect response model. A physiologically based indirect response model was first used to describe the anticoagulant effect of warfarin by relating changes in pro-thrombin time, which are an apparently indirect effect of warfarin, with the synthesis rate of prothrombin complex activity, which was shown to be a direct effect of warfarin plasma concentrations.[12] Based on this concept, Dayneka et al.[13] elaborated a basic indirect response model, in which the measured response variable was governed by an input or production process (rate constant k_{in}) and an output or degradation process (rate constant k_{out}). The changes in response (R) per time are described by Equation 2.1,

$$dR/dt = k_{in} - k_{out} \times R \qquad (2.1)$$

Depending on whether k_{in} or k_{out} is either inhibited or stimulated by the drug, four different submodels were deduced, in which the drug effect is mediated by a modified E_{max} model. The resulting Equations 2.2 through 2.5 follow:

- inhibition of k_{in}:

$$dR/dt = k_{in} \times (1 - C/(EC_{50} + C)) - k_{out} \times R \qquad (2.2)$$

- inhibition of k_{out}:

$$dR/dt = k_{in} - k_{out} \times (1 - C/(EC_{50} + C)) \times R \qquad (2.3)$$

- stimulation of k_{in}:

$$dR/dt = k_{in} \times (1 + E_{max} \times C/(EC_{50} + C)) - k_{out} \times R \qquad (2.4)$$

- stimulation of k_{out}:

$$dR/dt = k_{in} - k_{out} \times (1 + E_{max} \times C/(EC_{50} + C)) \times R \qquad (2.5)$$

where C is the concentration of the drug, E_{max} the maximum effect, and EC_{50} the concentration that produces half of the maximum effect.

The concept of physiologically based indirect response models has been widely used in PK/PD modeling as far as the underlying physiological mechanism of drug action and observed effect are known.[14] However, such an approach would definitely require more detailed information about the involved indirect response mechanism to provide valid, physiologically meaningful results when applied to clinical data.

PK/PD models are always a simplification of the real physiological process. They either might be purely descriptive by simply combining the observed time course of concentration and effect neglecting the underlying physiological mechanism involved or might be mechanism-based by appreciating the physiological events involved in the elaboration of the observed effect. Since predictive, not descriptive modeling is the ultimate goal of PK/PD analysis, mechanism-based models should be preferred as they not only describe the observations but also offer some insight into the underlying biological processes involved, and hence provide flexibility in extrapolating the models to other clinical situations.

Indirect response (IDR) models may be relevant for (but not limited to) the following kinds of drugs:

- Anticoagulant agents[15,16]
- Corticosteroids[17,18]
- H_2-Antagonists[19]
- Antipyretic agents[20–22]
- 5α-Reductase inhibitors[23–25]
- Aldose reductase inhibitors[26]
- LH-RH antagonists[27,28]
- Thromboxane synthase inhibitors[29]
- Cholinesterase inhibitors[30]
- Diuretics[31]
- GABA inhibitors[32]
- Dopamine antagonists[33,34]
- Calcimimetic agent[35]
- Antihuman CD4 monoclonal antibody of immunoglobulin G4 (IgG4)[36]
- β-Agonists (hypokalemia)[37]
- Recombinant interleukin-2 (IL-2)[38]
- Recombinant interleukin-10 (IL-10)[39]
- GH-releasing peptide[40]

PK/PD MODEL CLASSIFICATION

Ideally, concentrations should be measured at the effect site, the site of action, or biophase, where the interaction with the respective biological receptor system takes place. Thus, concentrations in easily accessible body fluids like plasma or blood are frequently used to establish these relationships under the assumption that the pharmacologically active, unbound drug concentration at the effect site is directly related to the one in the respective body fluid and they are, under PK steady-state plasma concentrations, in equilibrium.

The PK/PD models can be classified by PK steady-state conditions into two types: PK/PD models for steady-state conditions and those for non-steady-state conditions.

PK/PD Models for Steady-State Situations

For steady-state conditions, the most commonly used PK/PD models are listed as follows:

Fixed Effect PK/PD Model

A fixed effect model, also known as a quantal effect model, relates a certain drug concentration with the statistical likelihood of a predefined, fixed effect to be present or absent.

The simplest fixed effect model is a threshold model, where the effect E_{fixed} occurs after reaching a certain drug threshold concentration $C_{threshold}$,

$$E = E_{fixed} \text{ if } C \geq C_{threshold} \tag{2.6}$$

where E is the measured effect and C is the measured drug concentration.

Since the threshold concentration will vary among patients, the probability of the effect to be present at a certain concentration will be a function of the threshold concentration distribution in the population. This approach may be useful in the clinical setting as an approximation of dose–response relationships but has major limitations for the prediction of complete effect–time profiles.[41]

Linear PK/PD Model

The linear model assumes drug concentration is directly proportional to the drug effect observed, as shown in the following equation:

$$E = E_0 + b \times C \tag{2.7}$$

where E_0 is the baseline effect in the absence of drug and b is a slope.[42]

Log-Linear PK/PD Model

A much more common situation than the linear model shown above is the log-linear model with the following expression:

$$E = a + b \times \log C \tag{2.8}$$

where a and b are the *intercept* and *slope*, respectively, in a plot of effect E vs. the logarithm of the drug concentration C. The symbol b is an empirical constant that has no physiological meaning, especially not that of a baseline value.

The log-linear model can be considered a special case of the E_{max} model, regarding the range between 20 and 80% of E_{max}, where effect E and logarithm of the concentration C follow a linear relationship. This model was successfully used to model the relationship of the synthesis rate of prothrombin complex activity to the plasma concentration of warfarin.[43]

The linear and log-linear models are often applied approximations for the description of concentration–effect relationships in human studies when (1) the occurrence of side effects prevents the administration of higher doses or (2) the drug has a long half-life, resulting in a narrow concentration range being studied. It needs to be kept in mind that the linear and log-linear models are useful only for interpolation, but not extrapolation.

E_{max} PK/PD Model

E_{max} model relates concentration C and effect E as

$$E = E_{max} \times C / (EC_{50} + C) \tag{2.9}$$

where E_{max} is the maximum effect possible and EC_{50} is the concentration that results in 50% of E_{max}. E_{max} refers to the intrinsic activity of a drug. EC_{50} refers to its potency. This equation can be derived from the equilibrium interaction of a drug with its site of action, e.g., a receptor, enzyme, or ion channel, producing the effect E.

The E_{max} model describes the concentration–effect relationship over a wide range of concentrations from zero effect in the absence of a drug to the maximum effect at concentrations much higher than EC_{50}. The E_{max} model successfully described the relationship between propranolol plasma concentrations and decrease in heart rate.[44]

In the presence of a baseline effect E_0, Equation 2.10 results.

$$E = E_0 + (E_{max} \times C)/(EC_{50} + C) \tag{2.10}$$

The E_{max} model presented above assumes an increase of the effect with increasing concentrations. On the contrary, inhibitory effects can be described by the following equation:

$$E = E_0 - (E_{max} \times C)/(EC_{50} + C)$$

Sigmoid E_{max} PK/PD Model

The sigmoid E_{max} model is an expansion of the E_{max} model. Effect (E) relates to the concentration (C) as follows,

$$E = E_{max} \times C^{\gamma}/(EC_{50}^{\gamma} + C^{\gamma}) \tag{2.11}$$

This relationship can be theoretically described based on the interaction between γ drug molecules and one common interaction site. However, in most cases γ only serves as a shaping factor to allow for a better data fit. Therefore, γ is not necessarily an integer value. The steepness of the concentration–effect curve depends on the magnitude of γ; the larger γ, the steeper the linear phase of the log-concentration–effect curve. The slope of the linear portion of the log-concentration–effect is equal to $\gamma \times E_{max}/4$. The E_{max} model can be considered a special case of the sigmoid E_{max} model with $\gamma = 1$. The sigmoid E_{max} model best fitted the concentration–EEG effect relationship of the intravenous anesthetic ketamine and its two enantiomers.[45]

PK/PD Models for Non-Steady-State Situations

Under non-steady-state conditions, time courses of plasma concentration and effect may dissociate. Thus, to characterize fully the time course of drug action under non-steady-state conditions, PK and PD have to be adequately linked to predict the relationship of PD effect vs. drug concentration in plasma. This link is provided by the following four attributes in the integrated PK/PD models.

Direct Link vs. Indirect Link

If equilibration between the drug concentrations in plasma and at the effect site occurs rapidly enough so that the temporal delay is negligible, then the concentrations can be directly proportional to each other at any time despite the non-steady-state conditions. In such situations, effect and plasma concentration lack any detectable hysteresis and may be directly linked as under steady-state conditions. However, in many cases, the concentration at the site of action may lag behind that in plasma; therefore, no direct link can be established. This is usually manifested by a pronounced counterclockwise hysteresis between plasma concentration and effect. The extent of hysteresis is dependent on the degree of delay between the concentrations in plasma and at the effect site. This dissociation between the time courses of plasma concentration and effect can be resolved by introduction of a hypothetical effect compartment representing active drug concentration at the effect site. This is realized by linking the effect compartment to the kinetic model by a first-order process with negligible mass of drug into the effect compartment.

Although the effect compartment concept achieved wide popularity in the PK/PD modeling field, it is not suitable for a fairly large number of different pharmacological effects, and it does not help to elucidate some underlined mechanisms of pharmacological effects.

Direct Response vs. Indirect Response

Direct response models are characterized by a direct correlation between the effect site concentration of the drug and the observed effect without time lag. Direct response models can comprise either a directly linked direct response model[46] or an indirectly linked indirect response model.[47] Indirect response models are models for drugs whose mechanism of action consists of either inhibition or stimulation of a physiological process involved in the elucidation of the clinical expression of the observed effect, as discussed in detail in the previous section.

Soft Link vs. Hard Link

In soft-link models, both observed PK and PD data are used to determine the characteristics of the link between them. The link then serves as a buffer and accounts for any misspecification between the PK and PD relationships describing the respective data sets. In contrast, in hard-link models, the PD data are not used in the characterization of the model. Instead, the PK data are combined with additional information from *in vitro* studies like binding affinities to receptors, enzymes, or ion channels to allow true predictions of the PD data. The additional *in vitro* information determines the link between the PK and PD data.[5]

Time Variant vs. Time Invariant

Most of the PK/PD models assume time invariant, i.e., the PD parameters, e.g., E_{max} and EC_{50} stay constant over time. However, when time-variant PD occurs, a specific model for the involved process of either tolerance or sensitization is required. PD tolerance is defined as a decrease in drug effect over time despite constant drug concentrations at the effect site and is characterized by a clockwise hysteresis loop in a plot of effect vs. drug concentration. In an E_{max} model, tolerance is usually modeled as a time-dependent decrease of E_{max} if receptor downregulation is involved, or as a time-dependent decrease of EC_{50} if receptor desensitization is assumed. PD tolerance may also occur due to accumulated and antagonistic acting drug metabolites or negative physiological feedback.[5,48]

Sensitization is defined as the opposite effect of tolerance, an increase in drug effect over time despite constant drug concentrations at the effect site, which is manifested by a counterclockwise hysteresis loop in a plot of effect vs. concentration.

SURROGATE MARKER, BIOMARKER SELECTION, APPLICATION, AND VALIDATION

DEFINITIONS

- Clinical outcome: The clinical outcome is the ultimate efficacy measure quantifying the direct benefit to a patient, for example, cure or decreased morbidity; it is, however, often difficult to quantify. Instead, clinical outcomes are often predicted from surrogate markers that can be observed readily and early in drug candidate process and can be quantified easily.
- Biomarker and surrogate marker: Biomarkers are measurable physiological or biochemical parameters that reflect some PD activity of the investigated drug, even if they are not directly related to clinical outcome. They may be useful to obtain insight into the overall PK/PD behavior of the compound of interest. However, only if changes in biomarkers are predictive of clinical outcome are they considered surrogate markers. Before surrogate end points can be used to predict outcomes, it is crucial to validate them in accordance with Good Clinical Practice (GCP) and Good Laboratory Practice (GLP) methods. This must include the proof of relevance, i.e., that the changes in the marker are correlated to changes in disease state or outcome, as well as criteria similar to those used in analytical method validation, namely, repeatability, reproducibility, and sensitivity.

SURROGATE MARKERS IN DIFFERENT STAGES OF DRUG DEVELOPMENT PROCESSES

- Preclinical studies: Both beneficial and adverse PD markers would be identified from anticipated efficacy end points and side effects, respectively, in the animal pharmacological and toxicological studies.
- Phase I/II clinical studies: The first-time-in-human PK and tolerability study likely will use the widest dose range and the highest dose ever in the clinical development. Therefore, it provides the unique opportunity to assess the clinical PK/PD data. A list of candidate surrogate markers both from preclinical studies or previous clinical experience should be established at this stage. The PK/PD information from the placebo controls should also be obtained because it is invaluable for future development. In addition, the interspecies differences in PK/PD responses should also be recognized. During Phase II development (especially proof-of-principle in Phase IIa), careful evaluation of PK/PD responses in both healthy subjects and patients should be performed to detect differences, if any.
- Phase II/III clinical studies: The PK/PD model(s) in population studies during Phase II/III will be tested to assess variation due to differences from genetics, demographics (age, sex, ethnicity, etc.), diet, smoking, alcohol intake, disease conditions, concomitant drugs, hepatic or renal insufficiency, and other altered metabolic conditions. Table 2.1 represents a list of some potential biomarkers used in the clinical trials.

As surrogate markers evolve from proof-of-concept to prediction of outcomes, validation takes on increasing importance for both regulatory authorities and sponsoring companies. If the clinical experiences support the theoretical and preclinical experiences, the marker continues to be a valid marker. Differences between markers used for decision making and those used for regulatory approval must be acknowledged and delineated. When surrogate markers are related to the therapeutic and toxic end points, the decision-making process in drug development can be facilitated by PK/PD modeling.

The Food and Drug Administration (FDA) may grant marketing approval for a new chemical entity on the basis of adequate and well-controlled clinical trials establishing that the drug product has an effect on a surrogate end point that is reasonably likely, based on epidemiological, therapeutic, pathophysiological, or other evidence, to predict clinical benefit or on the basis of an effect on a clinical end point other than survival or irreversible morbidity.[49]

PK/PD APPLICATIONS IN DIFFERENT THERAPEUTIC AREAS

Some examples of a general characterization of types of PK/PD relationships in different therapeutic areas are summarized below.

- Reversible PK/PD model: Central nervous system (CNS), cardiovascular, muscle relaxation, and enzyme activity
- Indirect PK/PD model: Synthesis, secretion, mediator flux, cell trafficking, and enzyme inductions
- Irreversible PK/PD model: Chemotherapeutic, enzyme inactivation, reactive metabolites, and carcinogenicity

CENTRAL NERVOUS SYSTEM

Despite extensive research, there are major experimental design difficulties in quantitatively correlating the time course relationship between drug plasma concentrations and PD effects of CNS agents. Suitable PD effects, which should possess most of the following characteristics, are listed below:

- The concentration–effect relationship is curvelinear.
- The effects are reproducible within and between subjects and repeatedly over time.
- The effects are amenable to careful experimental design to control for tolerance or learning effects.
- The PD measure is meaningful either as an indicator of therapeutic effect and/or toxic effects.

Benzodiazepines have been extensively used in the treatment of anxiety. However, the untoward CNS impairment effects of benzodiazepines are much more readily quantified using cognitive neuromotor testing than measures of anxiolytic effects due to lack of effective PD markers. Psychometric tests (hand–eye coordination, memory, body sway, and psychomotor speed), anesthetic effects (sleeping time), eye movement, and neurophysiological effects have been used as the PD end points to model the PK/PD relationships of benzodiazepines.

The time course changes in the CNS impairment on psychomotor and cognitive skills related changes to the plasma lorazepam concentrations in a PK/PD model in healthy subjects. The PD end points included computerized continuous tracking, body sway with eyes opened, and digit symbol substitution (DSS) tests.[50]

Our knowledge of PD in epilepsy is limited partially due to lack of suitable measures of anticonvulsant effect intensity. Because of the intermittent nature of the clinical manifestations of epilepsy, studies on the PD of antiepileptic drugs in humans are extremely difficult. The following measures have been used as PD markers in PK/PD modeling in variety of antiepileptic drugs: effects on the frequency of interictal spikes in the electroencephalogram (EEG) (midazolam,[51] phenytoin,[52] and lamotrigine[53]), effects on seizure activity induced by photic stimulation (loreclezole[54] and β-carboline ZK 95962[55]), and effects on hyperventilation-induced EEG changes.[56]

Several complicating factors could affect the determination of PK/PD relationships for analgesics. These factors include technical challenges in measuring pain, the episodic nature of pain, large intersubject variability, tolerance development, existence of active metabolite (agonistic or antagonistic), presence of endogenous analgesic, etc. The three most commonly applied approaches to assess pain intensity

TABLE 2.1
List of Potential Biomarkers

Drug Class	Desirable Markers	Undesirable Markers
Asthma	Serum IgE, T-cell markers, non-T-cell markers	
Angiotensin II antagonists	Angiotensin II, angiotensin I, PRA, aldosterone, urine Na^+/K^+, ACE	
ACE inhibitors	Plasma and tissue ACE, angiotensin II, PRA, angiotensin I, urine Na^+/K^+, aldosterone	
Neutral endopeptidase inhibitors	Atrial natriuretic peptide, cGMP, aldosterone, urine Na^+/K^+	
HMG-CoA reductase inhibitors	Cholesterol, HMG-CoA reductase, triglycerides, LDL, HDL, HDL-apolipoprotein, cortisol, ACTH, testosterone, urine 17-OH-corticosteroids	Bilirubin, AGT, AST
Neurodengenerative disease therapeutics	Acetylcholinesterase, MAOa, MAOb, ACTH, cGMP, phenylethylamine, catecholamines, neuroimaging (MRI, MR spectroscopy, magnetization transfer imaging, T2 relaxation analysis, SPECT, PET, fMRI)	
Mood disorders	FMRI, sleep electroencephalograph, 24-h cortisol patterns, dexamethasone suppression tests, ACTH	
Antidiabetics	Insulin, glucose, C-peptide, cAMP, proinsulin, amylin, glucagon, lactate	Anti-insulin
Alcohol and drug abuse disorders	Computerized cognitive testing, glucose metabolism by PET, CBF by PET and fMRI	
Steroid hormones	Estradiol, estrone, FSH, GnRH, LHRH, LH	
Anti-inflammatory agents	Glucocorticoids, histamine, glucagon, cortisol, leukotrienes	

Immunosuppressants	TNFα, IL-1α, IL-2β, eicosanoids, creatinine, IgE, β2-microglobulin	Antibodies
Anticoagulants	Coagulation cofactors, other cofactors, clotting time	
Anti-HIV/AIDS	Viral burden/PCR, CD-4/CD-8, HIV Ab titer, p-24 antigen, neopterin	Cell toxicity
H_2 antagonists, proton pump inhibitors	Gastric acidity	
H_1 antagonists	Skin wheal and flare	Sedation
Antiglaucoma	Intraocular pressure	
Oncology		Thrombocytopenia, leukopenia, neutropenia

Abbreviations:

IgE = immunoglobulin E; PRA = plasma renin activity; ACE = angiotensin-converting enzyme; AGT = alanine glyoxylate aminotransferase; AST = aspartate aminotransferase; LDL = low-density lipoprotein cholesterol; HDL = high-density lipoprotein cholesterol; ACTH = adrenocorticotropin; MAO = monoamine oxidases; cGMP = cyclic guanosine 3', 5'-monophosphate; MRI = magnetic resonance imaging; SPECT = single photon emission computed tomography; PET = positron emission tomography; fMRI = functional magnetic resonance imaging; FMRI = ^{19}F-magnetic resonance imaging; cAMP = cyclic adenosine monophosphate; CBF = cerebral blood flow; FSH = follicle-stimulating hormone; GnRH = gonadotropin-releasing hormone; LHRH = luteinizing hormone-releasing hormone; LH = luteinizing hormone; TNFα = tumor necrosis factor-α; IL-1α = interleukin-1α; IL-1β = interleukin-1β; PCR = polymerase chain reaction; CD-4 = cluster of differentiation-4; CD-8 = cluster of differentiation-8; HIV Ab = human immunodeficiency virus antibody; AIDS = acquired immunodeficiency syndrome.

are the visual analogue scale (VAS), the verbal rating scale, and the numerical rating scale. VAS scales have been found to be reliable, feasible, and sensitive. "Pain" can be regarded as a quantal (all-or-none) response, but "pain relief" may be a graded response because it can be described on a continuous scale. In spite of the challenges, many successful PK/PD models have been developed for analgesics. The relationship between morphine plasma concentration and analgesic effect (a cumulative frequency of patients attaining a certain comfort score) in neonates was successfully modeled.[57] The pharmacological responses (pain relief, sedation, and pupillary size) were linked to methadone plasma concentration in patients with cancer pain by PK/PD modeling.[58] The slowing of the EEG by fentanyl was related to the plasma concentrations by using PK/PD modeling with a sigmoid E_{max} model that incorporates the hypothetical effect compartment.[59] Using PK/PD modeling and subsequent simulations, optimum dosing regimens have been designed for fentanyl, alfentanil, and sufentanil.[60]

CARDIOVASCULAR

Heart rate reduction (bradycardia) at rest and during exercise has been considered an important mechanism for the antianginal and anti-ischemic effects of β-adrenergic blocking agents and some calcium channel blockers. However, these agents may have side effects, such as negative inotropic and hypotensive effects. Ivabradine is a selective bradycardic effect without vasodilatory or inotropic side effects. Its active metabolite has the similar pharmacological effect. After oral administration of ivabradine, an indirect relationship (a counterclockwise hysteresis loop) between the bradycardiac effect and plasma concentration of both parent compound and its active metabolite was observed. A population PK/PD model (a link model with an effect compartment) showed that the active metabolite of ivabradine contributes in part to the overall activity of ivabradine. The active metabolite is responsible for the initial bradycardiac effect, whereas the parent compound is responsible for the duration of action.[61]

Angiotensin-converting enzyme (ACE) inhibitors (e.g., captopril) are effective agents in treating hypertension and congestive heart failure. ACE not only converts angiotensin I into angiotensin II but also inactivates bradykinin, the latter of which is a potent vasodilator causing the release of nitric oxide (NO), etc. In the human, dorsal hand veins have been shown to be responsive to angiotensin I and II as well as to bradykinin. The dorsal hand vein technique was therefore used to investigate the effect of oral single-dose captopril or enalaprilat infusion on bradykinin-induced venodilation in the dorsal hand vein preconstricted with phenylephrine.[62] Dose–response curves to bradykinin were constructed after preconstriction of the vein with phenylephrine by infusion of the bradykinin at increasing infusion rates. A second dose–response curve to bradykinin was repeated after ACE inhibition. The dose–response curves to bradykinin with or without ACE inhibitors were described by a four-parameter logistic equation, with which the E_{max} and ED_{50} could be estimated. The PK/PD (dose–response) model by using the dorsal hand vein technique indicated that the sensitivity to bradykinin after ACE inhibition in human veins should be augmented.

NO, an important mediator of vasodilatation and inhibition of platelet aggregation via increased formation of cyclic GMP, is synthesized in the vascular endothelium from the terminal guanidino nitrogen of L-arginine (L-Arg) by a constitutive isoform of NO synthase (NOS). The inhibition of NOS by NOS inhibitors, such as N^G-monomethyl-L-arginine (L-NMMA), can result in changes in hemodynamic responses. The relationship of plasma levels of L-NMMA with changes in PD end points (cardiac output, pulsatile blood flow, and exhaled NO) could be satisfactorily described by direct E_{max} models. Based on the PK/PD models, L-NMMA could significantly decrease cardiac output, exhaled NO, and pulsatile blood flow in the absence of a systemic hypertensive response. The EC_{50} value was equivalent for exhaled NO and cardiac output, but markedly higher for pulsatile blood flow.[63]

Because L-Arg is the physiological precursor of NO, the relationship of the temporal pattern of L-Arg plasma concentration and that of its vasodilator effect, i.e., the reduction in total peripheral resistance was investigated. The PD effect was directly linked to L-Arg plasma concentration by a linear effect model with lack of hysteresis loop. This result suggests that L-Arg has a direct vasodilator action within the vasculature.[64]

ONCOLOGY

Most of the acute toxicity of cancer chemotherapy is due to growth inhibition of rapidly dividing normal tissues, such as the bone marrow and gastrointestinal (GI) mucosa. To date, most clinical PD studies in oncology have focused on modeling the extent of myelosuppression.[65] Myelosuppression (thrombocytopenia, leukopenia, neutropenia, etc.) is easily quantified, occurs at predictable times, and is usually readily reversible. Clinical oncologists face the dilemma that levels of systemic exposure necessary to achieve more effective antitumor therapy are often similar to those associated with greater toxicity. The goal is to define a level of systemic exposure that maximizes the likelihood of therapeutic response while producing minimal (acceptable) toxicity, then use PK principles to select a patient-specific dosage that will yield a level of systemic exposure within this narrow range.

The PD parameters (e.g., nadir count or the surviving fraction, i.e., the nadir count divided by the pretreatment count) are usually correlated to PK parameters, such as the integral of total drug exposure over time (total area under the concentration–time curve, or AUC), steady-state plasma concentrations during continuous infusion (C_{ss}), or the time above the threshold concentration using linear, log-linear, or E_{max} models. Table 2.2 lists PD studies of hematological toxicity.

Significant relationships between the PK (e.g., C_{ss} or AUC) of selected anticancer drugs and their clinical effects (e.g., toxicity) have been demonstrated. These data suggest that individualized dosages of certain anticancer drugs, based on PK/PD, could increase the probability of response, decrease the probability of toxicity, or both.

Several models have been developed to describe the time course of leukopenia after chemotherapy, but their structures were not mechanistic. Dose instead of a concentration–time curve was used as input to their models. Therefore, the PK variability was ignored during the model development.

TABLE 2.2
PD Studies of Hematological Toxicity or GI Toxicity

Agent	PK Parameter	Hematological Toxicity
Carboplatin	AUC	Thrombocytopenia
	AUC	Leukopenia
Doxorubicin	C_{ss}	Leukopenia
Etoposide	C_{ss}	Leukopenia
Fluorouracil	C_{ss}	Leukopenia
Iproplatin	AUC	Thrombocytopenia
Hexamethylene bisacetamide	C_{ss}/AUC	Thrombocytopenia/leukopenia
Menogaril	AUC	Leukopenia/neutropenia
SAM486A	AUC	Neutropenia
Teniposide	Free AUC	Leukopenia
	C_{ss}	Mucositis
ThioTEPA	AUC	Leukopenia/thrombocytopenia/neutropenia)
Thymidine	C_{ss}	Thrombocytopenia
Trimetrexate	AUC/C_{ss}	Thrombocytopenia
Vinblastine	C_{ss}	Leukopenia

More mechanistically modeling efforts have recently been focused on the prediction of the entire time course of myelosuppression.[66] A physiological indirect-response PD model was developed to describe the entire time course of leukopenia or neutropenia after treatment with anticancer drugs (paclitaxel and etoposide). The model successfully related the drug concentration vs. time curve to the time course of leukopenia. The model consisted of two compartments corresponding to leukocyte pools in bone marrow and peripheral blood, because anticancer drugs work on myeloid cells in bone marrow, whereas leukocyte counts are measured in peripheral blood. A concentration–time curve (in terms of PK model parameters) of an anticancer drug was used as input to the model. It was assumed that only mitotic components consisting of myeloblasts to myelocytes are sensitive to chemotherapeutic agents. It was hypothesized that the inhibition of leukocyte production in bone marrow was controlled by the exposure of myeloid cells to anticancer drugs when they are in the sensitive cell stages. With the model mechanistically developed, the entire time course of leukopenia or neutropenia after chemotherapy with anticancer drugs could be successfully described, and it may provide a platform for physiological analysis of PD of anticancer drugs.

GASTROINTESTINAL

It is believed that the presence of a high intragastric concentration of H[+] ions and pepsin appear to be a prerequisite for the development of GI bleeding from stress ulcers. Therefore, a reduction of gastric acidity may prevent these lesions.[67]

Intravenous ranitidine is widely used as a prophylactic agent for stress ulcer-related bleeding in patients in intensive care units. The development of an integrated

PK/PD model for H_2-receptor blockers can certainly establish dosage guidelines for this indication. It has been recognized that there is presence of a delay between drug concentration of H_2-receptor blockers and the effect on intragastric pH. Both the effect compartment model and the indirect response model have been used to describe the delay.[68] The effect compartment approach assumes that the delay is due to the slow distribution of the drug to the effect site, whereas the indirect response model assumes the delay may be largely determined by a gradually developing physiological effect.

Theoretically, for acid inhibitory agents, the delay between concentration and onset of drug effect could be dependent on the acid present in the stomach at the time of initiation of therapy and the transport of gastric juice to the duodenum. Therefore, the effect of acid inhibitory agents on gastric pH may be characterized more appropriately by a physiological indirect response PK/PD model rather than an empirical effect compartment model. Both models have been challenged by both single and multiple doses of ranitidine with either an intermittent bolus or a continuous infusion scheme. The physiological indirect response model seems to be more meaningful because of its direct relevance to the physiology of gastric acid production and elimination, its ability to predict multiple-dose PK/PD data, its better quantification of tolerance development, and *in vitro* and *in vivo* correlation.

SELECTED EXAMPLES OF PK/PD APPLICATION IN DRUG DEVELOPMENT

Example 1

Octreotide acetate is a synthetic octapeptide that suppresses secretion of peptide hormones and neurotransmitters in a fashion similar to the endogenous somatostatins. Subcutaneous injection of octreotide acetate leads to symptomatic control in patients with acromegaly and some gastro-entero-pancreatic tumors. However, to be fully effective, a regimen of three times daily dosing is warranted. To improve patient convenience and compliance, octreotide acetate has been reformulated into microspheres of a biodegradable polymer (poly(DL-lactidecoglycolide) D-(+) glucose). Drug release occurs slowly as cleavage of the polymer ester linkage takes place primarily through tissue fluid hydrolysis. This formulation allows for a single monthly intragluteal injection with sustained release of octreotide.

Following a single intragluteal intramuscular injection of microencapsulated octreotide acetate in healthy cholecystectomized subjects, the pharmacokinetic profile of octreotide was characterized by an initial peak of octreotide followed by a sustained release of drug. Plateau concentration was sustained up to day 70, and gradually declined to below the detection limit by day 112. A population covariate pharmacokinetic model was constructed that consisted of two absorption processes, one for immediate-release portion from the surface, unencapsulated drug and another for sustained-release portion from the microencapsulated drug.[69]

It has been hypothesized that growth hormone (GH) may facilitate the progression of various disease states. These pathological effects may occur either by direct GH stimulation or indirectly by the elevation of insulin-like growth factor-1 (IGF-1). For more than 30 years, the importance of GH in the development and progression of diabetic retinopathy has been recognized. Recently, there was a growing interest in IGF-1 in proliferative diseases and the potential of altering the course of the disease by reducing IGF-1 concentrations. A population covariate PK/PD model (inhibitory E_{max} model with baseline) could adequately describe the inverse correlation between the IGF-1 concentration and octreotide concentration in healthy cholecystectomized subjects following a single intragluteal intramuscular injection of microencapsulated octreotide acetate.[69] Computer simulations based on the developed PK/PD model recommended the dosing regimen of 30 mg microencapsulated octreotide acetate intramuscularly every 4 weeks can achieve adequate suppression of IGF-1 concentration in at least 70% of the population, which is considered sufficient for therapeutic purposes.[69]

Example 2

SAM486A is a potent inhibitor of S-adenosylmethionine decarboxylase (SAMDC), which has been selected as potential therapeutic target for cancer chemotherapy. Three Phase I ascending-dose tolerability and pharmacokinetic studies with different dosing schedules have been performed in 112 patients with refractory malignancies.[70] Hematological toxicity manifested by dose-dependent neutropenia has been observed. Study 1 was an open-label, noncomparative, multicenter, between-patient, dose-escalating trial (39 patients). The starting dose, 3 mg/m^2 per cycle, was selected as approximately one third of the human equivalent of the no-adverse-effect level (0.06 mg/kg) in the most sensitive species, the rat, with daily dosing for 3 months. This corresponded to 5 days of continuous infusion (120 hours) of 0.6 mg/m^2 per day, repeated every 4 weeks. The dose was escalated to 700 mg/m^2/cycle based initially on a modified Fibonacci scheme. Later, the protocol was amended to permit a toxicity-driven escalation. The maximum tolerated dose (MTD) not associated with dose-limiting toxicity was defined as 400 mg/m^2/cycle. Study 2 was an open-label, nonrandomized, noncomparative, dose-escalating trial (50 patients). SAM486A was administered every week for 4 weeks, followed by a 2-week rest period or after recovery from the previous cycle, whichever came later. The dose regimens comprised 10-min to 3-h intravenous infusions once a week with doses ranging from 1.25 to 325 mg/m^2/week. MTD was defined as 270 mg/m^2/week. Study 3 was an open-label, noncomparative, single-center, dose-escalating trial (23 patients). SAM486A was given by a 1-h intravenous infusion daily for 5 days every 21 days. The starting dose for this trial was 3.6 mg/m^2/day. The dose was escalated up to 202.8 mg/m^2/day using a modified continuous reassessment method. MTD was defined as 102.4 mg/m^2/day.

The pharmacokinetics of SAM486A after intravenous infusions with different durations, dose ranges, and dosing regimens in these three Phase I studies was fully characterized using a population modeling approach.[70] The pharmacokinetic/toxicodynamic relationships between toxicodynamic effects (neutrophil count nadir, percent decrease in neutrophil count from baseline to nadir, or neutropenia grade) and pharmacokinetic predictors (maximum plasma concentration, cumulative AUC before the occurrence of neutrophil count nadir, or duration of exposure above threshold levels) were described by an inhibitory sigmoid E_{max} model, simple E_{max} model, or logistic regression, respectively. Cumulative AUC was found the best pharmacokinetic parameter to predict grade 4 and grades 3 + 4 neutropenia by logistic regression models. Based on the logistic regression models, the Study 3 dosing regimen of 100 mg/m² 1-h intravenous infusion daily for 5 days showed the lowest occurrence of grades 3 + 4 as well as grade 4 neutropenia with predicted probabilities of <0.4% and <0.2%, respectively. Therefore, this regimen was selected for further clinical trials.[70]

PK/PD AND COMPUTER-AIDED CLINICAL TRIAL DESIGN

Computer simulations have been widely applied and demonstrated to be of great value for the product development process in other than the pharmaceutical industries (e.g., aerospace, automobile). Computer-aided clinical trial design (CATD) has recently attracted more and more attention and recognition from the pharmaceutical industry, academia, and regulatory agencies as a means of achieving greater efficiency and improving the cost-effectiveness and timeliness of the drug development process. CATD not only aids in the identification of optimal dosage regimens, but it also facilitates evaluation of effective PK/PD models, placebo response, disease progression, adverse event development, and, ultimately, the expedition of the decision-making, regulatory review, and commercialization of new drug processes.

PK/PD models form the backbone of any mathematical model in CATD. It can be expected that the wide application of CATD will produce a distinct shift from empirical to mechanistic PK/PD modeling. PK/PD models obtained in CATD process not only will present concise but informative knowledge about the drug, but will also serve as a communication platform for discussions among members from the many scientific disciplines involved in a given drug development project.[71]

A number of input factors in CATD can be classified into two groups: controllable factors and uncontrollable factors. Some major controllable ("protocol") factors include:

- Number of subjects and type of patient populations
- Overall study design
- Dosing regimen
- Sampling schemes of PK/PD evaluation
- Inclusion/exclusion criteria

Some major uncontrollable factors include:

- PK/PD model
- Compliance, dropout
- Adherence to sampling schedules
- Residual errors in proposed PK/PD model (e.g., assay errors, model misspecification)
- Country, center effects

Once a valid population PK/PD model has been selected (the most time-consuming step in CATD), the simulation outcome of a clinical trial only depends on the specification of controllable design factors and on assumptions of uncontrollable design features. It is not a trivial task to successfully build simulation. It requires overall understanding of drug development, physiology, and clinical practice, knowledge of PK/PD, and access to adequate simulation software. Simulation unavoidably forces study designers to state their assumptions explicitly, thereby facilitating the understanding among members of a clinical development team and therefore moving to a more PK/PD-oriented rational drug development process.

SUMMARY

In summary, the following general considerations should be taken into account before embarking on PD studies.[72]

- Determination whether the pharmacological effect is direct, reversible, indirect, irreversible
- Selection of a pharmacological response that is clinically relevant
- Validation of the measurement of pharmacological response (or surrogate end points)
- Establishment of the number of subjects needed to ensure adequate statistical power
- Determination of the optimal design and sampling strategy
- Consideration of the complicating factors in PK/PD modeling[73] (e.g., lack of understanding of underlining mechanism, tolerance, rebound, diurnal variation, sensitization, active metabolite, protein binding, multiple receptors, pathophysiological changes, drug interactions, nonstationarity, enantiomer, transduction, etc.)
- Equilibration delays between sampling and effect compartment for non-steady-state investigations
- Baseline or placebo effects
- Presence of endogenous substances
- Selection of a predictive and parsimonious PK/PD model
- Optimal parameter estimation, statistical inference, and model predictions

FUTURE PERSPECTIVE OF PK/PD

As the regulatory climate changes and science evolves, PK/PD modeling may take on different forms, but its importance in drug development should not diminish. A thorough and sound understanding of the dose–concentration–effect relationship should prevent marketing a drug at doses later recognized to be inappropriate. The FDA has a major interest in encouraging good practices in conducting, analyzing, and reporting the results of PK/PD studies. Guidance for population PK/PD studies is currently being developed. It can be expected that more and more scientifically sound PK/PD studies serving a rigorous scientific framework for development will emerge in the near future.

REFERENCES

1. Levy, G., Gibaldi, M., and Jusko, W. J., Multicompartment pharmacokinetic models and pharmacologic effects, *J. Pharm. Sci.*, 58:422–424, 1969.
2. Levy, G., Kinetics of drug action in man, *Acta Pharmacol. Toxicol. Copenh.*, 29 (Suppl. 3):203–210, 1971.
3. Gibaldi, M., Levy, G., and Weintraub, H., Drug distribution and pharmacologic effects, *Clin. Pharmacol. Ther.*, 12:734–742, 1971.
4. Wagner, J., Kinetics of pharmacologic responses: I. Proposed relationship between response and drug concentration in the intact animal and man, *J. Theor. Biol.*, 20:173–201, 1968.
5. Meibohm, B. and Derendorf, H., Basic concepts of pharmacokinetic/pharmacodynamic (PK/PD) modeling, *Int. J. Clin. Pharmacol. Ther.*, 35(10):401–413, 1997.
6. Sheiner, L. B. et al., Simultaneous modeling of pharmacokinetics and pharmacodynamics: application to d-tubocurarine, *Clin. Pharmacol. Ther.*, 35:1–8, 1979.
7. Holford, N. H. G. and Sheiner, L. B., Kinetics of pharmacological response, *Pharmacol. Ther.*, 16:143–166, 1982.
8. Segre, G., Kinetics of interaction between drugs and biological systems, *Farmaco Ed. Sci.*, 23:907–918, 1968.
9. Levy, G., Mechanism-based pharmacodynamic modeling, *Clin. Pharmacol. Ther.*, 56:356–358, 1994.
10. Lieberman, R. and Nelson, R., Dose-response and concentration-response relationships: clinical and regulatory perspectives, *Ther. Drug Monit.*, 15:498–502, 1993.
11. Tracewell, W., Ludden, T., and Owen, J., Early clinical development: population pharmacokinetics and pharmacodynamics, *Appl. Clin. Trials*, October:28–35, 1997.
12. Nagashima, R., O'Reilly, R. A., and Levy, G., Kinetics of pharmacologic effects in man: the anticoagulant action of warfarin, *Clin. Pharmacol. Ther.*, 10:22–35, 1969.
13. Dayneka, N. L., Garg, V., and Jusko, W. J., Comparison of four basic models of indirect pharmacodynamic responses, *J. Pharmacokinet. Biopharm.*, 21:457–478, 1993.
14. Jusko, W. J. and Ko, H. C., Physiologic indirect response models characterize diverse types of pharmacodynamic effects, *Clin. Pharmacol. Ther.*, 56:406–419, 1994.
15. Piotrovsky, V. K., Indirect pharmacodynamic response model do not require any parametric pharmacokinetic model to be fitted to effect-time data, *Methods Find. Exp. Clin. Pharmacol.*, 19(10):723–729, 1997.
16. Chan, E. et al., Stereochemical agents of warfarin drug interactions: use of a combined pharmacokinetic-pharmacodynamic model, *Clin. Pharmacol. Ther.*, 56(3):286–294, 1994.
17. Krzydanski, W., Chakraborty, A., and Jusko, W. J., Algorithm for application of Fourier analysis for biorhythmic baseline of pharmacodynamic indirect models, *Chronobiol. Int.*, 17(1):77–93, 2000.
18. Sun, Y. N. et al., Fourth-generation model for cortiocosteroid pharmacodynamics: a model for methylprednisolone effects on receptor/gene-mediated glucocorticoid receptor down-regulation and tyrosine aminotransferase induction in rat liver, *J. Pharmacokinet. Biopharm.*, 26(3):289–317, 1998.
19. Mathot, R. A. and Geus, W. P., Pharmacodynamic modeling of the acid inhibitory effect of ranitidine in patients in an intensive care unit during prolonged dosing: characterization of tolerance, *Clin. Pharmacol. Ther.*, 66(2):140–151, 1999.
20. Flores-Murrieta, F. J. et al., Pharmacokinetic-pharmacodynamic modeling of tolmetin antinociceptive effect in the rat using an indirect response model: a population approach, *J. Pharmacokinet. Biopharm.*, 26(5):547–557, 1998.

21. Boni, J. et al., Pharmacokinetic and pharmacodynamic action of etodolac in patients after oral surgery, *J. Clin Pharmacol.*, 39(7):729–737, 1999.

22. Troconiz, I. F. et al., Pharmacokinetic-pharmacodynamic modeling of the antipyretic effect of two oral formulations of ibuprofen, *Clin. Pharmacokinet.*, 38(6):505–518, 2000.

23. Olsson Gisleskog, P. et al., A model for the turnover of dihydrotestosterone in the presence of the irreversible 5α-reductase inhibitors GI198745 and finasteride, *Clin. Pharmacol. Ther.*, 64:636–647, 1998.

24. Olsson Gisleskog, P. et al., Validation of a population pharmacokinetic/pharmacodynamic model for 5α-reductase inhibitors, *Eur. J. Pharm. Sci.*, 8:291–299, 1999.

25. Ko, H. C. and Jusko, W. J., Pharmacodynamic modeling of finasteride, a 5α-reductase inhibitor, *Pharmacotherapy*, 15(4):509–511, 1995.

26. Van Griensven, J. M. et al., Tolrestat pharmacokinetic and pharmacodynamic effects on red blood cell sorbitol levels in normal volunteers and in patients with insulin-dependent diabetes, *Clin. Pharmacol. Ther.*, 58(6):631–640, 1995.

27. Schwahn, M., Nagaraja, N. V., and Cohen, A. F., Population pharmacokinetic/pharmacodynamic modeling of cetrorelix, a novel LH-RH antagonist, and testosterone in rats and dogs, *Pharm. Res.*, 17(3):328–335, 2000.

28. Pechstein, B. et al., Pharmacokinetic-pharmacodynamic modeling of testosterone and luteinizing hormone suppression by cetrorelix in healthy volunteers, *J. Clin. Pharmacol.*, 40(3):266–274, 2000.

29. Zheng, N. X. et al., Pharmacokinetic-pharmacodynamic modeling of DP-1904, a novel thromboxane synthetase inhibitor in rabbits, based on an indirect response model, *Eur. J. Drug Metab. Pharmacokinet.*, 21(4):285–293, 1996.

30. Abbas, R. and Hayton, W. L., A physiologically based pharmacokinetic and pharmacodynamic model for paraoxon in rainbow trout, *Toxicol. Appl. Pharmacol.*, 145(1):192–201, 1997.

31. Wakelkamp, M. et al., Pharmacodynamic modeling of furosemide tolerance after multiple intravenous administration, *Clin. Pharmacol. Ther.*, 60(1):75–88, 1996.

32. Cleton, A. et al., Application of a combined "effect compartment/indirect response model" to the central nervous system effects of tiagabine in the rat, *J. Pharmacokinet. Biopharm.*, 27(3):301–323, 1999.

33. Movin-Osswald, G. and Hammarlund-Udenaes, M., Prolactin release after remoxipride by an integrated pharmacokinetic-pharmacodynamic model with intra- and interindividual aspects, *J. Pharmacol. Exp. Ther.*, 274(2):921–927, 1995.

34. Bagli, M. et al., Pharmacokinetic-pharmacodynamic modeling of tolerance to the prolactin-secreting effect of chlorprothixene after different modes of drug administration, *J. Pharmacol. Exp. Ther.*, 291(2):547–554, 1999.

35. Lalonde, R. et al., Mixed-effects modeling of the pharmacodynamic response to the calcimimetic agent R-568, *Clin. Pharmacol. Ther.*, 65:40–49, 1999.

36. Mould, D. et al., A population pharmacokinetic-pharmacodynamic analysis of single doses of clenoliximab in patients with rheumatoid arthritis, *Clin. Pharmacol. Ther.*, 66:246–257, 1999.

37. Bouillon, T. et al., Concentration-effect relationship of the positive chronotropic and hypokalaemic effects of fenoterol in healthy women of childbearing age, *Eur. J. Clin. Pharmacol.*, 51(2):153–160, 1996.

38. Piscitelli, S. et al., Pharmacokinetic modeling of recombinant interleukin-2 in patients with human immunodeficiency virus infection, *Clin. Pharmacol. Ther.*, 64:492–498, 1998.

39. Radwanski, E. et al., Pharmacokinetics and leukocyte responses of recombinant human interleukin-10, *Pharm. Res.*, 15:1895–1901, 1998.

40. Gobburu, J. et al., Pharmacokinetic-pharmacodynamic modeling of ipamorelin, a growth hormone releasing peptide, in human volunteers, *Pharm. Res.*, 16:1412–1416, 1999.

41. Mawer, G. et al., Prescribing aids for gentamicin, *Br. J. Clin. Pharmacol.*, 1:45–50, 1974.

42. Weaver, M. L., Tanzer, J. M., and Kramer, P. A., Pilocarpine disposition and salivary flow response following intravenous administration to dogs, *Pharm. Res.*, 9:1061–1069, 1992.

43. Ngashima, R., O'Reilly, R. A., and Levy, G., Kinetics of pharmacologic effects in man: the anticoagulant action of warfarin, *Clin. Pharmacol. Ther.*, 10:22–35, 1969.

44. Lalonde, R. et al., Propranolol pharmacodynamic modeling using unbound and total concentrations in healthy volunteers, *J. Pharmacokinet. Biopharm.*, 15:569–582, 1987.

45. Schüttler, J. et al., Pharmacokinetic modeling of the EEG effects of ketamine and its enantiomers in man, *J. Pharmacokinet. Biopharm.*, 15:241–253, 1987.

46. Schaefer, H. G. et al., Pharmacokinetic-pharmacodynamic modeling as a tool to evaluate the clinical relevance of a drug-food interaction for a nisoldipine controlled-release dosage form, *Eur. J. Clin. Pharmacol.*, 51:473–480, 1997.

47. Suri, A., Grundy, B., and Derendorf, H., Pharmacokinetics and pharmacodynamics of enatiomers of ibuprofen and flurbiprofen after oral administration, *Int. J. Clin. Pharmacol. Ther.*, 35:1–8, 1997.

48. Derendorf, H. et al., Receptor-based pharmacokinetic-pharmacodynamic analysis of corticosteroids, *J. Clin. Pharmacol.*, 33:115–123, 1993.

49. Downing, G. J., Ed., Biomarkers and surrogate endpoints: clinical research and applications, in *Proceedings of the NIH-FDA Conference*, 15–16 April 1999, Bethesda, MD, Elsevier, 2000.

50. Gupta, S. K. et al., Simultaneous modeling of the pharmacokinetic and pharmacodynamic properties of benzodiazepines, I. Lorazepam, *J. Pharmacokinet. Biopharm.*, 18:89–102, 1990.

51. Jawad, S. et al., A pharmacodynamic evaluation of midazolam as an antiepileptic compound, *J. Neurol. Neurosurg. Psychiatr.*, 49:1050–1054, 1986.

52. Milligan, N., Oxley, J., and Richens, A., Acute effects of intravenous phenytoin on the frequency of interictal spikes in man, *Br. J. Clin. Pharmacol.*, 16:285–289, 1983.

53. Jawad, S. et al., The effect of lamotrigine, a novel anticonvulsant, on interictal spikes in patients with epilepsy, *Br. J. Clin. Pharmacol.*, 22:191–193, 1986.

54. Overweg, J. and De Beukelaar, F., Single dose efficacy evaluation of loreclezole in patients with photosensitive epilepsy, *Epilepsy Res.*, 6:227–233, 1990.

55. Möller, A. et al., Inhibition of photosensitive seizures in man by the β-carboline ZK 95962, a selective benzodiazepine receptor agonist, *Epilepsy Res.*, 5: 155–159, 1990.

56. Rockstroh, B., Hyperventilation induced EEG changes in humans and their modulation by an anticonvulsant drug, *Epilepsy Res.*, 7:146–154, 1990.

57. Chay, P. C. W., Duffy, B. J., and Walker, J. S., Pharmacokinetic-pharmacodynamic relationships of morphine in neonates, *Clin. Pharmacol. Ther.*, 51:334–342, 1992.

58. Inturrisi, C. E. et al., Pharmacokinetics and pharmacodynamics of methadone in patients with chronic pain, *Clin. Pharmacol. Ther.*, 41:392–401, 1987.

59. Scott, J. C., Ponganis, K. V., and Stanski, D. R., EEG quantitation of narcotic effect: the comparative pharmacodynamics of fentanyl and alfentanil, *Anesthesiology*, 62:234–241, 1985.

60. Shafer, S. L. and Varvel, J. R., Pharmacokinetics, pharmacodynamics, and rational opioid selection, *Anesthesiology*, 74:53–63, 1991.

61. Ragueneau, I. et al., Pharmacokinetic-pharmacodynamic modeling of the effects of ivabradine, a direct sinus node inhibitor, on heart rate in healthy volunteers, *Clin. Pharmacol. Ther.*, 64:192–203, 1998.

62. Chalon, S. et al., Inhibition of angiotensin-converting enzyme in human hand veins, *Clin. Pharmacol. Ther.*, 65:58–65, 1999.

63. Mayer, B. et al., Pharmacokinetic-pharmacodynamic profile of systemic nitric oxide-synthase inhibition with L-NMMA in humans, *Br. J. Clin. Pharmacol.*, 47:539–544, 1999.

64. Bode-Böger, S. et al., L-Arginine-induced vasodilation in healthy humans: pharmacokinetic-pharmacodynamic relationship, *Br. J. Clin. Pharmacol.*, 46:489–497, 1998.

65. Ratain, M. J. et al., Pharmacodynamics in cancer therapy, *J. Clin. Oncol.*, 8:1739–1753, 1990.

66. Minami, H. et al., Indirect-response model for the time course of leukopenia with anticancer drugs, *Clin. Pharmacol. Ther.*, 64:511–521, 1998.

67. Peura, D. A., Stress-related mucosal damage: an overview, *Am. J. Med.*, 83:3–7, 1987.

68. Mathôt, R. and Geus, W., Pharmacodynamic modeling of the acid inhibitory effect of ranitidine in patients in an intensive care unit during prolonged dosing: characterization of tolerance, *Clin. Pharmacol. Ther.*, 66:140–151, 1999.

69. Zhou, H. et al., Population PK and PK/PD modelling of microencapsulated octreotide acetate in healthy subjects, *Br. J. Clin. Pharmacol.*, 50:543–552, 2000.

70. Zhou, H. et al., Population pharmacokinetics/toxicodynamics (PK/TD) relationship of SAM486A in phase I studies in patients with advanced cancers, *J. Clin. Pharmacol.*, 40:275–283, 2000.

71. Gieschke, R., Reigner, B. G., and Steimer, J. L., Exploring clinical study design by computer simulation based on pharmacokinetic/pharmacodynamic modelling, *Int. J. Clin. Pharmacol. Ther.*, 35:469–474, 1997.

72. Lalonde, R. L., Pharmacodynamics, in *Applied Pharmacokinetics: Principles of Therapeutic Drug Monitoring*, 3rd ed., Evans, W. E., Schentag, J. J., and Jusko, W. J., Eds., Applied Therapeutics, Vancouver, 1992, chap. 4.

73. Course on Pharmacokinetic/Pharmacodynamic Modeling — Concept and Applications, Sponsored by School of Pharmacy, State University of New York at Buffalo and Leiden/Amsterdam, Center for Drug Research, Leiden University, the Netherlands, May 14–17, 2000.

3 Pharmacokinetics and Metabolism in Drug Discovery and Preclinical Development

Nishit B. Modi

CONTENTS

INTRODUCTION

Pharmacokinetics and drug metabolism continue to play an increasingly important role in drug development, starting with drug discovery and lead optimization, pharmacology and safety evaluation, continuing into clinical development, and finally helping to position the product in the marketplace. As outlined in Figure 3.1, the successful development of a drug candidate depends on a collaborative interaction of pharmacokinetics and metabolism with a number of disciplines, with careful consideration given to the appropriate use of limited resources and to the timely review of results. This chapter provides an overview of the role of pharmacokinetics and metabolism in drug discovery and in the preclinical development of drugs. The goals of *in vitro* and nonclinical studies in experimental animals are to demonstrate, directly or indirectly, the biological activity against the targeted disease, to provide data for toxicology and safety evaluation, and to provide pharmacokinetic and

FIGURE 3.1 Schematic of the interaction between pharmacokinetics and metabolism and other disciplines during drug development.

pharmacodynamic data that may be helpful in developing dosing regimens and dose escalation strategies in clinical trials.[1] During preclinical development, pharmacokinetics can also help support the choice of an animal species and the dosing regimen (amount and dose frequency) and can facilitate interpretation of findings in toxicology studies to allow rapid entry into clinical trials.

Figure 3.2 presents a schematic of the sequence of studies that are conducted during various stages of development for a typical development program.

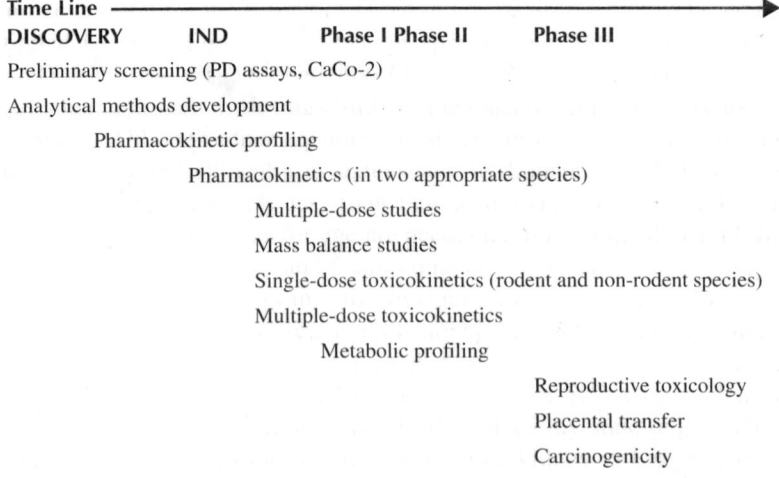

FIGURE 3.2 Schematic representation of the role of pharmacokinetics and metabolism in drug development.

Most people agree that pharmacokinetic and metabolism studies conducted during drug development should be based on scientific principles rather than on a rote approach of checking off items on a checklist. It is also appreciated that different approaches can be used by various groups to reach essentially similar objectives of getting a drug approved in an expeditious and safe manner. The sequence of studies depends on precedence, corporate culture, and the level of risk aversion within a company. As a result of a collaborative effort between the regulatory agencies and the pharmaceutical industry, great strides have been made to rationalize the development process through the International Conference on Harmonization (ICH). It is advisable to consult some of the regulatory guidance documents that have been issued as a result of these efforts prior to embarking on formal development studies. Table 3.1 presents a list of the current guidance documents applicable to the use of nonclinical pharmacokinetics and pharmacodynamics in development.

TABLE 3.1
ICH Guidance Documents Referring to the Application of Nonclinical Pharmacokinetics in Safety Evaluation and Drug Development

Stage	Document Title
S1C	Dose selection for carcinogenicity studies of pharmaceuticals
S1C(R)	Guidance on dose selection for carcinogenicity studies of pharmaceuticals: addendum on a limit dose and related notes
S3A	Toxicokinetics: assessment of systemic exposure in toxicity studies
S3B	Pharmacokinetics: guidance for repeated dose tissue distribution studies
S5A	Detection of toxicity to reproduction for medicinal products
S6	Preclinical safety evaluation of biotechnology-derived pharmaceuticals
M3	Nonclinical safety studies for the conduct of human clinical trials for pharmaceuticals

Source: http://www.fda.gov and http://www.eudra.org.

A number of *in vitro* and animal models are generally employed during the discovery and preclinical development stages to characterize pharmacokinetics, metabolism, and pharmacodynamics and to help identify agents with desirable pharmacological and biological properties. Typical information that is gathered during preclinical drug development is presented in Table 3.2.

Analytical methods that quantitate the parent and metabolite(s) in relevant biological fluids are a prerequisite for conducting robust pharmacokinetic studies. During the early discovery and development of a novel compound, analytical methods may not always be available, and it may not be prudent to invest a lot of resources to develop specific analytical methods. At this stage, limited pharmacokinetic data may be obtained using biological assays or radiolabeled material. However, results from such studies should be interpreted carefully since these methods often tend to be nonspecific. Analytical methods based on a biological property may not distinguish the parent drug from active metabolites. There have even been instances when the estimated bioavailability has exceeded 100%.[2] In addition, while radiolabeled studies using amino acid labeling methods are more common with recombinant proteins and peptides, the cost and resources required

TABLE 3.2
Typical Information Obtained during Preclinical Research to Facilitate
Understanding of the Pharmacokinetics and Metabolism of Drugs

Physicochemical properties (e.g., solubility, pK_a, partition coefficient)
Analytical methods for parent drug and metabolites
In vitro metabolism (e.g., enzyme induction, drug interaction)
In vivo metabolic profiling
Pharmacokinetics in rodent and non-rodent species (single and multiple dose)
Tissue distribution and whole-body autoradiography
Reproductive studies, placental and fetal transfer
Protein binding

to radiochemically synthesize chemical entities is generally not justifiable until the potential drug candidate is much further in development.

ABSORPTION

Various *in vitro* and *in vivo* methods have been used to predict drug absorption including Caco-2 cells, *in situ* intestinal permeability, whole-animal studies, and more recently chromatographic methods. Compared with *in vivo* absorption studies, evaluation of intestinal permeability *in vitro* requires less compound; is relatively easy to study, often avoiding complicated surgery; is rapid; and can allow a wider variety of variables to be controlled.[3]

IN VITRO CULTURED CELLS

Three major cell models obtained from human colon cancers have been used: HT-29, and T84, and Caco-2. HT-29 is a cell line capable of secreting mucin; it has been used to study the effect of mucin on drug absorption. The T84 cell line does not express biochemical differentiation markers and is not particularly useful for studying drug transport. However, the T84 cell line does express *P*-glycoprotein and can be used to study the role of this efflux pump.[4]

The most widely used cell line in drug transport studies is the Caco-2 cell line. This is an immortal cell line derived from human colon carcinoma that can be grown to monolayer on porous support. Functionally, this cell line models the colon more than the small intestine. The Caco-2 model allows characterization of both mucosal-to-serosal and serosal-to-mucosal transport and can also be used to study transcellular and paracellular transport. In addition to expressing small intestinal brush border enzymes, Caco-2 cells also express Phase I and Phase II enzymes and can be used to evaluate metabolism of compounds during transport across the intestinal barrier.[5-7]

USSING CHAMBERS

In this method, a small section of intestinal mucosa is sandwiched between two chambers containing buffer.[8-10] The compound of interest is introduced into the donor chamber and transport across the mucosa into the receiver chamber is measured as a function of time. Both mucosal-to-serosal and serosal-to-mucosal transport can be characterized by this method. The method also allows characterization of drug transport at different locations of the intestinal tract.

EVERTED GUT SAC

This model was first described by Wilson and Wiseman.[11] A segment of intestine is everted on a glass rod. One end is ligated and the other end is cannulated with a polyethylene tubing. The sac is filled with buffer solution and placed in a flask filled with buffer solution containing the drug. Both the mucosal and serosal solutions are maintained at 37°C and aerated. Samples may be taken from both the serosal and mucosal sides. At the end of the experiment, the intestinal segment can be dissolved in 1 M NaOH and the protein content measured to allow transport rate to be normalized. The method is simple, quick, reproducible, and inexpensive. The inverted sac method can allow study of intestinal metabolism and of the mechanism of drug transport, including the role of P-glycoprotein and absorption enhancers. One disadvantage of the everted sac method is that it is only possible to maintain the integrity of the tissue for short periods of time.

IN SITU STUDIES

Several techniques are available to bridge *in vitro* studies and studies in conscious animals. The common feature of these methods is that the gut remains *in situ* in an anesthetized rat, minimizing perturbation of normal function. However, data from *in situ* studies may not always correlate with those from studies in conscious animals. For example, it is known that splanchnic blood flow may be affected by anesthesia.[12] In addition, *in situ* methods do not adequately reproduce the multitude of environments encountered by a drug as it traverses the gastrointestinal tract.

   In situ methods of drug absorption include the Doluioso method (static method)[13] and the single- and continuous-loop perfusions (dynamic methods).[14] The Doluiosio method involves cannulation of a segment (~10 cm) of gut in an anesthetized rat. Syringes are attached via cannulas at either end to allow drug administration into and sampling of the luminal contents. The method is relatively simple, requiring only commonly available equipment and involving minimal surgery. Limitations of the method are that absorption is characterized by noting disappearance of drug. The method does not discriminate between intestinal metabolism, binding to intestinal wall, or precipitation of drug in the gut.

INTACT ANIMALS

Ultimately, there is no substitute for *in vivo* data. Studies in intact animals allow drug administration in the intestinal tract with sampling of the portal and systemic circulation to differentiate fully absorption from presystemic metabolism. Conducting studies in animals is expectedly slow and may not be viewed favorably during early development when the medicinal chemists need basic information in a very rapid fashion to help understand structure–activity relationships. The biggest impediments to conducting studies in animals are the time and resources required to develop analytical methods that are sensitive and specific enough to support *in vivo* studies. This investment is generally not justifiable during the early discovery phase. However, once a manageable number of leads have been generated, studies in appropriate animal models can provide valuable information for compound selection and modification. Understanding of the physiological, anatomical, and biochemical differences in the gastrointestinal tract of different animal species compared with humans can help select an appropriate animal model to mimic the bioavailability of drug in humans.[15]

To alleviate the analytical bottleneck, several groups have also investigated cocktail dosing (also referred to as "cassette dosing" or "N-in-one" dosing) where several drugs are injected simultaneously into animals and the ability of mass spectrometry to distinguish subtle differences in mass ion peaks is used to characterize the pharmacokinetics.[16,17] In these approaches, to reduce the potential for interactions between the agents tested, the dose of each individual compound is reduced, and a reference compound with known pharmacokinetic profile is usually included in the mixture.[18] Such approaches do not allow characterization of metabolism, and pharmacodynamic data cannot be obtained because it may not be possible to attribute pharmacological activity to a particular compound. The advantage of these methods is that more compounds can be dosed in a shorter time, fewer animals are used, and the analytical workload is decreased.

METABOLISM

Understanding the metabolic fate of a drug and routes of elimination early in drug development can provide valuable information for further development. Such information can help identify the potential for drug interactions in humans, explain interindividual and ethnic differences, and indicate the need for further pharmacological characterization or structure modification. Rapid progress in the inclusion of drug metabolism studies in development is attributable to three factors:[19]

1. Advances in our understanding of human drug metabolizing systems and the availability of *in vitro* models
2. The ability to use these absorption and metabolism models in conjunction with conventional *in vivo* studies
3. Development of appropriate analytical methods that can be coupled with automated sample handling systems

The cytochrome P-450 family is the most important class of enzymes involved in drug metabolism. The nomenclature uses an Arabic numeral to denote the family, followed by a capital letter to designate a subfamily, followed by another Arabic numeral to denote an individual gene/enzyme. In humans five CYP450 isoforms are commonly involved in hepatic drug metabolism: 1A, 2C, 2D, 2E, and 3A. Table 3.3 presents a summary of substrate specificities and inhibitors for the common isoforms.

The FDA has recognized the value of data from *in vitro* drug metabolism studies and has developed a guidance document for the pharmaceutical industry titled "Drug Metabolism/Drug Interaction Studies in the Drug Development Process: Studies *in*

TABLE 3.3
Cytochrome P-450 Isoforms

Cytochrome P-450	Polymorphism	Substrate	Inhibitor	Inducer
3A3/4	Not established	Acetaminophen Carbamazepine Cyclosporine Diazepam Diltiazem Erythromycin Ethinyl Estradiol Etoposide Imipramine Loratadine Midazolam Nifedipine Omeprazole Quinidine Rapamycin Steroids Taxol Terfenadine Theophylline Triazolam Verapamil Warfarin	Cyclosporine Fluoxetine Fluvoxamine Grapefruit Indinavir Itraconazole Low concentrations of ketoconazole Nefazodone Sertraline Troleandomycin Zafirlukast	Carbamazepine Dexamethasone Phenobarbital Phenytoin Rifampin Sulfathiazole Troleandomycin
2C19	20% poor metabolizers in Asians, 4–8% in Blacks, and 3% in Caucasians	Citalopram Diazepam Hexobarbital Imipramine Omeprazole Propranolol	Fluconazole Fluoxetine Fluvoxamine Tranylcypromine	Rifampin

(continued)

TABLE 3.3 (CONTINUED)
Cytochrome P-450 Isoforms

Cytochrome P-450	Polymorphism	Substrate	Inhibitor	Inducer
2D6	8% poor metabolizers in Caucasians and <1% in Asians	Amitriptyline, Captopril, Citalopram, Clozapine, Codeine, Debrisoquine, Desipramine, Dextromethorphan, Fluoxetine, Imipramine, Metoprolol, Nortriptyline, Paroxetine, Propranolol, Risperidone, Thioridazine, Tramadol, Venlafaxine	Cimetidine, Fluoxetine, Quinidine, Ritonavir, Sertraline, Yohimbine	Cigarette smoke
1A2	Not established	Acetaminophen, Caffeine, Clozapine, Estradiol, Imipramine, Phenacetin, Tacrine, Theophylline, Warfarin	Cimetidine, Ciprofloxacin, Fluvoxamine, Mexiletine, Zileuton, Paroxetine	Charcoal-broiled beef, Cigarette smoke, Cruciferous vegetables, Omeprazole

Adapted from Parkinson, 1996[43] and Michalets, 1998.[44]

TABLE 3.4

Application of *in Vitro* Methods of Drug Metabolism in Development

Generation and identification of metabolites

Development of analytical methods

Comparison of metabolic pathways across species which can in turn be used to explain pharmacologic
  or toxicologic differences between species

Justification of species for preclinical safety and pharmacokinetics studies

Help identify enzyme systems involved in metabolic clearance

Support clinical development

Vitro" that encourages routine, thorough evaluation of metabolism and interactions in *in vitro* studies whenever feasible and appropriate. Factors that have fostered the use of *in vitro* methodologies to characterize drug metabolism include the efficient and relatively inexpensive nature of the methods and the ready availability of human and animal tissues and enzyme systems. Table 3.4 outlines the value of *in vitro* metabolism in the development of drugs.

In vitro methods of drug metabolism can be divided into two classes:

1. Cellular systems, which include primary hepatocytes in suspension or monolayer culture, liver slices, and hepatocyte-derived cell lines
2. Enzyme systems, which include subcellular fractions (microsomal or cytosolic) and isolated enzyme preparations (purified or cDNA expressed)

Among the cellular systems, primary hepatocytes have the advantage that all the metabolic pathways are intact and available. Nuclei and mitochondria are present, which are known to contain enzymes that occasionally participate in metabolic pathways. Cellular systems have the disadvantage that they cannot be readily cryopreserved.

Liver slices have all the advantages of primary hepatocytes. In addition, the complete hepatic architecture is maintained so that sequential pathways that potentially involve several cells can also be studied. The limitation with liver slices is that they cannot be stored for long and not much work has been done to validate their use in enzyme induction. It is also very important to study a range of samples from different livers to obtain representative results. Human liver samples should be characterized as fully as possible with regard to gender, age, medical and drug history, and metabolic activity.[20] Liver tissue samples should be obtained as soon as possible after surgery and stored rapidly at −80°C.

The most widely used *in vitro* enzyme systems are the subcellular fractions comprising the S9 fraction, microsomes, and cytosolic fraction. Enzyme preparations have the advantage that they are easy to prepare and can be obtained from small amounts of tissue. The level of drug-metabolizing enzymes can be readily determined. In addition, their origin can be well characterized, allowing evaluation of the effect of disease or genetic factors on metabolism. However, all cofactors required for functioning must be added. A further limitation is that enzymes that work sequentially may be located in different subcellular fractions and such pathways may not be easily recognized with enzyme preparations.

TOXICOKINETICS

Toxicokinetics is defined as the study of pharmacokinetic data, either as an integral component in the conduct of nonclinical toxicity studies or in specially designed supportive studies, to assess systemic exposure (ICH Guidance document S3A). It is no longer sufficient to conduct toxicology studies on a dose–response basis.[21] Regulatory agencies are increasingly requiring information on toxicokinetics and the relationship between exposure and safety findings. Integration of pharmacokinetics in toxicology studies can contribute to study design and to the interpretation of study findings and can lead to the more rational use of animals and technical resources.

The objectives of a toxicokinetic program in toxicology studies are outlined in Table 3.5. These objectives may be met by characterizing one or more pharmacokinetic parameters from measurements made at appropriate time points in a toxicology study. The measurements are typically concentrations of the parent drug or metabolite in plasma or serum. However, in some cases it may be more appropriate to sample some other matrix or to quantify tissue levels. The pharmacokinetic parameters are usually area under the concentration–time curve (AUC), and maximum concentration (C_{max}), or the duration that the concentration is above a given threshold (C_t). The choice of the pharmacokinetic parameters should be made on a case-by-case basis.

Analytical methods used in toxicokinetic studies should be specific for the species of interest and of sufficient accuracy and precision. If the drug produces a metabolite that also circulates in the body and could contribute to the toxicology profile, assays may be needed for both the parent and metabolite. For chiral drugs, enantiospecific assays may be required so that exposure to both enantiomers can be quantitated. The sampling should be sufficiently frequent but no more than 10% of the blood volume can be removed without affecting homoeostasis[22] or compromising the integrity of the toxicology study. In some instances it may be preferable to conduct toxicokinetics in a satellite group of animals that are treated similarly to the main group so as not to compromise the safety assessments. The use of a satellite group can still provide valuable information on exposure and allow detection of nonlinear pharmacokinetics or saturable absorption. However, this approach does not allow explanation of unexpected findings in individual animals.

Table 3.6 presents AUC data from a 6-month dose-ranging toxicokinetic study in rats for Compound X. The data demonstrate a small increase in the AUC values with time, indicating some accumulation upon repeated dosing, but importantly, the

---

TABLE 3.5
Goals of Toxicokinetics in Nonclinical Toxicology Studies

To describe the system exposure and to characterize the relationship between dose and exposure

To relate exposure data to toxicological findings

To assess the findings in safety studies relevant to clinical studies

To support the choice of animal species and treatment regimen

To help in the design of future animal studies

TABLE 3.6
Summary of AUC Values from a Dose-Ranging Toxicokinetic Study for Compound X Demonstrating a Less Than Proportional Increase in AUC with Increasing Dose

Dose (mg/kg/day)	Week					
	1	5	9	13	19	26
5	1.4	1.4	2.3	1.6	2.6	2.9
50	9.8	15.0	15.2	19.6	18.2	23.3
500	50.4	64.0	53.3	63.2	58.5	68.3

data indicate saturation in the absorption with increasing dose. Although the AUC values increase dose-proportionally between 5 and 50 mg/kg/day, there is a less-than-proportional increase in the AUC values between 50 and 500 mg/kg/day. Toxicokinetic data such as these can be very valuable in choosing appropriate doses for subsequent toxicology studies.

There has also been an increase in the use of population pharmacokinetics in nonclinical studies.[23–26] These approaches allow the estimation of pharmacokinetic parameters with only a few samples (usually one to three) per animal. With appropriate attention paid to study design and sampling schedule, population methods can allow pharmacokinetic data to be obtained using limited sampling from animals included in toxicology studies rather than from a satellite group of animals.

ALLOMETRIC SCALING

Allometry is the study of size and its consequences. It is often of interest to predict pharmacokinetic behavior in humans based on data in animals.[27,28] The simplest allometric pharmacokinetic equation that is often used is the power relationship,

$$Y = aX^b \tag{3.1}$$

where Y is the dependent variable (e.g., half-life, clearance, etc.), X is the independent variable (usually species body weight), and a and b are allometric parameters. This equation may be linearized following log transformation:

$$\log Y = \log a + b \log X \tag{3.2}$$

A plot of data from appropriate animal species as shown above can provide estimates of the allometric parameters, allowing prediction of the pharmacokinetic parameters in other species including humans. This approach has been applied successfully for both chemical entities[29–32] and biologicals.[29,33,34]

Figure 3.3 presents the results of allometric scaling of a recombinant monoclonal antibody, rhuMAb VEGF.[35] Pharmacokinetic studies were conducted in mice, rats,

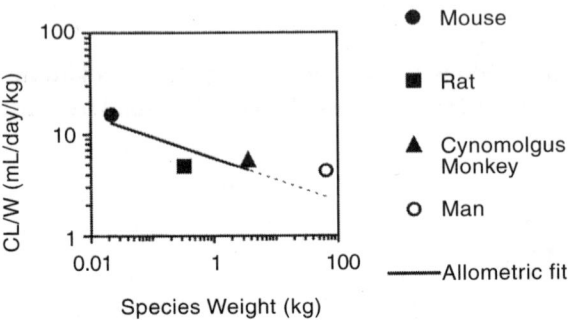

FIGURE 3.3 Allometric scaling of rhuMAb VEGF. (Adapted from Lin et al.[35])

and cynomolgus monkeys following intravenous dosing. The clearance data and species weight were plotted on logarithmic axes, and the allometric coefficients were estimated. The predicted clearance in humans based on the allometric equation (2.4 ml/kg/day) was comparable to the observed value in patients with cancer (4.3 ml/kg/day). A robust allometric relationship can help in the choice of the initial dose in clinical trials and can help support the dose escalation scheme.

Appropriate studies in animals can also allow a better understanding of the relationship between drug concentration and drug effects. Pharmacokinetic/pharmacodynamic studies in animals have been applied to compare dosing regimens,[36,37] to investigate drug interactions,[38,39] and to better understand the relationship between drug concentration and drug effect.[40,41]

Mordenti et al.[42] demonstrated the use of animal studies to evaluate the concentration–response relationship of a murine monoclonal antibody directed against an angiogenic growth factor (VEGF). This information was used to predict the efficacious plasma concentrations of a recombinant humanized version of the antibody in humans. In this panel of studies, the authors characterized the relationship between inhibition of tumor growth and serum drug concentration from a blood sample obtained at terminal sacrifice. The efficacy data were complemented with pharmacokinetic data obtained from a separate study, allowing the integration of pharmacokinetics and pharmacodynamics to predict a proposed dosing regimen in humans. Although the authors did not present human data to allow comparison with the predicted values, this study helps illustrate the value of appropriately designed pharmacokinetic and pharmacodynamic studies in animals.

Extrapolation of pharmacokinetic data from animals to humans is not without limitations and should be done carefully, keeping in mind the assumptions that are necessary for such extrapolation. It is important to select an appropriate *in vitro* or animal model that has relevance to humans. Extrapolation of animal data to humans is not likely to be successful in instances when the pharmacokinetics (absorption or disposition) is species-dependent, when active processes are involved, when drug action (either beneficial or adverse effects) is due to a poorly understood mechanism, or when species dependent mechanisms of tolerance or enzyme induction are present.[43,44]

REFERENCES

1. Peck, C. C. et al., Opportunities for integration of pharmacokinetics, pharmacodynamics, and toxicokinetics in rational drug development, *J. Clin. Pharmacol.*, 34:111–119, 1994.

2. Wills, R. J. and Ferraiolo, B. L., The role of pharmacokinetics in the development of biotechnologically derived agents, *Clin. Pharmacokinet.*, 23:406–414, 1992.

3. Smith, P. L., Methods for evaluating intestinal permeability and metabolism *in vitro*, in *Models for Assessing Drug Absorption and Metabolism*, Borchardt, R. T., Smith, P. L., and Wilson, G., Eds., Plenum Press, New York, 1996, 13–34.

4. Hunter, J., Hirst, B. H., and Simmons, N. L., Epithelial secretion of vinblastine by human intestinal adenocarcinoma cell (HCT-8 and T84) layers expressing *P*-glycoprotein, *Br. J. Cancer*, 64:437–444, 1991.

5. Gan, L. S. et al., Use of Caco-2 cells as an *in vivo* intestinal absorption and metabolism model, *Drug Dev. Ind. Pharm.*, 20:615–631, 1994.

6. Gan, L. S. I. and Thakker, D. R., Applications of the Caco-2 model in the design and the development of orally active drugs: elucidation of biochemical and physical barriers posed by the intestinal epithelium, *Adv. Drug Deliv. Res.*, 23:77–98, 1997.

7. Barthe, L., Woodley, J., and Houin, G., Gastrointestinal absorption of drugs: methods and studies, *Fundam. Clin. Pharmacol.*,13:154–168, 1999.

8. Ussing, H. H., Transport of ions across cellular membranes, *Physiol. Rev.*, 29:127–155,1949.

9. Hegel, U. et al., Bovine and porcine large intestine as model epithelia in a student lab course, *Adv. Phys. Ed.*, 10:S10–S19, 1993.

10. Field, M., Fromm, D., and McColl, I., Ion transport in rabbit ileal mucosa. I. Na and Cl fluxes and short-circuit current, *Am. J. Physiol.*, 220:1388–1396, 1971.

11. Wilson, T. H. and Wiseman, G., The use of sacs of everted small intestine for the study of the transference of substance from the mucosal to the serosal surface, *J. Physiol.*, 123:116–125, 1954.

12. Yuasa, H., Matsuda, K., and Watanabe, J., Influence of anesthetic regimens on intestinal absorption in rats, *Pharm. Res.*, 10:884–888, 1993.

13. Doluisio, J. T. et al., Drug absorption I: an *in situ* gut technique yielding realistic absorption rates, *J. Pharm. Sci.*, 58:1196–1200, 1969.

14. Lewis, L. D. and Fordtran, J. S., Effect of perfusion rate on absorption, surface area, unstirred water layer thickness, permeability and intraluminal pressure in the rat ileum *in vivo*, *Gastroenterology*, 68:1509–1516, 1975.

15. Kararli, T. T., Comparison of the gastrointestinal anatomy, physiology, and biochemistry of humans and commonly used laboratory animals, *Biopharm. Drug Dispos.*, 16:351–380, 1995.

16. Allen, M. C., Shah, T. S., and Day, W. W., Rapid determination of oral pharmacokinetics and plasma free fraction using cocktail approaches: methods and applications, *Pharm. Res.*, 15(1):93–97, 1998.

17. Shaffer, J. E. et al., Use of "N-in One" dosing to create an *in vivo* pharmacokinetic database for use in developing structure–pharmacokinetic relationships, *J. Pharm. Sci.*, 88:313–318, 1999.

18. Hop, C. E. C. A. et al., Plasma-pooling methods to increase throughput for *in vivo* pharmacokinetic screening, *J. Pharm. Sci.*, 87:901–903, 1998.

19. Rodrigues, A. D., Preclinical drug metabolism in the age of high-throughput screening: an industrial perspective, *Pharm. Res.*, 14:1504–1510, 1997.

20. Tucker, G. T., The *in vivo* assessment of human hepatic drug metabolism, in *Handbook of Phase I/II Clinical Drug Trials*, O'Grady, J. and Joubert, P. H., Eds., CRC Press, Boca Raton, FL, 1997, 51–63.

21. Welling, P. G., Role of pharmacokinetics in drug discovery and development, in *Pharmacokinetics of Drugs*, Welling, P. G. and Balant, L. P., Eds., Springer-Verlag, Heidelberg, 1994, 3–19.

22. McGuill, M. W. and Rowan, A. N., Biological effects of blood loss: implications for sampling volumes and techniques, *ILAR News*, 31:5–20, 1989.

23. Bouzom, F. et al., Use of nonlinear mixed effect modeling for the meta-analysis of preclinical pharmacokinetic data: application to S 20342 in the rat, *J. Pharm. Sci.*, 89:603–613, 2000.

24. Gething, P. A. and Daley-Yates, P. T., A sparse sampling, mixed effects approach to toxicokinetic analyses, *Drug Inf. J.*, 31:521–527, 1997.

25. Burtin, P. et al., Sparse sampling for assessment of drug exposure in toxicological studies, *Eur. J. Drug Metab. Pharmacokinet.*, 21:105–111, 1996.

26. Pai, S. M. et al., Characterization of AUCs from sparsely sampled populations in toxicology studies, *Pharm. Res.*, 13:1283–1290, 1996.

27. Mordenti, J. and Chappell, W., The use of interspecies scaling in toxicokinetics, in *Toxicokinetics and New Drug Development*, Yacobi, A., Skelly, J. P., and Batra, V. K., Eds., Pergamon Press, Elmsford, NY, 1989, 42–96.

28. Chappell, W. R. and Mordenti, J., Extrapolation of toxicological and pharmacological data from animals to humans, *Adv. Drug Res.*, 20:1–116, 1991.

29. Lave, T. et al., Interspecies scaling of interferon disposition and comparison of allometric scaling with concentration-time transformations, *J. Pharm. Sci.*, 84:1285–1290, 1995.

30. Lave, T. et al., Animal pharmacokinetics and interspecies scaling from animals to man of lamifiban, a new platelet aggregation inhibitor, *J. Pharm. Pharmacol.*, 48:573–577, 1996.

31. Lave, T. et al., Interspecies pharmacokinetic comparisons and allometric scaling of napsagartran, a low molecular weight thrombin inhibitor, *J. Pharm. Pharmacol.*, 51:85–91, 1999.

32. Feng, M. R. et al., Allometric pharmacokinetic scaling: towards the prediction of human oral pharmacokinetics, *Pharm. Res.*, 17:410–418, 2000.

33. Mordenti, J. et al., Interspecies scaling of clearance and volume of distribution data for five therapeutic proteins, *Pharm. Res.*, 8:1351–1358, 1991.

34. Khor, S. P. et al., Pharmacokinetics, pharmacodynamics, allometry, and dose selection of rPSGL-Ig for Phase I trial, *J. Pharmacol. Exp. Ther.*, 293:618–624, 2000.

35. Lin, Y. S. et al., Preclinical pharmacokinetics, interspecies scaling and tissue distribution of a humanized monoclonal antibody against vascular endothelial growth factor, *J. Pharmacol. Exp. Ther.*, 288:371–378, 1999.

36. Modi, N. B. et al., Pharmacokinetics and pharmacodynamics of TP-9201, a gpIIbIIIa antagonist, in rats and dogs, *J. Cardiovasc. Pharmacol.*, 25:888–897, 1995.

37. Eppler, S. et al., Pharmacokinetics and pharmacodynamics of recombinant tissue-type plasminogen activator following intravenous administration in rabbits: comparison of three dosing regimens, *Biopharm. Drug Dispos.*, 19:31–38, 1998.

38. Modi, N. B. et al., Pharmacokinetics and pharmacodynamics of TP-9201, a gpIIbIIIa antagonist, administered in combination with Activase® t-PA, heparin, and aspirin in beagles, *J. Cardiovasc. Pharmacol.*, 27:105–112, 1996.

39. Modi, N. B. et al., Pharmacokinetics and pharmacodynamics of sibrafiban (Ro 48-3657), an orally active IIbIIIa antagonist, administered alone or in combination with heparin, aspirin and recombinant tissue-type plasminogen activator in beagles, *J. Cardiovasc. Pharmacol.*, 32:397–405, 1998.

40. Refino, C. J. et al., Pharmacokinetics, pharmacodynamics, and tolerability of a potent, non-peptidic, GP IIb/IIIa receptor antagonist following multiple oral administration of a prodrug form, *Thromb. Haemost*, 79:169–176, 1998.

41. Veng-Pedersen, P. et al., A system approach to pharmacodynamics. Plasma iron mobilization by endogenous erythropoietin in the sheep fetus; evidence of threshold response in spontaneous hypoxemia, *J. Pharm. Sci.*, 82:804–807, 1993.

42. Mordenti, J. et al., Efficacy and concentration-response of murine anti-VEGF monoclonal antibody in tumor-bearing mice and extrapolation to humans, *Toxicol. Pathol.*, 27:14–21, 1999.

43. Parkinson, A., Biotransformation of xenobiotics, in *Casarett & Doull's Toxicology, The Basic Science of Poisons*, 5th ed., Klaassen, C. D., Ed., McGraw-Hill, New York, 1996, 113–186.

44. Michalets, E. L., Update: clinically significant cytochrome P-450 drug interactions, *Pharmacotherapy*, 18:84–112, 1998.

4 Phase I, II, and III FDA Submissions

Mark G. Eller

CONTENTS

1-56676-973-6/02/$0.00+$1.50
© 2002 by CRC Press LLC

INTRODUCTION

From the point of view of the industry sponsor, the goal of drug development is marketing approval so that drug candidates can be commercialized and profits made. From the view of the public, the goal is valuable new medicines. Between the two stand the regulatory agencies, which try to ensure that only truly beneficial drug candidates are approved and marketed. The ultimate regulatory barrier to approval is empirical proof (to the required statistical precision) that a new drug is both safe and effective in the proposed patient population when used as intended.

This proof of safety and efficacy takes the form of adequate and well-controlled clinical trials (Phase III studies), which can require thousands of patients, are time-consuming, and expensive. The cost of a Phase III study alone can easily exceed $10 million and can take several years.[1-3]

Because preclinical models for drug safety and efficacy have not yet demonstrated predictive ability to the point where risks of beginning the clinical phase of development with these large-scale pivotal studies are acceptable, clinical drug development occurs as a process. Small, well-defined groups of patients or healthy volunteers are intensively studied under highly controlled conditions to answer fundamental questions about the drug candidate. If the answers validate the basic premises upon which the decision was made to develop the drug, clinical studies are expanded, hypotheses are refined, and more data are collected until the point at which the pivotal clinical trials are undertaken. More often, this process leads to a decision not to conduct the pivotal studies. Only 20 to 30% of drug candidates entering clinical testing are ultimately approved, and the vast majority are eliminated before Phase III.[1,2] Early elimination is desirable, as the number of patient volunteers required for Phase III, as well as the time and expense involved, can easily exceed the totals for all the preceding clinical testing combined.[1-3]

The role of pharmacokinetics and pharmacodynamics in the decision-making process in drug development and their contribution to the assessment of safety and efficacy of investigational new drugs have been well established.[4-6] In this chapter, the role of pharmacokinetics and pharmacodynamics is described in terms of current approaches to typical questions or problems, that is, what is done and why. The details of how it is done are constantly evolving and are beyond the scope of this chapter.

GOAL OF PHARMACOKINETICS IN INDUSTRY

The objective of clinical studies in the pharmaceutical industry is to assess whether a drug candidate will be effective in the treatment of a disease or condition, its risks of unwanted effects, and the relative relationship of these assessments (benefit/risk assessment). These studies must be conducted so that the participating volunteers (healthy subjects or patients) are exposed to the least possible risk consistent with the potential benefit. In this context, clinical pharmacokinetics and pharmacodynamics in drug development is focused on three main objectives:

1. Weeding out drug candidates that have a low potential to become successful products as rapidly as possible
2. Expediting the development process for high-potential drug candidates
3. Providing the pharmacokinetic information to allow physicians and pharmacists to use the drug to the best advantage for potential patients

The main way these objectives are achieved is by providing a continuing assessment of the dose–concentration–effect relationship throughout the course of drug development. Effects include pharmacodynamic end points, surrogate markers for safety and efficacy, adverse events, and measures of efficacy. To be useful, the assessment must be quantitative and predictive. It must also include exploring and determining sources of variability. Choosing the right doses and dosage intervals to examine in each study to optimize the chance of success is the major advantage of incorporating pharmacokinetics into the decision-making process for clinical drug development. Identifying sources of variability to determine if dosage regimen adjustments should be recommended in labeling for certain patient subpopulations is the major contribution of industrial pharmacokinetics to clinicians. Thus, the contribution of pharmacokinetics in industry is analogous to the therapeutic setting, where a major contribution is dosage regimen selection and adjustment for individual patients.

PHASE I

The term *Phase I* is commonly used to refer to a type of study (any clinical pharmacology study during the course of development) as well as the beginning stages of clinical development, which involve the first human testing. For this discussion the focus is on the very first human studies: single and multiple dose-rising tolerance studies. These studies represent the beginning of the transition of a drug candidate from research to development. Before starting Phase I, the goals of the development program as a whole should be clearly defined. At a minimum, this should include answers to the following questions:

- What is the target patient population and indication for the drug candidate?
- What is the potential advantage over existing therapies?
- What is the desired route of administration and dosing frequency?
- What is the projected timeline to critical decision points?

PURPOSE OF PHASE I

The primary goal of Phase I is to determine the safety of the drug candidate. These studies use the highest doses that will ever be administered over the course of development and are the first opportunity to determine pharmacokinetics and, if possible, pharmacodynamics in humans. Thus, these trials address questions related to whether the candidate has any serious or unexpected toxicity, and whether it has desirable pharmacokinetic properties. Specific goals are to:

- Efficiently determine the safety and tolerance for acute and subchronic exposure.
- Identify and characterize intensity and duration of drug-related adverse events and establish the maximum tolerated dose (MTD), the largest dose that can be administered without causing harm.
- Obtain preliminary characterization of the pharmacokinetic profile. This includes preliminary assessment of:
  - Dose–concentration relationship (dose proportionality or linearity)
  - Variability in pharmacokinetic parameters between subjects
  - Dose–concentration–toxicity relationships (plasma concentrations associated with dose-limiting effects)
- If possible, obtain preliminary characterization of the pharmacodynamic profile.
- Determine dose and dose frequency for Phase IIa.

PREREQUISITES FOR PHASE I

The clinical pharmacokineticist on a drug development team does not simply analyze data at the end of the study. Typically, he or she is intimately involved in the design of the study, either making or contributing to the decision regarding the first dose to be given to humans, the dose escalation scheme, and potentially the decision to escalate doses. Often the pharmacokineticist will have to justify the sensitivity limits for the bioanalytical methods proposed for measuring drug concentrations in human plasma. To facilitate this, a thorough understanding of preclinical models for activity (including target drug concentrations), ADME data, pharmacokinetics, and toxicokinetics is required. This data can be characterized in the following ways.

Useful Preclinical Pharmacology Data
- Does the proposed pharmacological mechanism suggest that there should be a direct correlation between drug concentration and pharmacological effect, or is the effect more likely to be indirect?
- What is the IC_{50}/EC_{50} needed to produce the desired effect?
- What is the IC_{50}/EC_{50} that produces a toxic effect?
- Is the toxic effect an extension of the pharmacological effect?
- What is the duration of pharmacological effect?
- Is there an *in vitro/ex vivo* correlation to the therapeutic or pharmacological effect?
- Is there a surrogate marker for the therapeutic effect and is it validated?

Useful Preclinical Pharmacokinetics Data
- Pharmacokinetic parameters from several species, especially species used for toxicology studies
- Probable site of absorption, and evidence of saturable absorption
- Routes of elimination, evidence of first-pass metabolism, nonlinear pharmacokinetics, and enzyme induction or inhibition
- Active/toxic metabolites, metabolic paths
- Probable isozymes involved in metabolism from *in vitro* P-450 profiling, and uniqueness of metabolizing enzymes between species
- Protein binding, and evidence of saturable binding
- Penetration of drug into tissues/fluids of interest
- Drug/metabolites bioanalytical methodology

Useful Preclinical Toxicology and Toxicokinetic Data
- Maximally tolerated dose and resulting plasma concentration in two species (rodent and nonrodent)
- Dose and plasma concentrations that produced the first untoward effect
- Mechanism of toxic effect
- Toxicity profile and organs involved

Even a brief description of the regulatory requirements to initiate studies in humans is beyond the scope of this chapter. In the United States, the Investigational New Drug Application (IND) and, if desired, pre-IND meeting represent both the main regulatory barrier to initiate studies in humans and the first opportunity to educate regulators about the drug candidate.

Pre-IND meetings are not required but are sometimes requested by the sponsor and are conducted with the FDA division that would review the IND and, eventually, the NDA. Meetings at this early stage can be useful for open discussion of any scientific issue that may need to be resolved before IND submission. They are particularly useful if the drug candidate uses a new technology (e.g., gene therapy) or has a complicated history (e.g., developed or already marketed outside the United States, or a new indication for an existing drug). On the other hand, if the drug candidate is a new compound synthesized in the United States, with minimal preclinical and no clinical data, there is seldom a need to meet with the FDA to review data before submitting the IND.

STUDY DESIGN AND EVALUATION

The objectives of Phase I are achieved through an orderly sequence of studies of increasing dose and exposure. Clinical trials start with the acute (single-dose) administration of the drug candidate. After the first dose is observed to be safe, incrementally increased doses are given at intervals up to the maximum tolerated dose. Serial plasma and sometimes urine samples are obtained for assay of parent drug and metabolites. Response measures may be obtained for pharmacodynamic analysis. The information from single-dose administration trials gives the basic information for determination of the starting dose and dose frequency in subchronic (repeated dose) trials. These trials provide more safety data regarding long-term tolerability, drug accumulation, and elimination patterns.

Critical elements include study population and size, formulations, treatment assignments, starting dose, dose escalation scheme, safety assessments and determination of MTD, and pharmacokinetic and pharmacodynamic evaluations.

STUDY POPULATION AND SIZE

Except for drugs with known toxicity (e.g., cytotoxic anticancer drugs), the initial administration of an investigational new drug to humans is usually accomplished in healthy subjects rather than patients. A major goal is to identify drug-related adverse events, and this process is difficult in patients, since illness manifests itself as a collection of "adverse events." Healthy subjects are free from the abnormalities that would affect the subjects' sensitivity to the drug's toxic potential, or complicate the interpretation of the safety data. The health status of volunteers is controlled through rigorous inclusion and exclusion criteria that include prestudy clinical and laboratory screens of the prospective subjects. The number of subjects used in dose tolerance studies is relatively small, usually four to six subjects per dose group, to minimize exposure to investigational drugs, while still retaining the ability to identify common adverse events.

FORMULATIONS

Final formulations, like those used for marketed drugs, are generally not used for Phase I studies. Formulation development represents expenditure of significant resources, the approaches used may depend on the results of future bioavailability studies, and the success of the drug candidate is still an open question. Therefore, simple formulations are most often used. For orally administered drugs, solutions are ideal as they provide flexibility in dosing and minimize potential absorption or bioavailability problems. Capsules filled with neat drug or suspensions can be employed when solutions are not feasible.

TREATMENT ASSIGNMENTS

Phase I dose tolerance studies are typically double blind and placebo controlled. Various designs have been used; however, no single design for allocation of subjects to treatments has been accepted as the best approach.[7-10] The most common design is the "across subject ascending dose design." In this design, separate subjects are assigned to each dose level group, some receiving active drug, others placebo. Each dose level group has the same number of subjects, and the active/placebo ratio is fixed (e.g., six subjects on active drug and two on placebo per dose group). In this way, each subject is exposed to just one dose (active or placebo). Dosing groups are examined separately and sequentially (only one dose group at a time is in the clinic under evaluation). Other designs include "within-subject" designs, which require each subject receive more than one dose level, and pioneer designs in which two different active doses are administered during the same study period, with only a small group of "pioneer" subjects testing the higher dose.

STARTING DOSES

Ideally, the initial dose administered to humans would be selected to achieve peak plasma drug concentrations just below the target drug concentrations from pharmacology models for efficacy. However, this level of precision is not possible; a reasonable goal for selecting a starting dose is to identify a dose that is neither too high (in the toxic range) nor too low (requiring many escalations to reach MTD). As choice of dose is critical, detailed preclinical pharmacology, pharmacokinetic, toxicology, and toxicokinetic data must be available. A common approach is to use dosing safety factors of 10, 6, and 3 for rat, dog, and nonhuman primate, respectively, based on the species found to be most sensitive to the toxicologic effects of the drug.[8,11] For example, if rats were the most sensitive species, the starting dose for the human acute dose tolerance study would be one tenth of the largest no-observable-effect dose (mg/kg) from the chronic rat toxicity study.

Interspecies scaling based on physiological modeling or, more commonly, allometry, has been used to predict pharmacokinetics in humans from animal pharmacokinetic data.[12-14] This has been applied to choosing doses for acute dose tolerance studies, since dose selection can then be based on pharmacokinetic principles. However, this approach has recently been criticized as less predictive than usually envisioned and providing estimates of human pharmacokinetic parameters that are so imprecise that its prospective use to choose doses for first-in-human dose tolerance studies is not warranted.[15]

For subchronic dose tolerance studies, the starting dose is usually based on the first dose that exhibits any kind of pharmacological effect during the acute dose tolerance study. This dose should be administered as the total daily dose, with the dosing interval based on the half-life of the drug or duration of the pharmacodynamic response.

DOSE ESCALATION

The goal of dose escalation is to determine the maximum tolerated dose both efficiently and conservatively. Optimally, any scheme should not produce long and expensive Phase I studies, and at the same time should avoid the risks of overdosing and serious adverse events. One approach is to double doses with each escalation until a pharmacological response is observed, and proceed more conservatively with subsequent escalations, for example, calculating increases based upon a modified Fibonacci series (Table 4.1).

Perhaps a better approach than this arbitrary dose-driven escalation is pharmacokinetic or drug concentration-driven dose escalation. This requires flexibility in protocol design, rapid assay of plasma samples, and pharmacokinetic analysis of results between dosing escalations. However, there are many advantages with this approach. Issues related to bioavailability or nonlinear pharmacokinetics are immediately apparent and can be readily addressed. Selecting subsequent doses based on pharmacokinetic data from previous doses increases both safety and efficiency; fewer dose escalation steps are usually required. Another potential advantage is better correlation with safety and pharmacological effects.

TABLE 4.1
Dose Escalation Scheme for Dose Tolerance Trials

Doses (mg) to Administer before Onset of Pharmacological Activity[a]

N*	2N*	4N*	8N*	16N*	32N*	64N*

Possible Doses for Administration after Onset of Pharmacological Activity[b]

2N	4N	8N	16N	32N	64N	128N
3N	7N	13N	26N	52N	106N	211N
5N	10N	20N	40N	80N	160N	
7N	14N	28N	56N	112N	224N	
9N	18N	36N	72N	144N	288N	
12N	24N	48N	96N	192N		

[a] Doses are escalated across columns until a pharmacological response is observed.
[b] Escalation subsequent to pharmacological effect will proceed within a column.
Note: Starting dose (e.g., 1 mg); subsequent doses are expressed in multiples of the starting dose.

In either case, it is important to stress than under no circumstances should the next highest dose group start testing until sufficient exposure has occurred with the immediately preceding dose group to evaluate safety (see below).

SAFETY ASSESSMENT AND DETERMINATION OF MTD

Safety is monitored through basic laboratory and clinical observations (e.g., heart rate, blood pressure, respiration, electrocardiogram) made at frequent intervals throughout the expected duration of drug action. Clinical chemistry, hematology, and other appropriate laboratory observations are made at similar intervals. Special attention is paid to laboratory tests of organ function, especially critical target organs such as liver or kidney as well as the hematopoietic and reticuloendothelial systems. Because toxicity is often due to loss of pharmacological selectivity as doses and concentrations increase, the anticipated nonspecific receptor effects are monitored, if possible. Special tests to monitor the activity of particular organs of interest (e.g., pulmonary function tests) may also be used.

A dose-limiting adverse event is one that interferes with function (as assessed by either volunteer, investigator, or sponsor) but is not considered medically serious. Occurrence of a dose-limiting adverse event can be grounds for discontinuing a volunteer, but not terminating the study. The maximum tolerated dose is usually defined as one that causes dose limiting adverse events in some predefined fraction of the volunteers.[16,17] Because of the subjective nature of many adverse events, an additional criterion sometimes employed is that these adverse events must be reproducible when that same dose is repeated in a second group of volunteers. Any dose that produces *serious* adverse events (e.g., requiring hospitalization) in even one volunteer is beyond the MTD and its administration should be avoided by all means.

PHARMACOKINETIC EVALUATION

Pharmacokinetic evaluation typically includes noncompartmental analysis to characterize pharmacokinetics in terms of AUC or clearance, C_{max}, T_{max}, V_d, and half-life. This characterization should include an examination of pharmacokinetic parameters by dose level to look for obvious deviations from linearity. The high doses used in dose tolerance studies provide a unique opportunity to study linearity at concentrations that might only occur through overdosing in the clinical setting. Because of the relatively small number of subjects per dose level and because most designs involve separate groups of subjects receiving each dose, intersubject variability may mask some deviations from linearity. However, even with these constraints, dose tolerance studies can detect significant deviations from linearity. In subchronic studies, comparing steady state to single- or first-dose pharmacokinetics provides additional information about linearity and drug or metabolite accumulation. Pharmacokinetic evaluation should include an assessment of the variability in pharmacokinetic parameters. Characterizing variability is important because:

- Variability in pharmacokinetics can contribute to overall variability in therapeutic response.
- If safety data suggest a narrow therapeutic ratio, highly variable pharmacokinetics may be unacceptable and could ultimately result in a decision to discontinue the drug from development.
- An estimate of variability is required to optimize design of future studies, for example, to determine the number of subjects needed to achieve the desired statistical power in future pharmacokinetic studies, or to aid in selecting doses to produce the desired separation in plasma drug concentrations in Phase IIb studies.

Aside from trying to quantify variability, it may be possible to begin to identify sources of pharmacokinetic variability. The subject population for Phase I studies is by design small and relatively homogeneous in terms of demographics. Therefore, it will usually not be possible to investigate differences in pharmacokinetics due to age, gender, body size, etc. However, it may be possible to obtain preliminary information on variability due to nonlinear pharmacokinetic or saturable first-pass metabolism, diurnal variation, autoinduction, or genetic polymorphism.

Since dose tolerance studies usually produce adverse events, pharmacokinetic evaluation should include an assessment of the dose–concentration–toxicity relationship. In addition to regularly scheduled plasma samples obtained to calculate pharmacokinetic parameters, it is common to obtain a plasma sample at the time an adverse event is observed. Often it is possible to correlate acute toxicity with plasma concentrations as well as with dose, and occasionally even to develop a pharmacokinetic/pharmacodynamic model for acute or subchronic toxicity.

PHARMACODYNAMIC EVALUATION, PHARMACOKINETIC/PHARMACODYNAMIC MODELING

If the desired pharmacological effects are objective, quantitative, and the same as the therapeutic intent (e.g., increasing gastric pH to treat heartburn), then incorporating pharmacodynamic evaluations into Phase I studies can be extremely useful. If therapeutic effects are indirect or subjective (e.g., treatment of schizophrenia), there might not be a pharmacodynamic marker for the desired therapeutic effect, or it may only be meaningful or measurable in patients with the disease of interest. The challenge of Phase I pharmacodynamics is to carefully select the drug effect that best represents the desired therapeutic response, and to develop methods to measure it reproducibly with minimum variance. If this can be done, Phase I studies offer excellent opportunities for pharmacokinetic/pharmacodynamic modeling.[18-20] This modeling can be used to confirm or validate preclinical pharmacokinetic/pharmacodynamic models over a wide range of doses. Areas of uncertainty in the preclinical model can be explored, and confirmation of whether the dose or concentration–effect relationship is direct or indirect may be established. As with pharmacokinetics, variability in pharmacodynamics should be assessed. Once validated in Phase I, the pharmacokinetic/pharmacodynamic model can be useful in extrapolating beyond existing data to maximize the chance that effective and safe dosage regimens can be introduced as early as possible in Phase II studies.

PHASE II

Assuming the safety of the drug candidate was established in Phase I, the drug candidate will proceed to Phase II. Except for a few diseases like cancer, Phase II studies are the first to include patients with the disease intended for treatment. The introduction of disease represents a major step compared to testing in healthy subjects, and Phase II studies represent a major decision point for the continued development of a drug candidate. Aside from the Phase II studies themselves, there are a number of pharmacokinetic studies in healthy volunteers (Phase I clinical pharmacology studies) that are typically performed in parallel with Phase II. Again, this discussion focuses on the types of studies conducted while the drug candidate is "in Phase II," including these pharmacokinetic studies, as well as the Phase II studies themselves. Many of the basic concepts and practices previously described for Phase I apply to Phase II studies and thus are not repeated here.

GOALS OF PHASE II

Although safety data will continue to be collected and evaluated in Phase II, the emphasis of Phase II studies switches from safety and pharmacokinetics to efficacy and pharmacodynamics. Phase II studies are sometimes categorized as Phase IIa or Phase IIb depending on their goals:

- To prove the drug "works" in patients (Phase IIa)
- To determine the best dose, dose range, titration scheme, and dose interval (Phase IIb)

Phase IIa studies are sometimes called proof of concept or proof of principle studies. These studies are conducted to validate the principle upon which the drug was developed and provide unequivocal evidence of therapeutic effect. (For example, prove the drug inhibits enzyme x in patients as well as in the test tube and that this leads to a pharmacological effect resulting in clinical improvement or cure.) This involves addressing three related questions:

1. Proving the existence and functionality of the pharmacological or biochemical target in patients
2. Determining if the drug produces the intended pharmacological effect at clinically achievable concentrations
3. Determining if there is a therapeutic benefit from the pharmacological effect

Phase IIb studies explore the dose range and dose interval to determine the best dosing regimen to be used in Phase III. After Phase IIa demonstrates efficacy, it is sometimes tempting for sponsors to save time and expense by limiting Phase IIb, hoping that efficacy will be demonstrated in Phase III and that adverse events will be acceptable. Occasionally, this approach can be successful. For example, a mechanistic understanding of antibiotics has led to several being successfully developed with minimal Phase IIb programs: pharmacokinetics in healthy Phase I subjects may predict pharmacokinetics in patients, and *in vitro* activity may predict response. However, the risk of failing to define the dose–concentration–effect relationship in Phase IIb is a Phase III study containing the wrong dose and demonstrating unacceptable toxicity or marginal efficacy.

If Phase II studies do not demonstrate efficacy, then one must challenge the basic assumptions under which the drug was developed. In terms of pharmacokinetics, a drug candidate should not be considered ineffective in humans prior to achieving plasma concentrations that approach the effective concentrations in the animal species used for pharmacological testing.

PREREQUISITES FOR PHASE II

Preclinical prerequisites for Phase II are similar for Phase I, but with an emphasis placed on understanding preclinical models for efficacy. It is not possible to formulate a checklist for preclinical work required to support Phase II. Drugs that are new but have a known mechanism of action that has been validated in humans may require only minimal preclincal work. Drugs exploiting a known mechanism of action but with a new point of attack require more extensive pharmacological testing. New drugs with novel mechanisms of action will require extensive preclinical investigation. With truly innovative drugs based on new approaches, the validation of the preclinical model for efficacy will only come *after* the pivotal clinical studies. Therefore, it is vital that preclinical models provide pharmacodynamic measures that can also be applied in Phase II studies as markers of therapeutic efficacy.

STUDY DESIGN

Phase IIa studies are often conducted under optimal experimental conditions, using a relatively homogeneous group of uncomplicated patients (no concomitant illness, medication, or organ dysfunction) and doses close to MTD. These studies are closely monitored and tend to be short term. The number of subjects is relatively small, usually involving only one or two hospitals. Phase IIb studies are larger, involving more centers and a more diverse patient population. Phase II studies are typically double blind and, when possible, are "controlled," i.e., involve comparisons to a placebo and/or standard therapy. Elements of study design are similar to Phase I, but specifics will depend on the preclinical data and the pharmacokinetic, pharmacodynamic, and safety data obtained in Phase I, as well as the nature of the disease under investigation. If efficacy parameters are quantitative and objective, then one approach is to use a crossover design with the control agent(s) as treatment alternatives. If the disease has large natural intrapatient variability (e.g., arthritis) or subjective efficacy parameters (e.g., schizophrenia), then parallel designs with separate patient groups on active, placebo, or comparator drugs are common.

PHARMACOKINETIC/PHARMACODYNAMIC EVALUATION

Depending on the patient population, intensive plasma sampling schemes like those used in Phase I may or may not be feasible. If intensive sampling is not possible, population pharmacokinetic approaches, which require only a few samples per patient, can be used.[21]

Regardless of the techniques used, the goals should be to characterize the pharmacokinetics in the patient population, and to quantify the dose–concentration–effect relationship. Phase IIa studies allow the first characterization of pharmacokinetics in patients, but opportunities for pharmacodynamic evaluation or pharmacokinetic/pharmacodynamic modeling may be more limited than in Phase IIb. With Phase IIb data over several doses, it may be possible to define this relationship, or at least obtain estimates of the lowest useful concentration and the concentration beyond which additional response is not anticipated. Since Phase II patient populations tend to be larger and more diverse than the healthy volunteers used in Phase I studies, there are further opportunities to help identify sources of interindividual variability in pharmacokinetics and pharmacodynamics.

END OF PHASE II FDA CONFERENCE

The purpose of an "end of Phase II" meeting between the drug sponsor and FDA is to determine if it is appropriate to begin Phase III testing, and what this testing will entail. An information package containing the following elements should be submitted a month before the scheduled meeting:

- Summary results for all completed clinical studies, including statistical analysis of phase II data
- Summary of ongoing studies

- Summary of chemistry and preclinical data
- Phase III development plans, including protocols or protocol outlines of proposed pivotal Phase III studies, other remaining clinical studies (e.g., pharmacokinetic studies), and plans for any additional preclinical studies
- Proposed labeling claims or draft labeling
- List of specific questions for FDA reviewers

Agreement should be reached on the adequacy of the data submitted, plans to correct any deficiencies, the overall plans for Phase III, and on key elements of Phase III study design. With the potential exception of an advisory committee meeting, the end of Phase II meeting is perhaps the most important meeting with the FDA, because if properly conducted the sponsor should leave the meeting with a clear picture of what needs to be accomplished to gain approval of the drug candidate.

PK STUDIES IN PHASE II

After Phase I studies have established the safety of the drug candidate for continued development, there are numerous pharmacokinetic questions that need to be answered in addition to the basic questions of safety and efficacy addressed by Phase II and III clinical studies. These questions are answered with focused pharmacokinetic studies, the exact timing of which depends on the specifics of the drug and disease, as well as resources available and financial risks a sponsor is willing to take. Around the time of Phase II, pharmacokinetic studies are usually conducted to answer questions about metabolism in humans, pharmacokinetic linearity, and bioavailability.

HUMAN METABOLISM/MASS BALANCE STUDIES

Metabolism or mass balance studies help define the elimination characteristic of a drug in the body by investigating its biotransformation pathways and excretion pattern. This typically involves administration of a small quantity of ^{14}C-radiolabled drug. The information obtained, along with information from other pharmacokinetic, pharmacology, toxicology, and *in vitro* metabolism studies contributes to predictions about the safety of the drug in special patient populations, the reliability of animal toxicology data, and the design of future pharmacokinetic studies. Because of the fundamental nature of this information, human metabolism studies are often initiated soon after the Phase I dose tolerance studies have established the safety of the drug in healthy volunteers, in many cases before the initiation of the Phase II patient studies.

Specific goals for metabolism studies include:

- Determination of the elimination pathways (metabolism and excretion) of the drug
- Determination of the relative systemic exposure of parent drug and metabolites
- Determination of the distribution characteristics of the drug in blood

The cumulative recovery of >90% of the radioactive dose (urine + feces) is usually considered satisfactory to obtain information on the fate of the drug. If recovery is less than this, the reasons for the low recovery should be investigated. The excreted ^{14}C material can be profiled to identify structurally the major metabolites, and to determine their route of excretion (urine vs. feces). Together with *in vitro* studies, an elimination path can be determined, identifying the enzymes responsible for production of each metabolite and its route of excretion. By understanding which metabolic steps may be susceptible to induction or inhibition by other drugs, specific *in vitro* or clinical drug interaction studies may be conducted to determine the potential risks of drug–drug interactions. Similarly, an examination of excretion patterns can provide a preliminary indication of whether pharmacokinetics are likely to be altered in patients with renal or hepatic impairment.

The relative systemic exposure of the parent drug to the total drug-related substances (i.e., [^{14}C]-radioactivity) can be obtained by comparing the plasma AUC of the parent drug to that of total radioactivity. A large gap indicates significant metabolism, indicating the need for profiling to identify the structure of circulating metabolites. The definitive information on the extent of systemic exposure of each major metabolite should be obtained using a validated bioanalytical method. The relative systemic exposure to parent drug and metabolites, along with their pharmacological activities, will determine the most clinically relevant drug species to characterize in definitive pharmacokinetic studies.

The elimination profiles and relative exposure data in humans should be compared to that observed in animals. Differences in exposure to metabolites in humans and the animal species used for toxicology testing could limit the clinical relevance of toxicity data and suggest the need for additional testing. Similarly, differences in exposure to pharmacologically active metabolites could affect the interpretation of preclinical models for efficacy.

Binding of drug to plasma protein and the ratio of drug concentration in plasma to that in red blood cells should also be determined since this may influence the drug concentration available at the target site.

DEFINITIVE PHARMACOKINETICS AND DOSE PROPORTIONALITY

Crossover studies in healthy volunteers examining pharmacokinetic linearity, time dependency, and intra- and intersubject variability over the anticipated clinical dose range are generally required to ensure that the pharmacokinetic model developed for the drug candidate is suitable and predictive. Exceptions could include drugs with such low variability that definitive data on linearity are obtained from dose tolerance studies, or in cases where Phase IIb studies use crossover designs with extensive pharmacokinetic sampling.

At least three different doses should be tested, and the assessment of dose proportionality should be made from both single-dose and multiple-dose, steady-state data. These data can be obtained from one study if each treatment arm consists of a single dose followed by a washout and then multiple dosing to steady state. This design allows additional assessments of linearity by comparing pharmacokinetics after a single dose to that obtained over a dosing interval in the same subjects.

The time dependence of pharmacokinetic parameters and potential for metabolic induction can be also assessed.

BIOAVAILABILITY STUDIES

Phase II studies, especially Phase IIb studies, often require more-sophisticated formulations than those used in dose tolerance studies. However, it may be unwise to commit to a particular formulation strategy until after the clinical dose range is known. Therefore, formulations used in Phase II may not be as pharmaceutically elegant as marketed formulations. However, because bioavailability depends on formulation factors as well as intrinsic properties of the drug, bioavailability studies are conducted early in drug development to guide formulation development, establish the absolute or relative bioavailability of formulations used in clinical safety and efficacy studies, and determine the effect of food on bioavailability.

If an intravenous formulation can be developed, a crossover absolute bioavailability study in healthy volunteers can provide an unambiguous measure of clearance and volume of distribution as well as absolute bioavailability and absorption kinetics of the clinical formulation. Comparison of metabolite pharmacokinetics after oral and intravenous administration can provide information on first-pass metabolism.

If absolute bioavailability studies are not conducted, then the relative bioavailability of clinical formulations should be established by comparing them to a reference solution, ideally the same solution formulation used in dose tolerance studies. The use of common reference treatments in multiple studies facilitates comparison of data across studies. If the physical chemical properties of the drug suggest the potential for bioavailability problems, then it may be desirable to perform a bioavailability study to test alternate candidate formulations prospectively. Alternatively, if the drug is highly soluble and permeable, suggesting low risk for bioavailability problems, studies to document the bioavailability of clinical formulations can be done in parallel with (or after) the clinical studies.

Depending on how drug is to be administered in the Phase II and III studies, it may be useful to know the effect of food on absorption of drug from the clinical formulation prior to initiating these studies. This can often be examined in the same study used to characterize the bioavailability of clinical formulations.

The bioavailability studies discussed above are conducted primarily to manage the sponsor's risks in drug development (i.e., limit the possibility of a failed Phase IIb/III study due to poor or variable bioavailability). However, generating this same type of data on the final, to-be-marketed formulation is a regulatory requirement. If bioavailability is not anticipated to be a problem, and formulation development is not challenging, such that formulations used in Phase IIb and III trials represent the final, to-be-marketed formulation, then it may be possible to limit the bioavailability testing to the minimum regulatory requirements. More often than not, this represents unacceptable risks, clinical and final formulations are different, and bioavailability studies are conducted on both. The FDA has developed guidance documents (available on the Internet) that address study design and data analysis aspects of pivotal bioavailability, bioequivalence, and food effect studies. These should be consulted even if the sponsor anticipates repeating these studies with the final formulation at a later stage of development.

PHASE III

In Phase III clinical studies, the drug candidate is tested under conditions and in patients more closely resembling those who would be encountered if the drug were approved.

Phase III studies include the same type of blinded, randomized, controlled studies performed in Phase II, but in larger patient populations to provide statistical power to reject the null hypothesis of no treatment effect. The results of these pivotal trials are the primary factor determining the regulatory fate of the drug candidate. During Phase III, unblinded safety studies may also be conducted. These studies ensure that a sufficient number of patients are tested so that rare adverse events can be detected. Specific aspects of study design will depend on the drug, proposed patient population, and indication.

GOALS OF PHARMACOKINETICS IN PHASE III

Up until Phase III, the emphasis of pharmacokinetics has been on basic characterization, and on using this information to reduce drug development costs and time and to improve success rates by choosing the right doses and formulations to carry forward in the most well-designed studies possible. Once Phase III clinical studies are initiated, there are few drug development risks left to minimize. The focus now shifts to characterizing the remaining unknown sources of pharmacokinetic variability to identify subpopulations of patients who may have special risks or require dosage regimen adjustments. Two complementary approaches are used: characterizing pharmacokinetic variability in the patients participating in the Phase III clinical studies (population pharmacokinetics) and initiation of small, focused pharmacokinetic studies examining conditions that might be expected to affect pharmacokinetics.

POPULATION PHARMACOKINETICS

To characterize the pharmacokinetic and pharmacodynamic variability of a drug adequately, it needs to be studied in a representative population using relatively large numbers of patients. Considering all the volunteers participating in clinical studies, patients in Phase III studies are most representative of the population intended for treatment. Additionally, the patient population in Phase III is generally larger and more heterogeneous than that encountered in Phase I and II, so it becomes possible to examine the effects of various patient characteristics (e.g., age, body size, renal function, concomitant medications). Therefore, the Phase III patient population is an ideal target for pharmacokinetic and pharmacodynamic scrutiny. However, there are many practical reasons intensive plasma sampling, like that employed to characterize pharmacokinetics in Phase I, is not possible in Phase III studies. Sparse sampling, i.e., one to six plasma samples per patient obtained over the entire course of the study, is usually possible. Through the techniques of population pharmacokinetics, particularly nonlinear mixed effects modeling, these sparse data can be used to characterize pharmacokinetics, quantify variability, and examine its sources.[22,23] Regulatory expectations for this type of analysis are outlined in the 1999 FDA guidance document on population pharmacokinetics (available on the Internet).

Study Design

The primary goal of Phase III studies is to establish safety and efficacy; study design will be driven by this objective and is unlikely to be compromised to obtain pharmacokinetic information. Therefore, the methods of obtaining the data for population pharmacokinetics must be as simple as possible and not interfere with the major objectives of the study. Because population pharmacokinetics is by its nature an exploratory exercise, the modeling cannot be prospectively planned to a great level of detail. However, the objectives of the analysis, details of the data collection methods (sampling scheme, etc.) analysis population, pharmacokinetic parameters to be estimated, covariates to be explored, proposed methods of model building, and model validation should be determined prospectively and stated in the protocol or data analysis plan. Optimizing the plasma sampling design is especially important when there are severe limitations on the number of subjects or samples per subject. Similarly, defining the data analysis population requires considerable attention. In multicenter studies, it may be possible to obtain more subjects, or more samples per subject, from certain centers. This can be useful, but should be optimized so that the results obtained are genuinely representative of the overall patient population, and not just one or two more cooperative sites.

Data Analysis

Population pharmacokinetics usually consists of three steps: exploratory data analysis, population pharmacokinetic model development, and model validation.

Exploratory data analysis uses graphical and statistical techniques to reveal patterns in the data, identify potential outliers, and provide a diagnostic tool for confirming assumptions or suggesting corrective action if assumptions are not met.[24] Population pharmacokinetic model development integrates the basic pharmacokinetic model, which should have already been established from previous studies (e.g., two-compartment open model) with a model for the relationship between fixed effects (age, body weight, gender, creatinine clearance, etc.) and pharmacokinetic parameters (Cl, V_d, etc.) and a statistical model for the residual inter- and intrapatient variability.[25–27] Validation evaluates the predictive ability of the model by testing it against a different data set, either data from another study or data from the study in question in which a portion (e.g., 20%) of the total data set has been set aside for just such purposes.[28]

The results of the data analysis should provide population estimates for important pharmacokinetic parameters (Cl, V_d), equations describing how these parameters change with changes in covariates that were identified as significant, and a quantitative estimate of the magnitude of the unexplained variability in the pharmacokinetic parameters. An assessment of the unexplained, random variability is important because the efficacy and safety of a drug may decrease as unexplained variability increases. According to a survey conducted by the FDA and cited in their guidance documents, data from population pharmacokinetic studies provided useful new labeling information in 83% of submissions containing population pharmacokinetic data.[29] In the other 17% of submissions, the population pharmacokinetic analysis confirmed findings from previous pharmacokinetic studies and did not result in modifications to labeling.

Special Population and Drug Interaction Studies

Although the patient population used in Phase III studies should be fairly representative of the target patient population, it is common to exclude patients with extreme conditions that could reasonably be expected to result in severely altered pharmacokinetics, such as diseases of the organs involved in drug elimination (renal and hepatic disease). Exceptions include studies with drugs used to treat these diseases, or when these diseases are associated with the disease being investigated (e.g., renal disease in studies with drugs to treat diabetes). Similarly, unless the drug is targeted specifically for diseases of children or elderly people, very old and very young patients are sometimes excluded. It is also common to place some restrictions on concomitant medication. These exclusions are made for practical reasons and also represent a compromise. Patients must be representative of the target population, but by including patients with conditions likely to alter pharmacokinetics or response, the variability in clinical response may increase, thereby reducing the power to reject the null hypothesis of no treatment effect. Instead, separate studies are usually conducted in parallel during Phase III to examine the effect of these factors on pharmacokinetics:

- Renal disease
- Hepatic disease
- Elderly or pediatric populations
- Drug–drug interactions

The renal and hepatic patients and elderly volunteers participating in these studies typically do *not* have the disease intended for treatment. Entry criteria are usually designed to ensure that patients with varying degrees of renal or hepatic impairment are enrolled. A reference or control group of healthy volunteers with demographics closely matching the target patient population should also be included. If females have been excluded from previous pharmacokinetic studies (not recommended), the study in elderly individuals should include females in both the elderly group and the younger adult reference group. Comparing pharmacokinetics in these special populations to that in the reference or Phase III population should provide data upon which to base dosing recommendations for these patients. The FDA has developed guidance documents (available on the Internet) that address regulatory expectations for these types of investigations, with detailed suggestions related to study design, data analysis, and development of dosing recommendations.

Drug–drug interactions are a potential concern whenever there is concomitant therapy, and clinical drug interaction studies are often conducted in parallel with Phase III clinical trials. However, not all relevant drug combinations can be evaluated clinically.

At a fundamental level, a determination must be made whether a biological mechanism for an interaction exists. The most common pharmacokinetic mechanisms include metabolic inhibition or induction, protein binding displacement, and interference with absorption. Potential pharmacodynamic synergy or antagonism must also be considered. Interactions must be considered from the perspective of both changes affecting the drug candidate and changes affecting other concomitantly administered medication. If there is a mechanistic basis for a pharmacokinetic drug interaction, the next level of consideration takes into account expected changes in plasma drug concentrations. For example, even if metabolism of a drug candidate by a certain isoenzyme is completely blocked by another, coadministered drug, there may not be any apparent change in pharmacokinetics if that isoenzyme only contributes a small fraction to the total metabolism, or if the drug is also cleared renally.

Assuming there is a basis for believing the pharmacokinetics of the compound will be altered, the final level of analysis attempts to integrate this information with what is known about the safety and efficacy of the compound. If there is a sufficiently wide difference between the effective and toxic concentrations, a modest perturbation of plasma concentrations can be tolerated without a clinically relevant change in outcome. Determining which clinical drug interaction studies to conduct requires integrating preclinical, *in vitro* metabolism and human pharmacokinetic and safety data with mechanistic understanding of the basis for various potential interactions.[30,31] In some cases it may be possible to reach conclusions regarding potential drug interactions solely from preclinical or *in vitro* data. In other instances, *in vitro* data, for example, on inhibition of specific cytochrome P-450 isoenzymes, can be combined with limited clinical data and the results extrapolated to clinical situations involving other inhibitors of the same isoenzyme. By using basic pharmacokinetic principles, quantitative predictions can be made.[32] The FDA has developed guidance documents (available on the Internet) that address both *in vitro* and *in vivo* drug interaction studies and the level of evidence required for appropriate labeling regarding drug–drug interactions.

These special population and drug–drug interaction studies represent typical types of pharmacokinetic studies conducted in parallel with Phase III clinical trials. However, it would be unusual if a few of these studies were not conducted earlier. The priority of a given study will depend on the specific pharmacokinetic properties of the drug, the nature of the disease, and the target patient population. For example, if preliminary data suggest the drug candidate may have a narrow therapeutic range, and the human metabolism study demonstrates substantial first-pass metabolism solely by cytochrome P-450 3A4, it might be reasonable to conduct a ketoconazole interaction study in Phase II, irrespective of whether keotoconazole is likely to be a frequently encountered concomitant medication. Ketoconazole is a relatively strong 3A4 inhibitor, and testing this interaction represents a "worst-case" scenario. If this interaction is not clinically significant, it is unlikely that other 3A4 inhibitors will produce important interactions, and the use of these agents need not necessarily be restricted in Phase III trials.

NDA SUBMISSION AND APPROVAL

The new drug application (NDA) is the document through which a sponsor requests that the FDA approve a drug candidate for marketing. Included are all the individual study reports as well as integrated analyses attempting to demonstrate that the drug candidate is safe and effective in its proposed use and that its benefits outweigh its risks. Also included is the specific labeling that the sponsor proposes.

NDA Preparation

The general format of an NDA is specified by federal regulations, and specific format requirements for the pharmacokinetic section of an NDA are given in FDA guidance documents. The format of an NDA proceeds from high-level summaries to more-detailed summaries to individual study reports and data listings. Pharmacokinetic data are presented in the application summary, in the Human Pharmacokinetics and Bioavailability section (section 6), and in the Clinical Pharmacology portion of the Clinical Data section (section 8.3). The pharmacokinetic and pharmacodynamic data collected in humans must be included in their entirety. The summary volumes should integrate the data to present a clear picture and support the proposed labeling. Generally, this will take some skill in data interpretation. Data from different sources may all contribute to making a particular point. For example, data collected in Phase III population pharmacokinetics may overlap with data from special population and healthy volunteer Phase I studies. Phase III studies might include some patients older than those used in Phase I, or with higher serum creatinine than permitted by typical entry criteria for Phase I studies, but not as old or with the severest degree of renal insufficiency as those examined in special population studies. All these sources of data will need to be considered when describing the effect of age or renal disease on pharmacokinetics, and when developing dosing recommendations. Knowing that NDA preparation will provide an ideal opportunity for this sort of analysis, it is best not to "overinterpret" results when preparing individual study reports, especially for studies conducted early in the development process when it can be anticipated that many other studies will be conducted before submission.

Pre-NDA Meetings and NDA Filing

Pre-NDA meetings are held primarily to discuss the presentation and format of data supporting the application. Prior to the meeting, the sponsor should provide a summary of clinical studies submitted in the NDA, the proposed format for organization of the submission, and any outstanding questions. Although the primary purpose is to familiarize FDA staff with presentation and organization of the data to facilitate review, it also provides a final opportunity to discuss any unresolved problems or issues.

After the FDA receives the NDA, it undergoes a technical screening generally referred to as a "filability" review. This evaluation ensures that sufficient data and information have been submitted in each area to justify "filing" the application and beginning formal review of the NDA.

Advisory Committee Meetings

The FDA may ask an advisory committee to provide independent expert scientific advice to aid in its evaluation of safety and efficacy. Advisory committees consist of individuals with recognized expertise in a specific therapeutic area who are capable of evaluating information objectively and interpreting its therapeutic significance. Each committee also has a consumer advocate to represent the consumer perspective. As the name implies, the committees are advisory in nature;

the FDA makes final decisions regarding approval. Sponsors as well as FDA staff may make presentations to the committee. Although expertise in the therapeutic area of interest is a given, the depth of each member's understanding of pharmacokinetics and pharmacodynamics will vary. Because of this, and because of time limitations, the sponsor's presentation of pharmacokinetic data usually focuses on aspects that are directly related to safety and efficacy or that support specific proposed labeling statements.

NEGOTIATING APPROVAL

During the course of reviewing an NDA, FDA staff usually communicates with the sponsor about scientific, medical, and procedural issues that arise during the review process. Usually these involve questions regarding the data or easily correctable deficiencies found during the review. The FDA also informs sponsors of the need for additional data or information, or for technical changes in the application needed to facilitate the review. This type of early communication does not ordinarily apply to major scientific issues, which require consideration of the entire pending application by agency final decision makers. Instead, major scientific issues are usually addressed in an action letter at the end of the initial review process.

At the conclusion of the review, there are three possible action letters that can be sent to the sponsor:

1. A "not approvable" letter listing the deficiencies in the NDA and explaining why it cannot be approved.
2. An "approvable" letter signaling that, ultimately, the drug can be approved. The letter may point out minor deficiencies that can be corrected, or request labeling changes or commitments to do postapproval studies.
3. An "approval" letter stating that the drug is approved. This may follow an approvable letter, but can also be issued directly.

When an NDA nears approval, FDA reviewers also evaluate draft package labeling. Each element, including indications, use instructions, and warnings, is evaluated in terms of conclusions drawn from animal and human testing. All claims, instructions, and precautions must accurately reflect submitted clinical results.

The labeling "negotiation process," through which the final approved labeling of a drug is agreed upon, can take a few weeks to many months. The length of the process depends upon the number of comments and the sponsor's willingness to reach agreement. Sometimes a sponsor will submit several revisions of labeling before agreement on labeling can be reached.

FUTURE DEVELOPMENTS

Today it appears that future trends in drug development and industrial pharmacokinetics will be driven by three factors: the emergence of genomics, advances in bioanalytical technology, and the need for rapid proof of concept and better surrogate markers for drug efficacy.

EMERGENCE OF GENOMICS

Although there have been some setbacks, the promise of genomics and gene therapy is still large. Small molecules that modulate gene behavior will likely be developed, as well as genes themselves developed as therapeutic agents. There are many important development and pharmacokinetic issues to consider. The delivery of the therapeutic agent to the appropriate site with the right vector is obviously important, as is the efficiency of the delivery. What will be the proper tissue distributions, which distributions will be effective, and which might be harmful? The development of these agents will require new models for drug development. The paradigm of extrapolation from animal data and models to human studies will need to be reexamined.

ADVANCES IN BIOANALYTICAL TECHNOLOGY

Advances in bioanalytical technology made clinical pharmacokinetics a feasible endeavor by the 1970s. Modern bioanalytical technologies are now sensitive to the point that minute concentrations of drugs in plasma can be determined from doses that are so small as to have no expectation of producing any pharmacological effect whatsoever. This development, along with a recent FDA guideline to allow single-dose human studies supported with only single-dose toxicology data, raises the possibility of using the first study in humans to perform pharmacokinetic screening. A "cocktail" approach similar to that already used preclinically, where small doses of several drugs are administered simultaneously, may be possible to screen drug candidates for absorption, metabolic stability, and disposition in humans.

NEED FOR RAPID PROOF OF CONCEPT AND BETTER SURROGATE MARKERS

It is no wonder that drug companies consider the most commercially attractive diseases the ones with accepted surrogate markers. Blood pressure, glucose concentrations, and lipid concentrations are good examples. The lack of surrogates, by contrast, forces industry sponsors into large, expensive trials with clinical end points for mere hypothesis testing. As a consequence, any useful surrogate marker for innovative drugs identified in preclinical research is highly welcome to increase predictability for clinical outcome. This supports pharmacokinetic and pharmacodynamic modeling for first-in-human studies and fosters evaluation of the link between the pharmacodynamic action and therapeutic efficacy. Aside from continuing efforts in basic preclinical research, the future may also see selection of the more sensitive patients by genotyping them for a fit to the new therapy.

REFERENCES

1. DiMasi, J. A. et al., Research and development costs for new drugs by therapeutic category. A study of the US pharmaceutical industry, *Pharmacoeconomics*, 7:152–169, 1995.
2. DiMasi, J. A., Grabowski, H. G., and Vernon, J., R&D costs, innovative outputs and firm size in the pharmaceutical industry, *Int. J. Econ. Bus.*, 2:201–219, 1995.

3. Mathiew, M. P., Ed., *Parexel's Pharmaceutical R&D Statistical Sourcebook 2000*, Parexel International Corporation, Waltham, MA, 2000, 65–76.
4. Salmonson, T. and Rane, A., Clinical pharmacokinetics in the drug regulatory process, *Clin. Pharmacokinet.*, 18:177–183, 1990.
5. Peck, C. C. et al., Opportunities for integration of pharmacokinetics, pharmacodynamics, and toxicokinetics in rational drug development, *J. Pharm. Sci.*, 81:605–610, 1992.
6. Reigner, B. G. et al., An evaluation of the integration of pharmacokinetic and pharmacodynamic principles in clinical drug development. Experience within Hoffmann La Roche, *Clin. Pharmacokinet.*, 33:142–152, 1997.
7. Metzler, C. M. and VanderLugt, J. T., Medical and statistical design issues in clinical pharmacology, *Drug Inf. J.*, 24:281–288, 1990.
8. Posvar, E. L. and Sedman, A. J., New drugs: first time in man, *J. Clin. Pharmacol.*, 29:961–966, 1989.
9. Skinner, J. B., Early exploration of safety and efficacy, *Drug Inf. J.*, 24:325–339, 1990.
10. Colburn, W. A., Phase I and II trials: designs and strategies, *Drug Inf. J.*, 30:573–582, 1996.
11. Boxenbaum, H. and DiLea, C., First-in-human dose selection: allometric thoughts and perspectives, *J. Clin. Pharmacol.*, 35:957–966, 1995.
12. Voisin, E. M. et al., Extrapolation of animal toxicity to humans: interspecies comparisons in drug development, *Regul. Toxicol. Pharmacol.*, 12:107–116, 1990.
13. Mahmood, I., Allometric issues in drug development, *J. Pharm. Sci.*, 88:1101–1106, 1999.
14. Lave, T., Coassolo, P., and Reigner, B., Prediction of hepatic metabolism clearance based on interspecies allometric scaling techniques and *in vitro – in vivo* correlations, *Clin. Pharmacokinet.*, 36:211–231, 1999.
15. Bonate, P. L. and Howard, D., Prospective allometric scaling: does the emperor have clothes? *J. Clin. Pharmacol.*, 40:335–346, 2000.
16. Cutler, N. R. et al., Defining the maximum tolerated dose: investigator, academic, industry, and regulatory perspectives, *J. Clin. Pharmacol.*, 37:767–783, 1997.
17. Culter, N. R. et al., Defining the maximum tolerated dose: an update, *J. Clin. Pharmacol.*, 40:1183–1204, 2000.
18. Breimer, D. and Danhof, M., Relevance of the application of pharmacokinetic-pharmacodynamic modeling concepts in drug development, *Clin. Pharmacokinet.*, 32:259–267, 1997.
19. Kroboth, P. D., Schmith, V. D., and Smith, R. B., Pharmacodynamic modeling, application to new drug development, *Clin. Pharmacokinet.*, 20:91–98, 1991.
20. Sheiner, L. B. and Steimer, L. J., Pharmacokinetic/pharmacodynamic modeling in drug development, *Annu. Rev. Pharmacol. Toxicol.*, 40:67–95, 2000.
21. Vozeh, S. et al., The use of population pharmacokinetics in drug development, *Clin. Pharmacokinet.*, 30:81–93, 1996.
22. Samara, E. and Granneman, R., Role of population pharmacokinetics in drug development, a pharmaceutical industry perspective, *Clin. Pharmacokinet.*, 32:294–312, 1997.
23. Jackson, K. A. and Rosenbaum, S. E., The application of population pharmacokinetics to the drug development process, *Drug Dev. Ind. Pharm.*, 24:1155–1162, 1998.
24. Ette, E. I. and Ludden, T. M., Population pharmacokinetic modeling: the importance of informative graphics, *Pharm. Res.*, 12:157–168, 1995.
25. Peck, C. C., Population approach in pharmacokinetics and pharmacodynamics: FDA view, in *Proceedings of the COST B1 Conference*, 1992, 157–168.

26. Mandema, J. W., Verotta, D., and Sheiner, L. B., Building population pharmacokinetic models I. Models for covariate effects, *J. Pharmacokinet. Biopharm.*, 20:511–528, 1992.

27. Mandema, J. W., Verotta, D., and Sheiner, L. B., Building population pharmacokinetic-pharmacodynamic models, in *Advanced Pharmacokinetic and Pharmacodynamic Systems Analysis*, Argenio, D. Z., Ed., Plenum Press, New York, 1995, 69–86.

28. Ette, E. I., Stability and performance of a population pharmacokinetic model, *J. Clin. Pharmacol.*, 37:486–495, 1997.

29. Ette, E. I. et al., The population approach: FDA experience, in *The Population Approach: Measuring and Managing Variability in Response, Concentration and Dose*, Balant, L. P. and Aarons, L., Eds., Commission of the European Communities, European Cooperation in the Field of Scientific and Technical Research, Brussels, Belgium, 1997.

30. Fuhr, U. et al., Systemic screening for pharmacokinetic interactions during drug development, *Int. J. Clin. Pharmacol. Ther.*, 34:139–151, 1996.

31. Thompson, T. M., A strategy for use of *in vitro* metabolism studies to evaluate potential drug-drug interactions: M100907 as a case study, in *Predicting Drug Metabolism*, Rudolph, N. S. and Tullock, M. H., Eds., AdvanceTech Monitor, Woburn, MA, 2000, 96–106.

32. Rowland, M. and Martin, S. B., Kinetics of drug-drug interaction, *J. Pharmacokinet. Biopharm.*, 1:553–567, 1973.

5 Bioavailability and Bioequivalence*

Veneeta Tandon

CONTENTS

---

* This chapter was written by Veneeta Tandon in her private capacity. No official support or endorsement by the Food and Drug Administration is intended or should be inferred.

1-56676-973-6/02/$0.00+$1.50
© 2002 by CRC Press LLC

INTRODUCTION

The U.S. Food and Drug Administration (FDA) has developed certain guidelines for pharmaceutical companies to ensure that the drug products have adequate bioavailability to produce a therapeutic effect. There are many drugs that are available in the market as generic products, in which case the generic product must produce a therapeutic effect equivalent to the brand name. To facilitate the decision of therapeutic equivalency between pharmaceutically equivalent drug products, the FDA requires bioequivalence testing for multisource drug products. Whereas clinical studies are conducted to determine the safety and efficacy of a drug product, bioavailability studies are performed to estimate the absorption, distribution, and elimination of the drug. The release of drug substance from the drug product, along with the permeability and presystemic effects on drug substance, contribute to systemic exposure patterns. Bioavailability studies are critical for efficient drug development.

Both bioavailability and bioequivalence studies are required by the regulations, depending on the type of application, i.e., Investigational New Drug Applications (INDs), New Drug Applications (NDAs), Abbreviated New Drug Applications (ANDAs), and their supplements. The regulatory requirements for the documentation of bioavailability and bioequivalence are provided in the Code of Federal Regulations (21 CFR 320), which contains two subparts. Subpart A covers general provisions and definitions, and subpart B covers procedures for determining bioavailability and bioequivalence of drug products.[1] In addition to the regulations provided in the 21 CFR 320, the FDA has developed various guidelines, which are published on the Internet (http://www.fda.gov). This chapter discusses the general principles and methods for bioavailability and bioequivalence testing.

DEFINITIONS

The following relevant definitions have been provided by the Code of Federal Regulations[1] to understand the principles behind bioavailability and bioequivalence testing:

- **Bioavailability:** The rate and extent to which the active ingredient or active moiety is absorbed from the drug product and becomes available at the site of action. For drug products that are not intended to be absorbed into the bloodstream, bioavailability may be assessed by measurements intended to reflect the rate and extent to which the active ingredient or active moiety becomes available to the site of action.
- **Pharmaceutical Equivalents:** Drug products that contain identical amounts of the identical active ingredient, i.e., the same salt or ester of the same therapeutic moiety, in identical dosage forms, but not necessarily containing the same inactive ingredients, and that meet the identical compendial or other applicable standard of identity, strength, quality, and purity, including potency and, where applicable, content uniformity, disintegration times, and/or dissolution rates.

- **Pharmaceutical Alternatives:** Drug products that contain the identical therapeutic moiety, or its precursor, but not necessarily in the same amount or dosage form or as the same salt or ester. Each drug product individually meets either the identical or its own respective compendial or other applicable standard of identity strength, quality, and purity, including potency and, where applicable, content uniformity, disintegration times, and/or dissolution rates.
- **Therapeutic Equivalents:** Drug products are considered to be therapeutic equivalents only if they are pharmaceutical equivalents and if they can be expected to have the same clinical effect and safety profile when administered to patients under the conditions specified in the labeling.
- **Bioequivalence:** The absence of a significant difference in the rate and extent to which the active ingredient or active moiety in pharmaceutical equivalents or pharmaceutical alternatives becomes available at the site of action when administered at the same molar dose under similar conditions in an appropriately designed study.

THE NEED FOR BIOAVAILABILITY AND BIOEQUIVALENCE STUDIES

Bioavailability and bioequivalence studies can provide useful information regarding the drug, such as:

- In the strict sense, bioavailability studies provide an estimate of the fraction of the orally administered dose that is absorbed into the systemic circulation when compared to the bioavailability for a solution, suspension, or intravenous dosage form that is completely available.
- Bioavailability studies provide other useful information that is important to establish dosage regimens and to support drug labeling, such as distribution and elimination characteristics of the drug.
- Bioavailability studies provide indirect information regarding the presystemic and systemic metabolism of the drug and the role of transporters such as p-glycoproteins.
- Bioavailability studies designed to study the food effect provide information on the effect of food and other nutrients on the absorption of the drug substance.
- Such studies when designed appropriately provide information on the linearity or nonlinearity in the pharmacokinetics of the drug and the dose proportionality.
- Bioavailability studies provide information regarding the performance of the formulation and subsequently are a means to document product quality.
- Bioequivalence studies provide a link between the pivotal and early clinical trial formulation, a link between formulations used in the pivotal clinical trial and the stability studies, the pivotal clinical trial and the to-be-marketed drug product, and other comparisons as appropriate.

- Bioequivalence studies are the basis for determination of the therapeutic equivalence between a pharmaceutically equivalent generic drug product and a corresponding reference listed drug. This list is provided in the book titled "Approved Drug Products with Therapeutic Equivalence Evaluations," commonly known as the Orange Book.[10]
- Bioequivalence studies provide information on product quality and performance when there are changes in components, composition, and method of manufacture after approval of the drug product. The FDA has provided guidance for the industry, such as SUPAC-IR[2] and SUPAC-MR,[3] to determine when changes in components and composition and/or method of manufacture of the drug product suggests a need to perform further *in vitro/in vivo* studies.

BIOAVAILABILITY AND BIOEQUIVALENCE TESTING RECOMMENDED BY THE FDA

Some of the situations when bioavailability and bioequivalence testing is essential for a drug are highlighted below:

- For all new molecular entities
- For new formulations of active drug ingredients.
- For a new dosage form of a drug
- For a new dosage strength or dosage regimen
- For a new salt or ester of a drug
- For a new indication
- For the administration in special patient populations, e.g., pediatrics
- For a change in the manufacturing process or site of the drug substance in which there is a change in the physical properties of a drug substance
- For a change in the manufacturing process of the drug or the drug product that produces variabilities beyond the specifications of approved applications

BIOAVAILABILITY ASSESSMENT METHODS

Bioavailability is the measurement of the rate and extent of drug that is systemically available. Hence, pharmacokinetic parameters that give information on the amount of drug reaching the systemic circulation (extent) and the time taken to reach the systemic circulation (rate) are used as measures for assessing bioavailability. Bioavailability can be measured by direct and indirect methods outlined below. The various methods used are presented in descending order of preference.

DIRECT MEASURES OF BIOAVAILABILITY

Based on Plasma Drug Concentrations

Drug concentrations in the blood and plasma are the most direct methods of determining the systemic availability of a drug. The pharmacokinetic parameters that describe the rate and extent of absorption and systemic exposure based on plasma drug concentration data are summarized below.

The area under the plasma drug concentration and time curve (AUC_t, AUC_∞, or AUC_τ)(units = $\mu g.h/ml$, $ng.h/ml$, etc.): AUC is the measure of the extent of drug bioavailability. This gives a measure of the total systemic exposure. The AUC can be calculated from Equation 5.1, which can be rearranged as Equation 5.2 to determine the fraction of the dose absorbed.

$$AUC_0^\infty = \frac{FD}{k_{el}V_d} = \frac{FD}{\text{Clearance}} \tag{5.1}$$

$$F = \frac{Cl(AUC)}{D_0} \tag{5.2}$$

where
F = fraction of dose absorbed
D = dose
k_{el} = elimination rate constant
V_d = volume of distribution
Cl = clearance

AUC can be obtained by a numerical integration method such as the trapezoidal rule. A recent recommendation by the FDA has been the use of early exposure as a measure of rate of systemic exposure.[5] This can be calculated as a partial AUC, where the area can be truncated at the population median of the t_{max} values. Measurement of early exposure may be useful when rapid onset of action is desirable (e.g., an analgesic effect) or if a slow input is required to achieve efficacy or safety. The FDA has recently proposed a shift away from the focus on rate and extent of absorption to the measurement of systemic exposure, which can be determined as total, peak, or early exposure (if needed). This is based on the understanding that these measures better reflect the rate and extent of absorption.[5-7]

The peak plasma drug concentration (C_{max}) (units = $\mu g/ml$ or ng/ml, etc.): The C_{max} is also a measure of the extent of bioavailability or peak exposure and indicates concentrations required for a therapeutic or toxic response. It relates to peak exposure of the drug. C_{max} is obtained directly from the plasma concentration time profile.

The time to peak plasma drug concentration (t_{max}) (units = hours, minutes, etc.): The t_{max} is a measure of the rate of drug absorption and is the time required to reach the maximum drug concentration after drug administration. The t_{max} is obtained directly from the plasma concentration time profile.

INDIRECT MEASURES OF BIOAVAILABILITY

Based on Urinary Excretion Data

This method can be used only if urinary excretion of unchanged drug is the main mechanism of elimination of the drug and urine samples have been collected in intervals as short as possible to measure the rate and amount of excretion as accurately as possible. Bioavailability can be calculated from Equation 5.3 using the urinary excretion data as follows.

$$F = \frac{(D_u^\infty)/\text{Dose}}{f} \qquad (5.3)$$

where F = fraction of the dose absorbed

D_u^∞ = cumulative amount of drug excreted in the urine

f = fraction of unchanged drug excreted in the urine

The pharmacokinetic parameters that describe the rate and extent of absorption and systemic exposure based on urinary excretion data are as follows:

- Cumulative amount of drug excreted in the urine (D_u^∞): This is directly related to the total amount of drug absorbed. When the plasma concentration approaches zero, then the maximum amount of drug excreted in the urine is obtained (i.e., D_u^∞). This is a measure of the extent of drug absorption.
- The rate of drug excretion in the urine (dD_u/dt)
- Time for maximum urinary excretion (t^∞)

Limitations of Using Urinary Data

- There is a high degree of variability associated with the cumulative amount of drug excreted in the urine, and the method is less reliable compared with the estimation of bioavailability from plasma concentration time profiles.
- Urinary data should be collected for a period of time equal to five times the half-life corresponding to the terminal phase of the drug concentration–time profile to achieve 97% recovery after a single dose.
- Urinary data are valid only if the excretion of the drug or metabolite is related to the bioavailable dose of the drug.
- Urinary data cannot be reliably used to determine bioequivalency, C_{max}, t_{max}, absorption rate, and duration. Theoretically, the data could be used for determination of bioequivalency, but practically they will not be reliable because of the high degree of variability that could be associated with the parameter estimation.

Based on Acute Pharmacodynamic Effect

This approach may be applicable when the drug is not intended to be delivered into the bloodstream for systemic availability. It is an indirect measure of bioavailability in cases where the analytical method for assessing drug concentrations in the plasma or other biological fluids cannot be developed. In such cases a dose–response relationship must be established. This method can be used only if the method is sensitive, accurate, and reproducible. The pharmacodynamic parameters evaluated to assess bioavailability are the following:

- Total area under the pharmacodynamic effect–time curve
- Peak pharmacodynamic effect
- Time to peak pharmacodynamic effect

From Well-Controlled Clinical Trials

Well-controlled clinical trials that establish safety and efficacy of a drug product, for purposes of establishing bioavailability, can be used. However, this approach is the least accurate, sensitive, and reproducible approach. This approach can be used when analytical methods cannot be developed for a particular drug.

From Dissolution Studies

In vitro dissolution studies are used to assess product quality. In ideal circumstances *in vitro* dissolution rate should correlate with *in vivo* bioavailability. A dosage form with a rapid dissolution rate is likely to have a rapid rate of drug bioavailability *in vivo*. However, bioavailability is not only dependent on the dissolution of the drug product, but also the permeability and solubility of the drug substance. When an *in vitro–in vivo* correlation is available, the *in vitro* test can serve as an indicator of how the product will perform *in vivo*.

ABSOLUTE AND RELATIVE BIOAVAILABILITY

ABSOLUTE BIOAVAILABILITY

Absolute bioavailability of a drug is the systemic availability of the drug after extravascular administration of the drug and is measured by comparing the area under the drug concentration–time curve after extravascular administration to that after IV administration, provided the k_{el} and V_D are independent of the route of administration. Extravascular administration of the drug comprises routes such as oral, rectal, subcutaneous, transdermal, nasal, etc.

Absolute bioavailability is denoted as F, which is also the fraction of the dose that is absorbed. After IV administration, the entire dose is placed into systemic circulation; therefore, the fraction of the dose absorbed (F) or the absolute bioavailability is equal to unity. For routes other than IV administration $F \leq 1$, absolute bioavailability is most commonly expressed as a percentage, where an F of 1 is 100% bioavailable or an F of 0.8 is 80% bioavailable.

Absolute bioavailability can be calculated from the following equations:

From Plasma Data

$$\text{Absolute Bioavailability} = F = \frac{[\text{AUC}]_{PO}/\text{dose}_{PO}}{[\text{AUC}]_{IV}/\text{dose}_{IV}} \qquad (5.4)$$

where PO refers to oral administration.

From Urinary Excretion Data

$$\text{Absolute Bioavailability} = F = \frac{[D_u]^{\infty}_{PO}/\text{dose}_{PO}}{[D_u]^{\infty}_{IV}/\text{dose}_{IV}} \qquad (5.5)$$

RELATIVE BIOAVAILABILITY

The relative bioavailability is the systemic availability of a drug from one drug product (*A*) compared to another drug product (*B*). Relative bioavailability can be calculated from Equations 5.6 and 5.7 using plasma concentration time data.

From Plasma Data

$$\text{Relative Bioavailability} = \frac{[\text{AUC}]_A}{[\text{AUC}]_B} \qquad (5.6)$$

When the doses are not equal, the dose-adjusted AUC values should be used as follows.

$$\text{Relative Bioavailability} = \frac{[\text{AUC}]_A/\text{dose}_A}{[\text{AUC}]_B/\text{dose}_B} \qquad (5.7)$$

From Urinary Excretion Data

Urinary excretion data can also be used to calculate relative bioavailability provided the urine collections are adequate. Equation 5.8 can be used to calculate relative bioavailability from the urine data.

$$\text{Relative Bioavailability} = \frac{[D_u]^{\infty}_A}{[D_u]^{\infty}_B} \qquad (5.8)$$

FACTORS AFFECTING BIOAVAILABILITY

Some of the important factors that affect bioavailability are outlined as follows:

- Gastric emptying — Although not true in all cases, increased gastric emptying generally enhances bioavailability of orally administered drugs. Gastric emptying depends on the following factors:
  - Volume of liquid intake
  - Volume of solid food intake and its fat content
  - Viscosity of stomach content

- pH of the stomach
- Intake of other drugs
- Age and weight of the patients
- Physical activity of the patients taking drug
- Emotional state of the patient
- Various disease states
- The variability seen in the absorption of orally administered drugs is mainly due to different rates of gastric emptying, which are affected by the various factors listed above. Hence, to minimize variability, bioavailability studies may be conducted under controlled conditions, such as healthy individuals of controlled weight and age under fasted conditions or with a controlled diet. The use of healthy subjects minimizes both inter- and intrasubject variability.
- Presystemic and systemic metabolism — Presystemic metabolism, which occurs during first-pass metabolism, can decrease the bioavailability of a drug. The following types of metabolism are commonly seen:
  - First-pass metabolism: First-pass metabolism occurs when an absorbed drug passes directly through the liver before reaching systemic circulation after oral administration.
  - Intestinal metabolism: Drug metabolizes in the intestine itself or during the passage through the intestinal wall.
  - Hydrolysis of the drug in the stomach fluids.
  - Transporters such as p-glycoprotein may influence the bioavailability of a drug.
- Complexation with other agents in the gastrointestinal tract.
- Formulation factors, such as may occur with inert ingredients, the manufacturing process and/or use of surfactants, etc.

DESIGN OF BIOAVAILABILITY AND BIOEQUIVALENCE STUDIES

Both bioavailability and bioequivalence focus on measuring the absorption of the drug into systemic circulation; hence, similar study design approaches are used to establish bioavailability of a drug or to assess bioequivalence. Bioavailability is a comparison of the drug product to an intravenous formulation, a solution, or a suspension, whereas bioequivalence is a more formal comparative test that uses specified criteria for comparisons with predetermined bioequivalence limits for evaluation.

The study design for bioequivalence mainly depends on the criteria for evaluation. Since July 1992, the Center for Drug Evaluation and Research (CDER) has recommended the use of the *Average Bioequivalence Criterion* as published in the guidance entitled "Statistical Procedures for Bioequivalence Using a Standard Two-Treatment Crossover Design."[8] This criterion calls for a conventional nonreplicate crossover study for evaluating bioequivalence. Recently, two new approaches have been described for evaluating bioequivalence, which are termed the *Individual*

Bioequivalence Criterion and the *Population Bioequivalence Criterion.* The Individual Bioequivalence Criterion calls for a replicate study design, whereas the Population Bioequivalence Criterion does not involve a replicate study design, but a replicate crossover design or parallel design, which can also be used for this criterion.[9] A replicate study design is one in which both the test and the reference drug products are administered to the same individuals on two separate occasions.

The general study design considerations for conducting bioavailability or bioequivalence studies are as follows:

- An initial pilot study with a smaller number of subjects to assess variability, optimize sample collection time (as suitable for the immediate release and modified release dosage forms), and other useful information.
- A conventional two-formulation, two-period, two-sequence nonreplicate crossover design. This design is used for the Average Bioequivalence Criterion. It is also used if a Population Criterion is chosen for bioequivalence comparisons. This study is recommended for a single-dose study. A single-dose bioequivalence study is generally more sensitive in assessing release of the drug substance from the drug product into systemic circulation for both conventional and modified release products. The nonreplicate design is recommended by the FDA for most orally administered immediate-release dosage forms.[4]
- A replicate-crossover study design with four periods, two sequences, and two formulations. The FDA recommends the use of the Average Bioequivalence Criterion for this study design as well,[4] although this study design is not necessary when an average approach is used to establish bioequivalence. Replicate crossover designs allow for estimation of within-subject variances for the Test (*T*) and Reference (*R*) measures and the subject-by-formulation interaction component. The same lot of the test and reference formulation should be used for the replicated administration. This design is desirable for modified-release dosage forms or highly variable drug products, and is suitable for an Individual Bioequivalence approach.
- A parallel design could also be used under special circumstances, for example, a drug with a long half-life.
- The reference standard in a bioequivalence study is a formulation currently marketed with an approved full NDA, for which there are valid scientific safety and efficacy data. The list of reference products is provided in the Orange Book ("Approved Drug Product with Therapeutic Equivalence Evaluations").[10] The reference product is usually the innovator's brand-name product. The total content of the active drug substance in the test product must be within 5% of the reference product. Usually similar routes of administration are used for the test and reference products unless an alternative route is needed to answer specific pharmacokinetic questions. In some cases the reference material could be a solution, suspension, IV product, or the clinical trial material containing the same quantity of active drug ingredient.

- Healthy subjects are preferred as the study population and should be ≥18 years of age. In some cases it may be useful to conduct the study in patients. A heterogeneous population would be preferable that includes males and females, young and elderly people, and subjects from different racial groups or the targeted age and gender if the drug product is to be specifically used in those populations.
- An adequate number of subjects should be enrolled to allow for dropouts. It is not desirable to replace dropouts. At least 12 subjects should be included in a study.
- The highest marketed strength should be used for evaluating bioequivalence.
- The test and reference product should be administered with 240 ml of water.
- The test and reference drug products should be administered under fasting conditions (overnight) and the fast should continue for up to 4 h after dosing. Subjects should abstain from alcohol for 48 h prior to each study period and until after the last sample from each period is collected. Subjects can be allowed water as desired except for 1 h before and after drug administration.
- An adequate washout period should separate each treatment.
- Plasma and blood samples are preferred over urine and other tissue samples for evaluating drug/metabolite concentrations. An adequate number of samples should be taken to characterize the absorption, distribution, and elimination phases of the drug/metabolite accurately.
- For bioavailability studies, the parent compound or the active moiety and the active metabolites should be measured if analytically feasible. For bioequivalence studies, the measurement of the parent compound is desirable, unless the parent drug levels in the plasma or serum are too low to allow reliable measurements. In addition to measuring the parent, the measurement of the metabolite is important when it contributes to either safety or efficacy of the drug product. The bioequivalence criterion is applied to the parent with supportive evidence from the metabolite measurements. Similarly, measurement of enantiomers or racemate may be necessary as appropriate.

PHARMACOKINETIC INFORMATION FOR EVALUATION OF BIOEQUIVALENCE STUDIES

The following pharmacokinetic information for the drug should be obtained for the evaluation of bioequivalence studies:

- Area under the plasma concentration–time curve from zero to time t (AUC_t) and its log transformation (ln AUC_t)
- Area under the plasma concentration–time curve from zero to infinity (AUC_∞) and its log transformation (ln AUC_∞)
- Peak drug concentration (C_{max}) and its log transformation (ln C_{max}) and time to peak drug concentration (t_{max})

- Elimination rate constant (k_{el}) and half-life of the drug ($t_{1/2}$)
- C_{min}, C_{ave}, and degree of fluctuation [$(C_{max} - C_{min})/C_{ave}$], swing [$(C_{max} - C_{min})/C_{min}$], and evidence of attainment of steady state, if steady-state studies are used.

BIOEQUIVALENCE EVALUTION CRITERIA

In the past 20 years the evaluation criteria for bioequivalence studies recommended by the FDA have evolved and been revised several times. For bioequivalence comparisons the new formulation or method of manufacture is the test product (T) and the prior formulation or method of manufacture is the reference (R) product. To establish bioequivalence, the difference between the bioavailability of the test product and the reference product must be within the prespecified bioequivalence limit as governed by the approach taken to assess bioequivalence, which is discussed in this section.

The first approach that was used by the FDA to evaluate bioequivalence was the *75/75/125 Rule*, which required that a test and reference ratio for 75% of the subjects should fall between the interval of 75 to 125%. In subsequent years this approach was replaced by the *Power Approach*, which utilized a standard *t*-test for testing equivalence. The Power Approach consisted of testing the hypothesis of no difference at a 0.05 level with an estimated power of 0.80 to detect a 20% difference in the means of the test and reference.

The current evaluation criteria are based on the two one-sided test approach, also commonly referred to as the *Confidence Interval Approach* or *Average Bioequivalence*, which determines whether the average values for the pharmacokinetic parameters measured after the administration of test and reference products are comparable. This approach involves the calculation of a 90% confidence interval about the ratio of the averages of T and R products for AUC and C_{max} values. To establish bioequivalence, the AUC and C_{max} of the T product should not be less than 0.80 (80%) or greater than 1.25 (125%) of the R product based on log-transformed data (i.e., a bioequivalence limit of 80 to 125%). For some time prior to the use of log-transformed data, the nontransformed data were used to assess bioequivalence. In 1989, it was realized that log transformation of the data enables a comparison based on the ratio of the two averages rather than the difference between the averages in an additive manner.[11] Moreover, most biological data correspond to a log-normal distribution rather than to a normal distribution.

More recent proposals discussed for evaluating bioequivalence are based on approaches termed *Individual Bioequivalence* and *Population Bioequivalence*. The average bioequivalence approach focuses only on the comparison of population averages (μ_T, μ_R) of a bioequivalence metric of interest and not on the variances of the metric for the T and R products. The individual bioequivalence approach not only compares the population averages (μ_T, μ_R), but also assesses the within-subject variability (σ_{WT}^2, σ_{WR}^2) as well as the subject-by-formulation interaction (σ_D^2). The population bioequivalence approach is designed to assess the total variability, i.e., within- and between-subject variability (σ_{TT}^2, σ_{TR}^2) of the pharmacokinetic parameter (metric) in the population. The individual and population bioequivalence approach allows the use of a mixed scaling, which takes into account the variability of the R product (termed as reference scaling) or a specified constant (termed as constant scaling). Reference scaling is used when the R product is highly variable; otherwise constant scaling is used.[9]

The bioequivalence for the evaluation of bioequivalence (BE) based on the average, individual, and population approaches are given in Equations 5.9 through 5.11. This criteria should be ≤BE limit (θ_A, θ_I, θ_p for average, individual, and population approaches, respectively) for each approach:

Average BE

$$(\mu_T - \mu_R)^2 \leq \theta_A^2 \tag{5.9}$$

Individual BE

$$\frac{(\mu_T - \mu_R)^2 + \sigma_D^2 + (\sigma_{WT}^2 - \sigma_{WR}^2)}{\sigma_{WR}^2} \leq -\theta_1 \tag{5.10}$$

Population BE

$$\frac{(\mu_T - \mu_R)^2 + (\sigma_{TT}^2 - \sigma_{TR}^2)}{\sigma_{TR}^2} \leq \theta_P \tag{5.11}$$

STATISTICAL MODELS TO ASSESS BIOEQUIVALENCE

Log-transformed data are used for comparisons to represent a normal distribution of the data. For AUC and C_{max}, the log of ratio (ln T/R or log T/R) between the test and reference are used for comparisons. The arithmetic mean for the test and reference products, geometric means, means of the logs, standard deviations of the logs, or coefficients of variation should be calculated for AUC and C_{max}. For population bioequivalence, the estimates of the total variance and for individual bioequivalence approach the subject-by-formulation interaction variance and the within-subject variance for the T and R product should be determined.

General linear model or mixed effect model procedures are performed on the pharmacokinetic parameters AUC and C_{max} to test the data for difference within and between test and reference groups. For a general linear model, the statistical model should include factors accounting for various sources of variability, such as sequence, subjects, study period, and treatment or formulation depending on the study design.

For the average bioequivalence approach, two one-sided tests of hypothesis at the 5% level of significance are carried out to construct the 90% confidence intervals. For the population and individual bioequivalence approach, an upper 95% confidence bound for the population or individual criterion is estimated, which should be less than or equal to the bioequivalence limit (i.e., θ_I, θ_p).

CRITERIA FOR WAIVER OF EVIDENCE OF *IN VIVO* BIOAVAILABILITY OR BIOEQUIVALENCE STUDIES

Under the following circumstances at the applicants' request, the FDA[1] may waive the requirement for *in vivo* bioavailability studies for a drug product if the drug product meets any of the following provisions:

- The following drug products that also meet the condition of containing the same active and inactive ingredients in the same concentration as a drug product that is the subject of a full approved NDA can receive a waiver for *in vivo* evidence of demonstrating bioavailability of the drug product:
  a. The drug product is a parenteral solution intended solely for administration by injection.
  b. The drug product is an ophthalmic or otic solution.
- The following drug products that also meet the condition of containing the same active ingredients in the same dosage form as a drug product that is the subject of a full approved NDA can receive a waiver for *in vivo* evidence of demonstrating bioavailability of the drug product:
  a. The drug product is administered by inhalation as a gas, e.g., a medicinal or inhalation anesthetic.
  b. The drug product is a solution for application to the skin.
  c. The drug product is an oral solution, elixir, syrup, tincture, or a similar solubilized form. These products should not contain any inactive ingredient that is known to significantly affect absorption of the active drug ingredient.
- If the drug product is in the same dosage form, but in a different lower strength and the following conditions have been met:
  a. The drug product is proportionally similar in its active and inactive ingredients to another product for which the same manufacturer has obtained approval by meeting the bioavailability requirements for a submission.
  b. Both drug products meet an appropriate *in vitro* test approved by the FDA.
  c. An *in vivo* study has been conducted on the highest strength.

These criteria could be used for immediate-release tablets or capsules.

- If the drug product is in the same dosage form, but in a higher strength, the waiver for *in vivo* bioavailability will depend on:
  a. Clinical safety or efficacy data
  b. Linear elimination kinetics over the dose range
  c. Higher strength being proportionally similar to the lower strength
  d. Similar dissolution profiles

These criteria could be used for immediate-release tablets or capsules.

- If the drug product is a modified-release dosage form, a lower strength could be waived if the following conditions are met:

a. For beaded capsules, the difference should only be in the amount of beads present and the lower-strength capsules should have similar dissolution profiles.

b. For tablets, the lower-strength tablet should be compositionally proportional to the higher strength; both should have the same drug release mechanism and similar dissolution profiles.

- The drug product shows an *in vitro–in vivo* correlation.
- The drug product is a reformulated product that is identical, except for a different color, flavor, or preservative that could not affect the bioavailability of the reformulated drug product, to another drug product for which the same manufacturer has demonstrated bioavailability and obtained approval and that has an FDA approved *in vitro* test.
- *In vivo* bioavailability requirements may be waived for a good cause that is compatible with the protection of public health.
- For a drug product that was approved prior to 1962 and is determined to be effective in at least one indication in a Drug Efficacy Study Implementation (DESI) notice and is listed not to have a potential bioequivalence problem.
- Recently the FDA has proposed the waiver of bioequivalence studies for immediate-release solid oral dosage forms for a Class 1 drug substance based on the Biopharmaceutics Classification System (highly soluble and highly permeable) and for a rapidly dissolving product.[12]

LIMITATIONS OF BIOAVAILABILITY AND BIOEQUIVALENCE STUDIES

For drugs and dosage forms with certain characteristics, systemic bioavailability and general bioequivalence studies cannot be designed and evaluated as conventional studies. Some of these situations are listed below.

- A crossover design may be difficult for drugs with a long elimination half-life. Three to four elimination half-lives may extensively prolong the duration of the study in a crossover design. In this situation a parallel design can be used for bioequivalence studies.
- Highly variable drugs may require a far greater number of subjects to meet the FDA bioequivalence criteria.[13] The variability seen in the performance of certain drug products may be due to the inherent characteristics of the drug or due to the drug formulation or both.
- Certain characteristics in the biotransformation of drugs make it difficult to evaluate the bioequivalence of such drugs. For example, for drugs that are stereoisomers with a different rate of biotransformation and a different pharmacodynamic response, the measurement of independent isomers may be difficult for analytical reasons.
- Drugs that are administered by routes other than the oral route or drugs/dosage forms that are intended for local effects have minimal systemic bioavailability. Some examples of such drug classes are the ophthalmics, dermals, intranasal, and inhalation drug products.

Bioequivalence assessments of drugs that are insignificantly absorbed into the systemic circulation are difficult. In some cases, for such drugs a biological marker has been established for the assessment of bioequivalence. Examples of biological markers used are skin blanching in the case of hydrocorticosteroids and neutralization of stomach acid for antacids. For certain cases a pharmacodynamic end point may be more appropriate for the assessment of bioequivalence.

REFERENCES

1. Code of Federal Regulation: Bioavailability and Bioequivalence Requirements, Vol. 21, Part 320, U.S. Government Printing Office, Washington, D.C., 2000.
2. Scale up and Post Approval Changes: Immediate Release Solid Oral Dosage Forms: Chemistry, Manufacturing and Controls, *in Vitro* Dissolution Testing and *in Vivo* Bioequivalence Determination, Guidance issued by FDA, November 1995.
3. Scale up and Post Approval Changes: Modified Release Solid Oral Dosage Forms: Chemistry, Manufacturing and Controls, *in Vitro* Dissolution Testing and *in Vivo* Bioequivalence Determination, Guidance issued by FDA, June 1997.
4. Bioavailability and Bioequivalence Studies of Orally Administered Drug Products— General Considerations, Guidance issued by FDA, October 2000.
5. Chen, M. L., An alternative approach for the assessment of rate of absorption in bioequivalence studies, *Pharm. Res.*, 9:1380–1385, 1992.
6. Bois, F. Y. et al., Bioequivalence: performance of several measures of rate of absorption, *Pharm. Res.*, 11:966–974, 1994.
7. Tozer, T. N. et al., Absorption rate vs. exposure: which is more useful for bioequivalence testing? *Pharm. Res.*, 13:453–456, 1996.
8. Statistical Procedures for Bioequivalence Studies Using Standard Two-Treatment Crossover Studies, Guidance Issued by FDA, June 1992.
9. Statistical Approaches to Establishing Bioequivalence, Guidance issued by FDA, January 2001.
10. Approved Drug Products with Therapeutic Equivalence Evaluations, 19th ed., U.S. Department of Health and Human Services, Washington, D.C., 1999.
11. Schuirmann, D. J., Treatment of bioequivalence data: log transformation, in *Proceedings of Bio-international '89 — Issues in the Evaluation of Bioavailability Data*, Toronto, Canada, October 1–4, 1989, 159–161.
12. Waiver of *in Vivo* Bioequivalence Studies for Immediate Release Solid Oral Dosage Forms Based on a Biopharmaceutics Classification System, Guidance issued by FDA, August 2000.
13. Shah, P. et al., Evaluation of orally administered highly variable drugs and drug formulations, *Pharm. Res.*, 13:1590–1595, 1996.

Section III

Clinical Applications

6 General Approaches to Clinical Pharmacokinetic Monitoring

Ronald A. Herman

CONTENTS

INTRODUCTION

Basic pharmacokinetics has focused on a detailed scientific study of what the body is doing to the drug to facilitate the following:

- Liberation of the drug from the dosage form
- Absorption of the compound into the systemic circulation
- Distribution to various sites of action
- Metabolism to active or inactive compounds
- Elimination from the body.

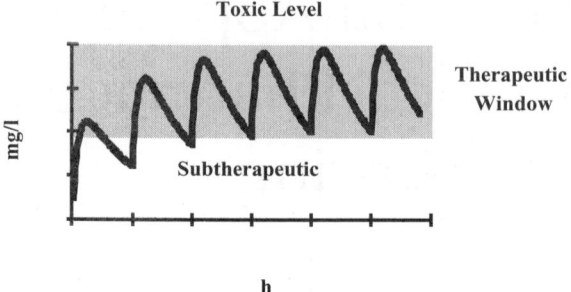

FIGURE 6.1 The therapeutic window.

Many medications being developed today must be used carefully because too much drug in the body may have toxic effects, or not enough drug can result in ineffective therapy because the threshold of the therapeutic response is not achieved. The application of basic pharmacokinetic knowledge to these medications that have this well-defined therapeutic window is referred to as therapeutic drug monitoring. Figure 6.1 illustrates this concept of a therapeutic window. Consequently, the area of pharmacokinetics that monitors therapeutic concentrations has evolved to another scientific subdiscipline of pharmacokinetics. Taking the basic pharmacokinetic information and combining that with a patient's actual physiological response to the medication is what clinical pharmacokinetics is all about. Gerhard Levy[1] has said, "Simply stated, clinical pharmacokinetics is a health science discipline that deals with the application of pharmacokinetics to optimize the pharmacotherapeutic management of individual patients." The terms *clinical pharmacokinetics*, *therapeutic drug monitoring* (TDM) and *applied pharmacokinetics* have all been used to describe the process of taking drug concentrations, basic pharmacokinetic principles, and the person's clinical response and combining them to optimize drug therapy for the patient. This is demonstrated in Figure 6.2.

Adjustment of drug dosage on the basis of individualized target drug concentrations can achieve better outcomes in therapy by preventing or minimizing adverse effects or reaching desired therapeutic outcomes more rapidly.[2–6] However, for this approach to be successful, there are three important things to consider: accurately determined drug concentrations, correctly selected pharmacokinetic equations, and

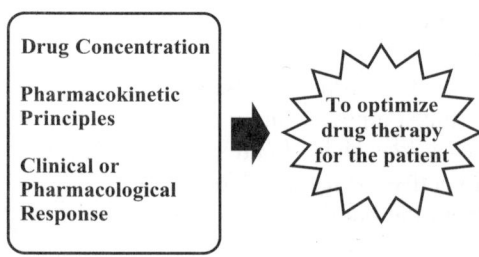

FIGURE 6.2 Clinical pharmacokinetics or TDM is applied pharmacokinetics.

a thorough understanding of the patient's medical conditions that can influence clinical and pharmacological response. They are all indicated in Figure 6.2.

DRUG CONCENTRATION

The clinical pharmacokineticist involved in making recommendations to adjust drug therapy must make sure that the blood level determinations that are being used to adjust therapy are properly collected and determined. Invalid starting information can result in inappropriate alterations in drug therapy. In general, blood level determinations are done when a medication has a narrow therapeutic index and monitoring levels of the drug in the sampling compartment can be helpful to keep the concentrations above the minimum threshold, and below any potential toxic levels.[7] If a compound has no documented toxicity in the range of typically used doses, there is no need to monitor levels.

WHY DO WE MEASURE DRUG LEVELS?

There are usually three reasons for monitoring drug levels:

1. Determine the best possible safe dosing regimen. When therapy is initiated it may be beneficial to measure levels to help design a dosing regimen that will keep levels within the desired therapeutic window.
2. If the organ systems that determine the rate at which the drug leaves the body are changing, then it is advisable to measure drug levels to maintain the drug at optimal levels.
3. Outpatients who have been stabilized on a dosing regimen sometimes benefit from having levels checked to confirm patient compliance or aid in the identification of noncompliance.

WHEN DO WE DRAW THEM?

Most Often at Steady State

Figure 6.1 illustrates that, if drug levels are measured before steady state is achieved, therapeutic adjustments could result in concentrations much higher than desired. Dosage adjustments can be made from non-steady-state levels, but the pharmacokineticist must make sure that the appropriate pharmacokinetic equations are chosen that include accumulation factors and that it is known how many doses have been given, and the timing of those doses. Many clinicians choose to wait until steady state is achieved before drug levels are measured to assess the current therapy so that simplified steady-state equations can be used.

Sometimes a Single Level

If the drug is given by a constant-rate infusion, or if the peak–trough fluctuation is minor, then it is sufficient to draw only a single steady-state level as shown in Figure 6.3.

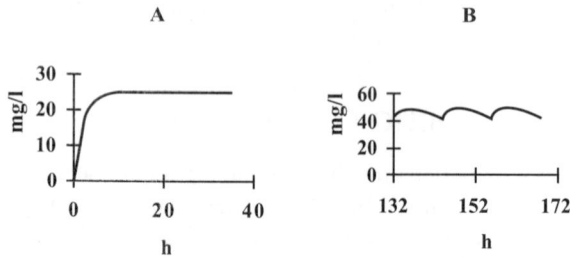

FIGURE 6.3 Plasma concentrations of constant rate infusion (A) and minimal peak–trough fluctuation (B).

Often a Peak and Trough

If the drug is given by extravascular administration or intermittent infusion and it demonstrates a significant difference in concentration before and after dosing, then generally a peak and a trough concentration will need to be measured. The semilog concentration vs. time plot of Figure 6.4 demonstrates this.

Optimal Sampling Times

Depending on the pharmacokinetic equations that are used to individualize therapy, the timing of samples may be very important.[8] The peak for an intravenous (IV) bolus dose would be obtained immediately after the dose is given. The peak concentration following an IV infusion of the drug will usually occur right after the infusion stops. If the concentration of the drug is plotted against time on a semilog plot and if the declining concentration following absorption or the infusion is not a straight line, then the drug exhibits a measurable distribution phase. If the drug demonstrates a significant distribution phase, as in Figure 6.4 and if the pharmacokinetic analysis will assume a one-exponential elimination, then the peak must be measured when distribution is complete.

Once distribution is complete, the slope of the log concentration vs. time curve from the measured peak down to the trough will be linear. A steady-state trough sample should be drawn immediately before the next dose is given after the peak

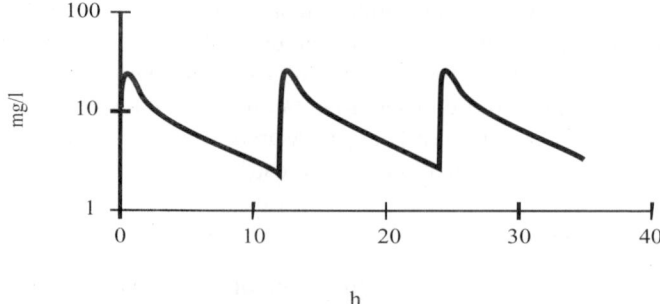

FIGURE 6.4 Often a peak and trough are necessary.

has been obtained. Generally, if the sample is drawn within 30 min of the dose being given, it can be treated as if it were drawn immediately before the dose was given. Ideally, the peak is measured and then the trough at the end of the dosing interval. Experience has shown that quite often those levels are done on two different nursing shifts. The more people involved in the collection of the blood samples, the greater the chance for error, like forgetting to draw the sample, drawing it at the wrong time, etc. Therefore, to be practical and to reduce errors, it is routine to draw a trough before a dose is given, then measure the peak at the appropriate time following the drug administration. Since the samples are being obtained at steady state, the trough obtained before a dose should be exactly the same as the trough following the peak. Therefore, that value is extrapolated to postdose trough and the pharmacokinetics can then be determined using this extrapolated value. Each drug that will be examined in subsequent chapters has optimal sampling times. Take note of these optimal sampling times. For example, gentamicin and tobramycin levels should be obtained at steady state immediately before a dose is given and 30 min after the intermittent infusion stops. Vancomycin levels should be obtained immediately before a dose is given and 60 min after the infusion stops.

How Are They Measured?

Normally, drug concentrations are obtained by venipuncture of venous blood. Arterial samples are much more difficult to obtain and in most cases arterial and venous concentrations of the drug are the same. Plasma is the fluid portion of whole blood that remains after the solid materials (red blood cells, white blood cells, and platelets) have been centrifuged off. Serum is the liquid that remains after blood has been allowed to clot. Thus, fibrin and the solid materials have been removed. Normally, serum and plasma concentrations of the drug are the same, because there is very little binding of most drugs to fibrin or other components of the blood clot.[9] Plasma samples are obtained by using an anticoagulant to prevent the blood from clotting. Heparin or a calcium-binding resin like EDTA, citrate, or fluoride can be used to prevent the clotting. However, some of these anticoagulants can interfere with the assay of some drugs. Consequently, as a general rule, drug concentrations are usually determined in serum.

There are many assay techniques that can be used to determine the drug concentrations that will be used to individualize drug therapy. Microbiological assays were one of the early assays used to determine the concentration of antibiotics. These assays tend a have a fairly high degree of error and they take a long time to achieve results (24 to 72 h). Radioimmunoassays tend to be able to detect reasonably low levels of compounds (sensitivity), but generally do not have high specificity (metabolites and other substances can easily interfere with the determination of the actual drug concentration). High-performance liquid chromatography (HPLC) has a reasonable sensitivity and specificity, tends to be inexpensive, and is generally very rapid, so it is a commonly used method of determining drug concentrations. Gas liquid chromatography (GLC) usually has a high specificity, is very sensitive, and is fast; however, it tends to be very expensive. Consequently, it is used frequently for research purposes, but not often for routine clinical monitoring of drug concentrations. Probably the most frequently used assay method for therapeutic drug monitoring is an enzyme multiplied immunoassay. Generally these are prepared in kits that have a very reasonable sensitivity and specificity and they tend to be very inexpensive and very rapid.

It is also very important to take note of the units in which assay results are reported. When the concentration information is used in the pharmacokinetic equations, it is important that the proper concentration units be used to obtain the correct units for volume of distribution and half-life. Once again, as individual drugs are reviewed, take note of the preferred assay technique, the proper sampling matrix, and the units of concentration reported.

BASIC PHARMACOKINETIC PRINCIPLES

Applied pharmacokinetics involves predicting steady-state responses. The three main approaches to pharmacokinetic analysis involve the use of physiological models, traditional compartmental models, or a model-independent (linear systems) approach. Physiological compartmental models are very complex and very good for describing in detail what the body is doing to the drug. However, often it is difficult to develop strong predictive relationships for drug response in the clinical environment because often there is limited information about all of the model parameters for the patient. Model-independent, or noncompartmental approaches, are suitable in some cases, but often dynamic clinical data without an extensive sampling scheme cannot effectively predict pharmacokinetic behavior. Consequently, compartmental models alone or sometimes in conjunction with a Bayesian forecasting approach are often used to optimize therapy.

Approval of new drug applications is only finalized when there is a thorough understanding of the basic pharmacokinetics of a drug. Pharmacokinetic studies have been conducted in animals in preclinical trials and in humans at each subsequent stage of the drug development process. For each drug, this basis of scientific knowledge of what the body is doing to the drug should lead to a concrete plan as the best way to monitor and individualize drug therapy for this compound. Drugs that have a linear response to dose and are eliminated in a mono-exponential fashion can use simple one-compartment model equations to predict response to future therapy. Drugs that exhibit a biphasic decline in concentration when the log concentration is plotted against time will require a more complex two-compartment model. If either absorption or elimination are dependent upon a saturable process, then a nonlinear pharmacokinetic model will be necessary to try to describe the drug response. Sometimes it is possible to take a model-independent approach and determine those parameters that are useful to use the drug wisely. Occasionally, it is even possible to take other information about the patient and use that to help forecast the person's response. Therefore, the clinical pharmacokineticist must be able to select the appropriate pharmacokinetic model to use when taking drug concentrations and using them to individualize therapy. It is necessary to choose a kinetic model that is descriptive of the time course of drug concentration and its relationships to therapeutic effect.

CLINICAL RESPONSE

The physiological processes that determine the rate at which a dr.ig leaves that body have a profound effect on the dose and drug concentration vs. time profile. In addition, many genetic characteristics and environmental factors, including diet and concomitant use of other drugs, can all impact this relationship. Physiological variables such as age, gender, body composition, disease, and pregnancy also affect the disposition of drugs and therefore can modify the relationship between the dose and therapeutic response. The clinical pharmacokineticist must not only ensure appropriate drug concentrations have been obtained and that the correct pharmacokinetic equations have been chosen, but also must know enough about the patient's health and well-being that adjustments can be made to therapy based on anticipated clearance changes the patient may exhibit.

THERAPEUTIC DRUG MONITORING

SINGLE-DRUG CONCENTRATIONS

Medications that exhibit a linear response or dose-independent response can have the therapy optimized by using a single blood level determination. If the drug is given by a constant-rate infusion or if it is an extravascular administration, the peak–trough fluctuations are minor, and the patient is at steady state, then a simple proportion can be used.

$$D_{new} = \frac{D_{current} \cdot C_{desired}}{C_{ss, measured}} \tag{6.1}$$

It is necessary to know the target, or desired, concentration and the current dose to make a modification in therapy. This approach can be used for lidocaine infusion, or for slow release theophylline, for digoxin, phenobarbital, and procainamide.

PEAK AND TROUGH CONCENTRATIONS

Individualization of dosage for patients receiving intermittent infusions or extravascular administration and for which a peak and trough concentration is required will need three types of information. The therapeutic target concentrations: the desired peak and trough concentrations must be known. Also, the patient's steady-state volume of distribution will be required as an estimate of the extent of drug distribution. Last, an estimate of the person's rate of drug elimination is required. The target concentrations will be chosen based upon information known about the patient's current health and the treatment goals. The steady volume of distribution and the rate of elimination can be estimated prospectively from population estimates for the patient or they can be determined retrospectively from measured peak and trough drug levels. (See illustration in Figure 6.5.)

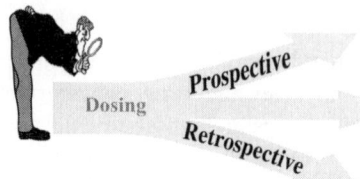

FIGURE 6.5 Therapeutic drug monitoring.

Prospective Population Estimates

An example of a population estimate for the steady-state volume of distribution of vancomycin may be 0.7 mg/kg of corrected body weight. Body weight, or perhaps body weight in conjunction with an estimate of the person's hydration status, is often used to estimate volume of distribution. If the drug is not lipophilic, then it will not distribute well into adipose tissue. Therefore, for this type of drug, an obese patient's actual body weight (ABW) is not a good estimate of steady volume of distribution. Because excess weight has significant interstitial fluid associated with it, one cannot exclude the excess weight and compute only lean body mass (LBW). As a result, a body weight correction is utilized if a person's actual body weight is more than 30% over his or her lean body weight.. The corrected body weight, or adjusted body weight that is used is 40% of the excess over lean body mass. Therefore, for hydrophilic drugs in obese patients, the dosing weight (DWT) or corrected body weight is

$$DWT = LBW + 0.4 \cdot (ABW - LBW) \tag{6.2}$$

The lean body weight or ideal body weight can be obtained from insurance tables or by using the following relationships:

$$
\begin{aligned}
LBW \; &= \; 50 \text{ kg} + 2.3 \text{ kg (for each inch over 5 ft)} && \text{for males} \\
&= \; 45.5 \text{ kg} + 2.3 \text{ kg (for each inch over 5 ft)} && \text{for females}
\end{aligned} \tag{6.3}
$$

The population estimate of rate of elimination may be a regression relationship based upon creatinine clearance (Cl_{cr}) or some other factor(s) that correlate well with the elimination of the drug. For example, an estimate of the elimination rate for vancomycin is $K_{el} = 0.000954 * Cl_{cr} + 0.0036$.[10] Measurement of creatinine clearance is usually unpleasant for patients and difficult to perform accurately, so often creatinine clearance is predicted. One of the commonly used methods of predicting Cl_{cr} is the Cockcroft and Gault method:[11]

$$Cl_{cr} = \frac{(140 - Age) \cdot LBW}{72 \cdot SrCr_{ss}} \cdot (0.85 + 0.15 \cdot Sex) \tag{6.4}$$

where Sex is 1 for a male and 0 for a female and $SrCr_{ss}$ is serum creatinine.

This equation estimates Cl_{cr} in ml/min, which is a patient-specific value. This is what would be used in the prospective equation to estimate the elimination rate. Some of the other methods of estimating creatinine clearance compute an estimate that is normalized to a standard body surface area, usually 1.73 m². These estimates would need to be un-normalized to estimate a patient-specific elimination rate. To normalize a Cl_{cr} estimate:

$$\text{NormCl}_{cr} = Cl_{cr} \cdot \frac{1.73}{\text{BSA}} \tag{6.5}$$

The body surface area (BSA) can be determined using a person's weight in kilograms and height in centimeters according to the following relationship, which has been shown to be useful for children and adults by Haycock and co-workers:[12]

$$\text{BSA} = \text{Wt}^{0.5378} \cdot \text{Ht}^{0.3964} \cdot 0.024265 \tag{6.6}$$

Population estimates of steady-state volume of distribution, elimination rate, and target concentrations are listed for each drug in subsequent chapters.

RETROSPECTIVE DETERMINATIONS

The measured peak and trough concentration can be plotted on a graph with the slope being the rate of elimination and the extrapolated intercept is the initial concentration. They are used to determine an estimate of the steady-state volume of distribution. However, the following equations, which Sawchuk and Zaske[13] originally published, can be easily used to obtain the same information with greater reliability.

• Calculate the elimination rate constant:

$$k_l = \frac{\ln C_1 - \ln C_2}{t_2 - t_1} = \frac{\ln C_{pk} - \ln C_{tr}}{t_{tr} - t_{pk}} = \frac{\ln(C_{pk}/C_{tr})}{\tau - t_{inf} - t_{pi}} \tag{6.7}$$

Here t_{pk} is the time the peak was measured relative to the start of the infusion and t_{tr} is the elapsed time of the trough relative to the start of the infusion that preceded the peak. τ is the length of the dosing interval, t_{inf} is the length of the infusion, and t_{pi} is the time from when the infusion stopped to when the peak is measured. All times are in hours.

• Calculate C_0:

$$C_0 = \frac{C_{pk}}{e^{-k_{el}(t_{pk} - t_{inf})}} \tag{6.8}$$

This is an extrapolation of the linear portion of the curve back to when the infusion stopped and is an approximation of the concentration at the end of the infusion. This corresponds to the maximum steady-state concentration.

- Calculate the half-life:

$$t_{1/2} = \frac{\ln 2}{k_{el}} \tag{6.9}$$

- Calculate the steady-state volume of distribution:

$$V_{ss} = \frac{R_0}{k_{el}} \cdot \frac{1 - e^{-k_{el}t_{inf}}}{(C_0 - C_{tr} \cdot e^{-k_{el}t_{inf}})} \tag{6.10}$$

Please note that R_0 is the zero-order rate of input, the infusion rate. It is the dose (in mg) divided by the length of the infusion (in hours). A very frequent error is to use the dose, but fail to divide by the length of infusion. Remember C_0 is not the measured peak; rather, it is the value extrapolated back to when the infusion stopped. In the denominator another frequent mistake is made when the trough concentration (C_{tr}) is subtracted from C_0 before the multiplication. The trough must be multiplied by the exponential term ($e^{-k_{el}t_{inf}}$) before it is subtracted from C_0.

- Calculate the dosing interval:

$$\tau = \frac{\ln(C_{\text{Max, desired}} / C_{\text{Min, desired}})}{k_{el}} + t_{inf} \tag{6.11}$$

Remember to select a practical dosing interval. Every 6, 8, 12, 24, 36, or 48 h would all be practical intervals.

- Calculate the new infusion rate:

$$R_0 = C_{\text{Max, desired}} \cdot k_{el} \cdot V_{ss} \cdot \frac{(1 - e^{-k_{el}\tau})}{(1 - e^{-k_{el}t_{inf}})} \tag{6.12}$$

This equation determines an infusion rate in mg/h. To compute a dose to be ordered for the patient, take R_0 times the length of infusion that will be used. Then, select a practical dose. For example, vancomycin is often rounded to the nearest 250 mg to allow for the most economical preparation of the dose to be delivered to the patient.

- Calculate the new peak using the practical dosing interval and the practical dose that have just been determined.

$$C_{ss, pk} = \frac{R_0}{V_{ss} \cdot k_{el}} \cdot \frac{(1 - e^{-k_{el}t_{inf}})}{(1 - e^{-k_{el}\tau})} \tag{6.13}$$

- Calculate the new trough.

$$C_{ss,\,tr} = C_{ss,\,pk} \cdot e^{-k_{el}(\tau - t_{inf})} \tag{6.14}$$

This approach to dosing patients retrospectively will allow the clinical pharmacokineticist to optimize therapy by maintaining concentrations within certain therapeutic guidelines.

SUMMARY

Properly collected drug concentrations can be very useful to optimize drug therapy for those agents that have a narrow therapeutic window. Prospectively and retrospectively monitoring therapeutic drug concentrations can improve patient care and reduce hospital costs.[2–7] The prospective and retrospective dosing relationships can be incorporated into spreadsheets that can be used in handheld personal computers at the patient's bedside. They can also be programmed into programmable calculators that are also useful for bedside consultations.

REFERENCES

1. Levy, G., An orientation to clinical pharmacokinetics, in *Clinical Pharmacokinetics — A Symposium*, Levy, G., Ed., American Pharmacology Association, Washington, D.C., 1974, 1–9.
2. Whiting, B. et al., Clinical pharmacokinetics: a comprehensive system for therapeutic drug monitoring and prescribing, *Br. Med. J. (Clin. Res. Ed.)*, 288(6416):541–545, 1984.
3. Mungall, D. et al., Individualizing theophylline therapy: the impact of clinical pharmacokinetics on patient outcomes, *Ther. Drug Monit.*, 5(1):95–101, 1983.
4. Ried, L. D., McKenna, D. A., and Horn, J. R., Meta-analysis of research on the effect of clinical pharmacokinetics services on therapeutic drug monitoring, *Am. J. Hosp. Pharm.*, 46(5):945–951, 1989.
5. Ried, L. D., Horn, J. R., and McKenna, D. A., Therapeutic drug monitoring reduces toxic drug reactions: a meta-analysis, *Ther. Drug Monit.*, 12(1):72–78, 1990.
6. Schumacher, G. E. and Barr, J. T., Making serum drug levels more meaningful, *Ther. Drug Monit.*, 11(5):580–584, 1989.
7. Schumacher, G. E. and Barr, J. T., Using population-based serum drug concentration cutoff values to predict toxicity: test performance and limitations compared with Bayesian interpretation, *Clin. Pharm.*, 9(10):788–796, 1990.
8. Schumacher, G. E., Choosing optimal sampling times for therapeutic drug monitoring, *Clin. Pharm.*, 4(1):84–92, 1985.
9. Sadee, W. and Beelen, G. C. M., *Drug Level Monitoring*, John Wiley & Sons, New York, 1980.
10. Matzke, G. R. et al., Pharmacokinetics of vancomycin in patients with various degrees of renal function, *Antimicrob. Agents Chemother.*, 25(4):433–437, 1984.
11. Cockcroft, D. W. and Gault, M. H., Prediction of creatinine clearance from serum creatinine, *Nephron*, 16(1):31–41, 1976.

12. Haycock, G. B., Schwartz, G. J., and Wisotsky, D. H., Geometric method for measuring body surface area: a height-weight formula validated in infants, children, and adults, *J. Pediatr.,* 93(1):62–66, 1978.

13. Sawchuk, R. J. and Zaske, D. E., Pharmacokinetics of dosing regimens which utilize multiple intravenous infusions: gentamicin in burn patients, *J. Pharmacokinet. Biopharm.,* 4(2):183–195, 1976.

7 Aminoglycosides and Other Antibiotics

Michael E. Klepser and Erika J. Ernst

CONTENTS

INTRODUCTION

The term *antimicrobial* is broadly used to describe any agent that produces a detrimental effect on the growth or reproduction cycle of a bacteria, fungus, or parasite. In order for an antimicrobial to be effective, it must reach its target site within an organism in sufficient quantities and remain there for an adequate length of time to interrupt the normal life cycle of the organism. Therefore, predictions regarding the activity of an antibiotic can be assessed by examining pharmacokinetic (how much drug gets to the active site) and pharmacodynamic (interaction between drug and active site and the desired biological effect) properties and comparing data with those of other agents against similar pathogens.

Discussion of the pharmacodynamic characteristics of antibiotics generally focuses on two properties:

1. Concentration–effect relationships
2. The postantibiotic effect (PAE)

Evaluation of antibacterial concentration–effect relationships integrates measures of drug exposure and inherent activity against specified pathogens. The peak concentration, rate of drug elimination, and area under the serum concentration vs. time curve (AUC) are parameters used to assess drug exposure (Figure 7.1). The

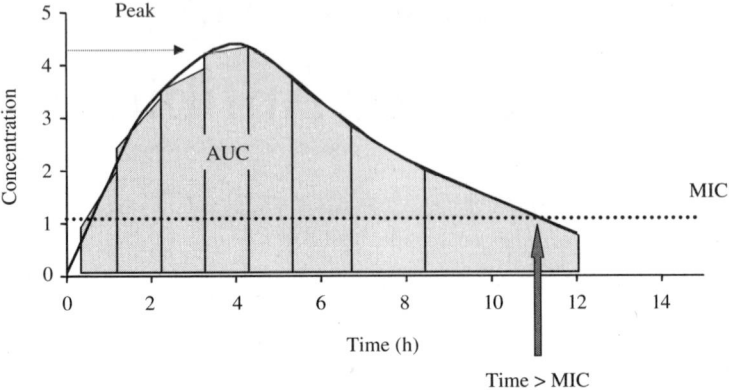

FIGURE 7.1 Schematic representation of pharmacodynamic components.

minimum inhibitory concentration (MIC) is the lowest concentration of a drug that arrests or inhibits the growth of a bacteria. The MIC is routinely used as a marker of antimicrobial activity. The concentration–effect profile of all antimicrobials can be described by a sigmoidal curve. Antimicrobials can be roughly classified as either exhibiting concentration-dependent or non-concentration-dependent activity depending on the characteristics of the concentration–effect curve (Figure 7.2).

Concentration-dependent activity

- The rate and/or extent of antibacterial activity is improved as the amount of drug at the site of infection increases.
- Parameters which incorporate magnitude of drug exposure (i.e., peak:MIC or AUC:MIC) correlate with the effect.

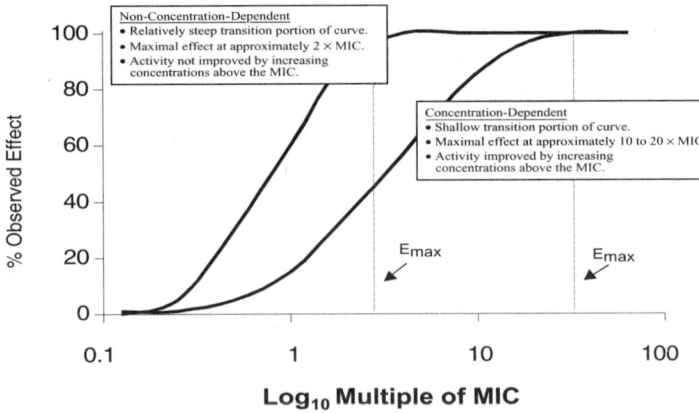

FIGURE 7.2 Description of concentration–effect relationships base on the characteristics of a dose–response curve.

- Transition portion of concentration-effect curve falls over clinically achievable concentration range.
- Maximum effect noted at concentrations equal to 8 to 20 × MIC of the organism.

Non-concentration-dependent activity

- The rate and/or extent of kill are not improved once concentrations exceed 2 to 4 × MIC of the organism.
- Activity is most closely linked with the duration of antimicrobial exposure, i.e., time that concentrations remain above the MIC (time > MIC) or AUC:MIC.
- Transition portion of curve is generally far exceeded by clinically achievable concentrations.

It is interesting to note, that because AUC is a function of drug concentration and time, it can serve as a marker of both magnitude and duration of drug exposure. Therefore, it is not surprising that AUC:MIC correlates well with the activity exhibited by agents possessing concentration-dependent and non-concentration-dependent activity. Table 7.1 categorizes various antimicrobials by their concentration–effect relationships and lists which parameters best correlate with killing/inhibitory activity.

The PAE is a persistent inhibitory effect measured after a brief exposure of an organism to an antimicrobial followed by removal of the antibiotic. The PAE is likely the result of nonlethal cellular damage, residual drug binding to a target site, or a combination of both. Agents exhibiting relatively long PAEs may be dosed at intervals longer that might otherwise be predicted based solely on half-life. In contrast, dosing of agents with shorter PAEs may require selection of dosing intervals that are tied more closely with the half-life of the agent. It should be pointed out that because drug concentrations *in vivo* do not immediately drop to zero but rather

---

TABLE 7.1
Pharmacokinetic and Pharmacodynamic Parameters Correlating with Antibacterial Activity

Concentration–Effect Relationship	Antimicrobial Agents	Pharmacodynamic Correlate(s)
Concentration-dependent activity	Aminoglycosides	Peak:MIC
	Fluoroquinolones	Peak:MIC; AUC:MIC
Non-concentration-dependent activity	β-Lactams	Percent time > MIC
	Macrolides	Percent time > MIC (erythromycin, clarithromycin)
		AUC:MIC (azithromycin)
	Vancomycin	AUC:MIC; Percent time > MIC

decline gradually with time, there are generally prolonged periods of time over which organisms are exposed to sub-MIC concentrations. Therefore, it may be more accurate to refer to the *in vivo* PAE as a sub-MIC effect.

CLINICAL APPLICATION
OF PHARMACODYNAMIC CONCEPTS

The goal of an antimicrobial dosing regimen should be to maximize the antibacterial effect of the drug while minimizing patient exposure to that agent. A secondary consideration in the development of a dosing regimen should be to prevent the emergence of resistance. If development of a dosing regimen is approached rationally and the pharmacokinetic and pharmacodynamic properties of the drug properly considered, all of these goals may be realized.

Agents such as the aminoglycosides and the fluoroquinolones exhibit concentration–dependent bactericidal activity. Owing to this concentration–effect relationship, the activities of these agents are improved as drug concentrations at the site of infection increase above the MIC of the target pathogen. It therefore is no surprise that dosing regimens that maximize the magnitude of a pathogen exposure to these agents result in optimal killing.[1-9] The peak drug concentration obtained is the most straightforward measure of the magnitude of antimicrobial exposure. Several investigators have reported that for agents exhibiting concentration-dependent killing such as the quinolones and the aminoglycosides the maximal killing effect is observed when concentrations at the site of infection of approximately 8 to 20 × MIC are achieved.[5,9,10] Unfortunately, as is often the case with the fluoroquinolones, these target peak:MIC are unachievable because of toxicity experienced at such high drug levels. Therefore, when an optimal peak:MIC is not obtained, the AUC:MIC becomes the pharmacodynamic parameter most closely associated with outcome.[10]

In contrast, the activity of agents that exhibit non-concentration-dependent activity does not improve appreciably as the organism is exposed to drug concentrations much higher than the MIC. For this reason, administration of doses larger than required to ensure that concentrations at the site of infection exceed the MIC of the microbe do not enhance antibacterial activity. For these agents, it is not the magnitude of exposure that influences activity; rather, it is the duration of exposure. Therefore, to maximize the extent of killing, one must dose these compounds so that the amount of time concentrations at the site of infection are optimized and remain greater than the MIC. This may be accomplished by administering relatively small doses at frequent intervals, using continuous infusion to maintain constant concentrations, administering an extended-release formulation, or selecting an agent with a long half-life relative to the proposed dosing interval. Regardless of the approach taken, effort should be given to lowering peak concentrations, as they do not contribute to efficacy and may only increase toxicity, and increasing the percent of the dosing interval over which serum concentrations remain above the MIC of the pathogen. Although maintaining drug concentration greater than the MIC for 100% of the dosing interval would appear to be optimal, it has been noted that a time greater than the MIC for as little as 40% is often associated with a positive outcome.[2-4]

Although our use of pharmacodynamic principles has not yet been translated into widespread individualized patient application, they have changed the way we think about antimicrobial selection, dosing, and formulary decisions.

AMINOGLYCOSIDES

The aminoglycosides are a group of semisynthetic antimicrobials that exhibit a broad spectra of activity against a variety of medically important Gram-positive and Gram-negative bacteria. These agents generally exhibit bactericidal activity secondary to inhibition of protein synthesis via irreversible binding to the bacterial 30S ribosomal subunit. Some general pharmacokinetic features of the aminoglycosides include:

- Poor absorption from the gastrointestinal tract
- Extensive distribution throughout the body ($V_d \sim 0.25$ l/kg), which is greatly dependent of the fluid status of the patient
- Low degree of protein binding
- Not metabolized and are excreted unchanged in the urine primarily by glomerular filtration
- Among patients with normal renal function, the plasma half-life of the aminoglycosides is approximately 2 to 4 h. Renal impairment will significantly prolong the observed half-life of these compounds.

Significant renal and ototoxicity are associated with the use of the aminoglycosides. Toxicity is the result of drug accumulation within target organ cells and has been linked to elevated trough concentrations (i.e., >2 µg/ml for gentamicin and tobramycin and >10 µg/ml for amikacin). As a result of these toxicities having been linked with drug concentrations, clinicians utilize serum drug levels to optimize aminoglycoside therapy. Currently, two dosing approaches are considered acceptable for monitoring and adjusting aminoglycoside dosing regimens.

1. Multiple daily dosing (traditional approach)
   - Dose to achieve target peak (i.e., 6 to 10 µg/ml, gentamicin and tobramycin; 20 to 30 µg/ml, amikacin) and trough concentrations (i.e., <2 µg/ml, gentamicin and tobramycin; <10 µg/ml, amikacin)
   - Dosing interval generally ranging from every 8 h to every 24 h
2. Extended interval dosing (also known as once-daily dosing)
   - Dose to optimize the peak:MIC ratio, target ratio approximately 10:1 to 20:1
   - Dosing interval usually every 24 h
   - Results in very low trough concentrations
   - Thought to reduce likelihood of toxicity

Despite the differences with respect to the desired serum concentration goals, evaluation of aminoglycoside levels and pharmacokinetic adjustment of dosing regimens are common to both approaches. Using a first-order pharmacokinetic model, patient-specific elimination and volume of distribution can be calculated based on patient serum drug concentrations. However, when large doses of the aminoglycosides are administered, one should wait until at least 6 h following the start of the infusion to draw the first drug level. This ensures that the drug has fully distributed throughout the body.

$$k_{el} = \frac{\ln(C_{peak}/C_{trough})}{\tau - (t_{peak} - t_{trough})} \tag{7.1}$$

assumes that the trough (C_{trough}) is obtained immediately before the dose for which the peak (C_{peak}) is drawn and

$$V_{SS} = \frac{(Dose/t')(1 - e^{-k_{el}t'})}{k_{el}[C_0 - (C_{trough} \cdot e^{-k_{el}t'})]} \tag{7.2}$$

where t' is the time of infusion, k is the first-order elimination rate constant, t_{peak} and t_{trough} are the times for the peak and trough concentrations, and C_0 is the extrapolated peak. By using these patient-specific values for elimination (k_{el}) and volume of distribution (V_d), the drug dose and schedule may be adjusted to obtain the desired serum concentrations.

$$\text{Maintenance Dose} = [k(V_d)(C_{peak})t'](1 - e^{-k_{el}\tau})/(1 - e^{-k_{el}t'}) \tag{7.3}$$

β-LACTAM ANTIMICROBIALS

The term *β-lactam* refers to several classes of antimicrobials including the penicillins and cephalosporins, which are related via possession of a common β-lactam ring structure. Although these compounds differ somewhat with respect to their spectra of activity, the β-lactams share many similar pharmacokinetic and pharmacodynamic characteristics.

- β-Lactams are widely distributed throughout the body into many tissues and fluids including synovial fluid, pleural fluid, and to a somewhat more variable degree the cerebrospinal fluid.
- The half-life of the various compounds in this class range from 0.4 h for penicillin to >10 h for ceftriaxone.
- This class of drugs follows linear pharmacokinetics and exhibits first-order elimination.
- The majority of β-lactams are primarily eliminated in the urine. Some notable exceptions, however, include nafcillin, which undergoes hepatic hydrolysis, and ceftriaxone, which is excreted in the bile.

As a class, the β-lactams are extremely well tolerated and have a large therapeutic window. For this reason, therapeutic monitoring of β-lactam serum concentrations is not warranted. Although individualized pharmacokinetic monitoring is not employed to direct therapy, knowledge of general pharmacokinetic and pharmaco-dynamic characteristics can help guide therapeutic and formulary decisions. The β-lactams exhibit non-concentration-dependent bactericidal activity. Additionally, it has been determined that the amount of time that serum concentrations remain above the MIC of the infecting pathogen is the pharmacodynamic parameter most closely associated with efficacy for these compounds. To account for various dosing sched-ules, this parameter is generally expressed as the percent of the dosing interval for which concentrations exceed the MIC of the organism (percent time > MIC). The percent time > MIC can be used to facilitate comparisons among agents and thus guide therapeutic and formulary decisions.

- Requires knowledge of population pharmacokinetic parameters for peak (C_{max}), half-life ($t_{1/2}$), and time to achieve peak concentration (t_{max}).
- Calculate the elimination rate constant (k_{el}).

$$k_{el} = 0.693/t_{1/2} \tag{7.4}$$

- Calculate the time (t_{MIC}) following the peak at which serum concentrations are equal or above the MIC (C_{MIC}).

$$C_{MIC} = C_{MAX}e^{-k_{el}(t_{MIC} - t_{MAX})} \tag{7.5}$$

- Divide t_{MIC} by the dosing interval and multiply by 100 to express as a percent.
- Gives a rough estimate for the percent time > MIC.

Although having serum concentrations remain above the MIC of the pathogen for 100% of the dosing interval would be optimal, data have demonstrated positive outcomes with percent time > MIC as low as 40%.[2-4]

The goal of dosing the β-lactams is to maximize the time > MIC. In an effort to obtain this goal, dosing regimens may be manipulated. Large doses may be administered that would result in increased time > MIC; however, this strategy would also result in higher peak concentrations, which do not contribute to efficacy. Smaller doses may be administered more frequently, therefore, not allowing concentrations to fall below the MIC of the organism. However, this approach can be inconvenient for patients and health-care providers. One method for optimizing the duration of drug exposure that has been proposed is the use of continuous infusions of parenteral β-lactams. Use of continuous infusion regimens would eliminate drug concentration fluctuations, thus allowing drug concentrations to be maintained above the MIC for 100% of the dosing interval. The daily dose to be administered via continuous infusion can be either the total daily dose usually administered by intermittent infusion or a calculated dose.

$$\text{Dose (mg/day)} = V_d \text{(l/kg) Weight (kg)} k_1 \text{(h}^{-1}) C_{desired} \text{(mg/l) 24 h} \tag{7.6}$$

By calculating a daily dose to be administered by continuous infusion, the amount of drug required over the course of 24 h can generally be reduced.

FLUOROQUINOLONES

Relegated to the management of urinary tract infections only two decades ago, the fluoroquinolones are now cornerstones in the management of a variety of inpatient and outpatient infections. Key pharmacokinetic characteristics of the quinolones include:

- Rapid absorption from the gastrointestinal tract (t_{max} ~1 to 2 h)
- Excellent oral bioavailability (70 to >95%)
- Large volume of distribution (3.5 to 4.5 l/kg)
- Protein binding ranges from 20 to 50%
- Most renally eliminated; moxifloxacin exhibits the largest degree of non-renal elimination (i.e., biliary secretion)

The quinolones exhibit concentration-dependent bactericidal activity against a variety of Gram-positive and Gram-negative bacteria. Several *in vitro* and *in vivo* studies have demonstrated that the activity of the fluoroquinolones is closely associated with the peak:MIC and AUC:MIC ratio.[1,3,5,10] If a peak:MIC of greater than 10:1 can be achieved, bactericidal activity is maximized and the likelihood of resistance developing is lessened. However, in many cases achieving the target peak:MIC is not clinically feasible because of dose-related toxicities noted with these agents. When a peak:MIC of greater than 10:1 is not attainable, then the AUC:MIC serves as the best correlate with activity. The optimal AUC:MIC target is somewhat controversial; however, a ratio of greater than 50:1 and 100:1 appears to be reasonable for Gram-positive and Gram-negative pathogens, respectfully.

Comparison of the relative activity of various quinolones can be accomplished by comparing pharmacodynamic parameters against key pathogens using population pharmacokinetic data and published or institutional susceptibility data. Evaluation of such data can assist in the selection of agents for formulary consideration.

MACROLIDES

Currently, azithromycin, clarithromycin, and erythromycin are the most commonly prescribed macrolide antimicrobial agents. The macrolides exhibit bacteriostatic/bactericidal activity against a variety of Gram-positive and to a more limited extent Gram-negative pathogens. The pharmacokinetic profiles of the macrolides vary considerably.

Erythromycin
- The rate and extent of absorption is highly dependent on the specific formulation used.
- Protein binding ranges from 70 to 90%.

- Elimination is primarily via biliary excretion.
- Half-life ranges from 1.5 to 3 h.

Clarithromycin
- Bioavailability is 50 to 55%.
- Exhibits nonlinear, dose-dependent pharmacokinetics.
- Undergoes significant hepatic metabolism to an active metabolite.
- Half-life ranges from 3 to 7 h.

Azithromycin
- Bioavailability is approximately 40% and is reduced by approximately 50% if administered with food.
- Primary route of elimination is via excretion in the bile.
- Exhibits a rapid distribution phase followed by a terminal half-life of greater than 60 h.

Despite some of the pharmacokinetic differences noted among these agents, they all exhibit wide distribution throughout the body. Of particular interest is the ability of these agents, especially azithromycin and clarithromycin, to accumulate within white blood cells.

Although the macrolides have been reported to exhibit non-concentration-dependent antibacterial activity, there remains some uncertainty with respect to the pharmacodynamic parameter that best correlates with activity. It has been suggested that the percent time > MIC best correlates with the activity of erythromycin and clarithromycin. However, owing to the extremely long half-life of azithromycin, time of exposure becomes a less sensitive predictor of activity. As a result, incorporation of a measure of the magnitude of exposure, i.e., AUC, may improve our ability to establish a correlation between drug concentrations and activity. Currently, no targets for percent time > MIC or AUC:MIC have been established to provide clinical guidance with the macrolides.

GLYCOPEPTIDES

Currently, vancomycin is the only approved glycopeptide available for use in the United States. For several decades, vancomycin has been considered the most potent antimicrobial against Gram-positive pathogens. The pharmacokinetic properties of vancomycin are listed below.

Vancomycin
- Poor absorption from the gastrointestinal tract
- Large volume of distribution (0.5 to 0.9 l/kg)
- 50 to 60% protein bound
- Rapid distribution phase followed by terminal elimination phase ($t_{1/2}$ = 4 to 6 h)
- Eliminated via glomerular filtration

Historically, vancomycin serum concentrations have been monitored and adjusted using first-order pharmacokinetic calculations (see aminoglycosides). However, difficulty obtaining a reliable peak concentration because of the initial distribution phase and lack of correlation between serum concentrations and efficacy and toxicity have led to a reappraisal of the value of pharmacokinetic monitoring with vancomycin.

Vancomycin exhibits non-concentration-dependent bactericidal activity; therefore, percent time > MIC or AUC:MIC should correlate with outcomes. As a result, investigators have attempted to optimize therapy by dosing vancomycin to maintain serum concentrations above the MIC of specific pathogens.[6-8] As a means to optimize exposure, some clinicians continue to monitor trough concentrations and attempt to maintain this concentration greater than the MIC of the pathogen.

SUMMARY

With the exception of the aminoglycosides, individualized pharmacokinetic monitoring does not occur routinely for the currently available antimicrobial agents. However, evaluation and application of pharmacokinetic and pharmacodynamic data do impact antimicrobial utilization practice on a more global scale. An understanding of these principles allows clinicians to compare various agents against a variety of pathogens and to make scientifically supported antimicrobial selections. Additionally, by understanding and applying pharmacokinetic and pharmacodynamic principles, antimicrobial activity of these agents may be maximized while patient exposure is minimized, thus allowing for safer, more efficacious, and more cost-effective utilization of antibiotics.

REFERENCES

1. Blaser, J et al., Comparative study with enoxacin and netilmicin in a pharmacodynamic model to determine importance of ratio of antibiotic peak concentration to MIC for bactericidal activity and emergence of resistance, *Antimicrob. Agents Chemother.*, 31(7):1054–1060, 1987.
2. Craig, W. A., Antimicrobial resistance issues of the future, *Diagn. Microbiol. Infect. Dis.*, 25(4):213–217, 1996.
3. Craig, W. A., Pharmacokinetic/pharmacodynamic parameters: rationale for antibacterial dosing of mice and men, *Clin. Infect. Dis.*, 26(1):1–10; quiz 11–2, 1998.
4. Craig, W. A. and Andes, D., Pharmacokinetics and pharmacodynamics of antibiotics in otitis media, *Pediatr. Infect. Dis. J.*, 15(3):255–259, 1996.
5. Drusano, G. L. et al., Pharmacodynamics of a fluoroquinolone antimicrobial agent in a neutropenic rat model of Pseudomonas sepsis, *Antimicrob. Agents Chemother.*, 37(3):483–490, 1993.
6. Duffull, S. B. et al., Efficacies of different vancomycin dosing regimens against *Staphylococcus aureus* determined with a dynamic *in vitro* model, *Antimicrob. Agents Chemother.*, 38(10):2480–2482, 1994.

7. James, J. K. et al., Comparison of conventional dosing versus continuous-infusion vancomycin therapy for patients with suspected or documented Gram-positive infections, *Antimicrob. Agents Chemother.*, 40(3):696–700, 1996.

8. Klepser, M. E. et al., Comparison of bactericidal activities of intermittent and continuous infusion dosing of vancomycin against methicillin-resistant *Staphylococcus aureus* and *Enterococcus faecalis*, *Pharmacotherapy*, 18(5):1069–1074, 1998.

9. Moore, R. D., Lietman, P. S., and Smith, C. R., Clinical response to aminoglycoside therapy: importance of the ratio of peak concentration to minimal inhibitory concentration, *J. Infect. Dis.*, 155(1):93–99, 1987.

10. Preston, S. L. et al., Pharmacodynamics of levofloxacin: a new paradigm for early clinical trials [see comments], *J. Am. Med. Assoc.*, 279(2):125–129, 1998.

8 Cardiovascular Agents

John M. Dopp and Bradley G. Phillips

CONTENTS

1-56676-973-6/02/$0.00+$1.50
© 2002 by CRC Press LLC

INTRODUCTION

This chapter reviews practical pharmacokinetic and pharmacodynamic consider-ations for the use of select cardiovascular drugs. In the clinical setting the majority of cardiovascular drugs are empirically dosed and adjusted based on efficacy and toxicity. For certain drugs, outlined in this chapter, concomitant disease states and drug interactions can impact pharmacokinetic and pharmacodynamic parameters necessitating dose adjustment and careful monitoring.

CLASSIFICATION

The Vaughan Williams classification (shown as types in Table 8.1) and therapeutic ranges for cardiovascular drugs are summarized in Table 8.1.[1]

ANTIARRHYTHMICS

PROARRHYTHMIA

The Cardiac Arrhythmia Suppression Trial (CAST) highlighted the importance and awareness of proarrhythmia. The main finding of CAST was that, despite elimination of complex ventricular ectopy after myocardial infarction, mortality was significantly higher in patients treated with encainide or flecainide.[2] Others have reported that the overall risk of cardiac mortality is higher, presumably due to proarrhythmia, in patients treated with Type 1a antiarrhythmics for atrial fibrillation who have con-

---

TABLE 8.1
Vaughan Williams Classification and Therapeutic Ranges of Cardiovascular Drugs

Type	Cardiovascular Drug	Therapeutic Range
1a	Quinidine[a]	2–5 µg/ml
	Procainamide[a]	4–10 µg/ml
	Disopyramide[a]	2–5 µg/ml
1b	Lidocaine[a]	2–6 µg/ml
	Tocainide	4–10 µg/ml
	Mexilitine	0.5–2 µg/ml
1c	Flecainide	0.2–1 µg/ml
	Encainide	—
	Propafenone	0.5–3 µg/ml
II	Beta blockers (see Table 8.12), e.g., atenolol, metoprolol, propranolol	—
III	Amiodarone[a]	1–2.5 µg/ml
	Dofetilide[a]	—
	Sotalol	—
	Bretylium	—
IV	Calcium channel blockers (see Table 8.13), e.g., amlodipine, diltiazem, verapamil	—

[a] Antiarrhythmics commonly dosed and monitored pharmacokinetically.

Source: Data from Vaughan Williams, E. M., *Clin. J. Pharmacol.*, 24:129, 1984.

---

comitant heart failure.[3,4] The potential for proarrhythmia, choice of anitarrhythmic, and factors that may predispose to proarrhythmia, including concomitant disease states, electrolyte abnormalities, and drug interactions, should be evaluated before and during antiarrhythmic therapy. Proarrhythmia should be expected if one of the following occurs:

- The existing arrhythmia becomes worse.
- The symptoms with an existing arrhythmia worsen.
- A new arrhythmia develops (e.g., torsades de pointes) during therapy.

Proarrhythmia and the advances made in nonpharmacological therapies, like radiofrequency catheter ablation and implantable defibrillators, have tempered the use of antiarrhythmic agents.[5]

SERUM CONCENTRATION MONITORING

Goal: Provide patients with an adequate dose of the drug to prevent recurrence of their arrhythmia, while minimizing the incidence of adverse effects. Clinical considerations include the following evaluations:[6]

- Decide if the patient has an arrhythmia that should be treated. Always evaluate the risk-to-benefit profile for treatment in each patient.
- Consider the type of arrhythmia and the patient's concomitant diseases.
- Choose an initial dose. Calculation of dose should include patient factors (age, diseases), drug pharmacokinetics, and the recommended therapeutic range of agent.
- Optimize short-term regimen. Adjust acutely to optimize therapy (subtherapeutic and toxic concentrations).
- Test the efficacy of the agent to predict long-term effectiveness. Tests to evaluate efficacy include serial electrophysiological (EP) drug testing, serial Holter monitoring, and electrocardiogram. If a drug is determined to be effective, a serum drug level may be obtained to document an effective or "target" concentration for that patient.
- Initiate long-term regimen. Continue therapy with a regimen designed to achieve and maintain an effective (target) serum concentration.
- Optimize long-term therapy. Evaluate the need to adjust dose if subtherapeutic or toxic concentrations are achieved.

RATIONALE FOR MONITORING ANTIARRHYTHMIC DRUGS[6]

- Exhibit considerable variability in dose–response between patients
- Have low toxic-to-therapeutic ratio
- Generally follow a predictable relationship between serum drug concentration and therapeutic response
- Therapeutic ranges that are not absolute targets

INTERPRETATION OF SERUM DRUG CONCENTRATIONS

Consider:

- The specific arrhythmia being treated
- The criteria for antiarrhythmic efficacy — suppression vs. decrease in the severity
- The method of antiarrhythmic monitoring (e.g., EP testing, Holter monitor)
- The degree of protein binding
- The presence of active metabolites
- Potential drug interactions

INDICATIONS FOR DETERMINING SERUM DRUG CONCENTRATIONS

- To document the effective concentration for an individual patient
- To determine the reason for antiarrhythmic drug failure
- To assess potential toxicity
- To assess compliance
- To assist in dose adjustment

ANTIARRHYTHMIC AGENTS

AMIODARONE

Amiodarone is a Type III antiarrhythmic agent. It has been shown in several clinical trials to be safe and effective in the treatment of supraventricular and ventricular arrhythmias in patients with cardiovascular disease.[7] Amiodarone is also associated with fewer proarrhythmic effects compared with other antiarrhythmic drugs. Amiodarone is empirically dosed and adjusted based on efficacy and toxicity.

Clinical Pharmacokinetic/Pharmacodynamic Considerations

- Amiodarone is safe and effective in patients with heart failure and myocardial infarction.
- Intravenous loading dose is given over 24 h; oral loading dose given over weeks.
- Elimination half-life is long and variable between patients.
- Peak effect may not be seen for 1 week to 5 months.
- Potential drug side effects require determination of chest, pulmonary, liver, and thyroid function prior to and during therapy.
- Dose-dependent side effects may include nausea, vomiting, loss of appetite, visual disturbances, and pulmonary toxicity.

- Lowest effective maintenance dose (200 to 400 mg/day) should be achieved.
- Potential drug interactions should be considered.
- Amiodarone effect may last up to 2 to 3 months after the drug is discontinued.

Absorption

The absorption of amiodarone is erratic and unpredictable ($F = 0.35$, range 0.2 to 0.7). After a single dose, peak concentrations are observed after 2 to 12 h (mean 4 to 5 h).[8,9]

Distribution

Amiodarone is extensively bound to plasma proteins and is widely distributed throughout the body. The mean volume of distribution is 66 l/kg and ranges from 7 to 148 l/kg. Steady-state amiodarone serum concentrations range between 1 and 2.5 µg/ml.[9,10]

Metabolism

Nearly 100% of amiodarone is eliminated by the liver (CL = 90 to 160 ml/h/kg). Desethyl-amiodarone is the main active metabolite and has EP activity similar to amiodarone (CL = 200 to 300 ml/h/kg).[10] The mean elimination half-life after a single oral dose ranges from 4.6 to 36 h.[11] During chronic therapy, elimination half-life is 40 to 55 days (range 26 to 107 days). The antiarrhythmic effect of amiodarone may last for 2 to 3 months after it is discontinued.[12]

Excretion

Amiodarone is not removed renally and is not removed by hemodialysis.

Drug Interactions

Amiodarone is a CYP3A3/4 enzyme substrate; CYP3A3/4, CYP2C9, and CYP2D6 enzyme inhibitor. Because of the long half-life of amiodarone, drug interactions may be observed after amiodarone is discontinued. The majority of clinically important drug interactions include the effect of amiodarone on other drugs, as shown in Table 8.2.

Dosing

Amiodarone is empirically dosed based on efficacy (termination or suppression of the arrhythmia) and toxicity. Baseline tests should be obtained prior to initiating therapy, including electrocardiogram, thyroid function test, chest radiograph, pulmonary function test, and liver enzymes.[13] The proposed therapeutic range for amiodarone is 1 to 2.5 µg/ml; levels are not monitored in the clinical setting for efficacy and toxicity.[6]

TABLE 8.2
Drug Interactions for Amiodarone

Drug	Interaction
β-Blockers and calcium channel blockers	Additive negative chronotropic effects
Cimetidine	May increase amiodarone serum levels
Cyclosporine	Increased cyclosporine serum levels
Digoxin	Increases digoxin serum levels; decrease digoxin dose by 50%
Dofetilide	Increased risk of dofetilide-induced proarrhythmia
Drugs that prolong QT interval (e.g., erythromycin, haloperidol, tricyclic antidepressants, sotalol, bepridil)	Increase the risk of torsades de pointes (proarrhythmia)
Other antiarrhythmics (e.g., quinidine, procainamide)	Increase quinidine and procainamide serum levels
Warfarin	Increases INR

Intravenous

A loading dose of 1 g of amiodarone is given over 24 h (see table below) and maintained if needed in the short term at 0.5 mg/min. If breakthrough arrhythmias occur during therapy, supplemental doses of 150 mg may be administered. Intravenous amiodarone should be switched to the oral route when possible.[8,14]

IV	Oral
150 mg over 10 min followed by 360 mg over 6 h then 540 mg over 18 h	800–1600 mg/day for 1 to 3 weeks 600–800 mg/day for 1 month
Bolus: 150 mg PRN for VF/VT[a]	
Maintenance 0.5 mg/min	Maintenance: 200 to 400 mg/day

[a] VF: ventricular fibrillation; VT: ventricular tachycardia

Oral

Amiodarone may be loaded orally. Typically, 800 to 1600 mg a day in divided doses is administered until a therapeutic response is observed. The dose is then tapered to 600 to 800 mg a day for 1 month. Maintenance therapy should be the lowest effective dose (200 to 400 mg/day). Patients with ventricular arrhythmias may require a higher loading and maintenance dose.

Conversion from IV to Oral Administration

For patients initiated on IV therapy the following can be used as a guide to switch to oral amiodarone.[8] Oral therapy should then be tapered to the lowest effective dose.

Duration of Infusion, weeks	Initial Daily Oral Dose, mg
<1	800–1600
1–3	600–800
>3	400

DOFETILIDE

Dofetilide is a Type III antiarrhythmic agent used in the treatment of chronic atrial fibrillation or atrial flutter. Physicians and hospital staff need to complete a dofetilide education program before the drug can be received by the hospital and prescribed. Dofetilide can cause life-threatening ventricular arrhythmias and should therefore be used in select patients.

Clinical Pharmacokinetic/Pharmacodynamic Considerations

- Dofetilide is safe and effective for the treatment of chronic atrial fibrillation in patients with structural heart disease.
- Initiation or reinitiation of therapy is completed in a controlled setting in a hospital.
- Initial dose is based on renal function.
- Subsequent doses are adjusted on QT or QTc changes on the electrocardiogram.
- Changes in QT interval are linear with changes in dofetilide serum concentrations.
- Assess for proarrhythmia (e.g., torsades de pointes).
- Assess for drug interactions that may interfere with metabolism or renal elimination of dofetilide predisposing to proarrhythmia.
- Avoid concomitant use of drugs that can prolong QT interval.

Absorption

The oral bioavailability of dofetilide is >90%. After a single dose, peak plasma concentrations are achieved in 2 to 3 h. Bioavailability is unaffected by food or antacids.[15]

Distribution

Approximately 70% of dofetilide binds to plasma proteins, independent of plasma dofetilide concentrations. Volume of distribution is 3 l/kg and elimination half-life is 10 h.[15]

Metabolism

A small portion (20%) of dofetilide is metabolized (N-dealkylation and N-oxidation) in the liver. Metabolites have no or little antiarrhythmic activity.

Elimination

The majority of dofetilide (80%) is excreted in the urine: 80% as unchanged dofetilide and 20% metabolites. Renal elimination involves both glomerular filtration and active tubular secretion.[16] Because QT interval and risk of ventricular arrhythmias are directly related to plasma concentrations of dofetilide, dose adjustment based on creatinine clearance is clinically very important.

Drug Interactions

Dofetilide has weak affinity for CYP3A4. Concomitant use of dofetilide with cimetidine, hydrochlorothiazide (including combination products), verapamil, trimethoprim, prochlorperazine, megestrol, or ketoconazole is contraindicated due to significant increases in dofetilide serum concentrations (see Table 8.3).

TABLE 8.3
Drug Interactions for Dofetilide

Drug	Interaction
Enzyme inhibitors (e.g., erythromycin, ketoconazole, amiodarone)	Increase dofetilide serum levels
Drugs that prolong QT interval (e.g., erythromycin, haloperidol, tricyclic antidepressants, sotolol, bepridil)	Increase the risk of torsades de pointes (proarrhythmia)
Type I and other Type III antiarrhythmic agents	Increase risk of proarrhythmia. Space drugs at least three half-lives apart. For amiodarone, wait at least 3 months after discontinuation or until amiodarone level is <0.3 μg/ml before initiating dofetilide.
Hydroclorothiazide (alone or in combination products)	Increase dofetilide serum levels and QT interval.

Dosing

Initial therapy, given twice a day, is administered in the hospital under controlled conditions that allow for continuous monitoring of the electrocardiogram for 3 days. The baseline QT and QTc interval must be ≤440 ms (500 ms in patients with ventricular conduction abnormalities); otherwise, dofetilide is contraindicated. The serum potassium should be in the normal range and at least 4 meq/l before and during acute and chronic dosing to minimize risk of proarrythmia. The initial dose is empirically based on renal function according to the following:

Cl_{cr} (ml/min)	First dose (μg)
>60	500
40–60	250
20–<40	125
<20	Contraindicated

At 2 to 3 h after administration of the first dose, the QT or QTc interval is determined. If the QT or QTc interval is not changed from baseline, then the initial dose, based on creatinine clearance, is administered every 12 h. However, if the QT or QTc has increased >15% from baseline or if it is >500 ms (550 in patients with ventricular conduction abnormalities), subsequent doses are as follows:

If the starting dose was	Adjust dose for QT prolongation
500 µg	250 µg twice daily
250 µg	125 µg twice daily
125 µg	125 µg once a day

The QT and QTc are monitored 2 to 3 h after each subsequent dose while the patient is in the hospital. If at any time the QT or QTc is >500 ms (550 in patients with ventricular conduction abnormalities), dofetilide should be discontinued. If this guide to dosing is followed, the incidence of torsades de pointes (proarrhythmia) is less than 1%.

DISOPYRAMIDE

Disopyramide is a Type Ia antiarrhythmic agent used to treat supraventricular and ventricular arrhythmias. Of the Type Ia agents, disopyramide possesses the greatest negative inotropic and anticholinergic activity.[17] These properties limit the use of this drug over other antiarrhythmics, particularly in patients with congestive heart failure.[7] Disopyramide is empirically dosed in the clinical setting and the dose is adjusted based on heart, liver, and renal function.

Clinical Pharmacokinetic/Pharmacodynamic Considerations

- The dosing regimen should be individualized because of a large variation in bioavailability, distribution, and elimination between patients.
- Disopyramide should be avoided in patients with heart failure due to negative inotropic properties.[7]
- Dose-dependent protein binding exists.
- Therapeutic index is narrow (2 to 5 µg/ml).
- Serum disopyramide concentrations are unreliable for monitoring, in part, because of high variability in protein binding. Therapeutic range for free disopyramide is not well established.[6]
- Assess for proarrhythmia (e.g., torsades de pointes).
- Adjust dose empirically on concomitant diseases: heart failure, liver disease, renal dysfunction.

Absorption

Oral absorption is variable for both immediate- and sustained-release formulations ($F = 0.8$; range 0.6 to 0.9). Following a single dose, T_{max} is reached between 0.5 and 3.5 h.[18]

Distribution

Disopyramide is highly protein bound (~90%) predominantly to α_1-acid glycoprotein. Conditions that increase α_1-acid glycoprotein (e.g., myocardial infarction, trauma, cardiac arrest) may dramatically alter the unbound fraction. Binding to α_1-acid glycoprotein is concentration dependent at therapeutic serum concentrations (i.e., 60% bound at 1 µg/ml and 14% bound at 8 µg/ml). As a result, pharmacokinetic parameters based upon total drug concentrations are unreliable. Unbound mean V_c = 0.39 l/kg (range 0.16 to 0.72): unbound mean V_{ss} = 1.3 to 1.7 l/kg.[18] The therapeutic range for free disopyramide is not well established.

Metabolism

Approximately 35 to 50% of disopyramide is metabolized in the liver to inactive metabolites (e.g., mono-N-dealkylated disopyramide). In normal subjects, unbound Cl is 1.5 to 4.5 ml/min/kg. The elimination half-life is variable; it ranges from 4 to 10 h and is prolonged in patients with hepatic or renal disease.[18]

Elimination

Up to 60% of disopyramide is excreted unchanged in the urine.

Drug Interactions

Disopyramide is a CYP3A3/4 enzyme substrate; its interations are shown in Table 8.4.

TABLE 8.4
Drug Interactions for Disopyramide

Drug	Interaction
Enzyme inducers: (e.g., phenytoin, phenobarbital, rifampin)	Decrease disopyramide serum levels
Enzyme inhibitors (e.g., erythromycin, ketoconazole)	Increase disopyramide serum levels
β-Blockers and calcium channel blockers	Additive negative inotropic effects
Drugs that prolong QT interval (e.g., erythromycin, haloperidol, tricyclic antidepressants, sotalol, bepridil)	Increase the risk of torsades de pointes (proarrhythmia)

Source: Data from Boyles, R. N. et al., *Clin. Pharmacol. Ther.*, 12:105, 1971.

Dosing, Empiric Dosing

Oral dose:

- Initial dosage: 100 mg every 6 h. The dose is titrated to a usual daily dose of 300 to 1200 mg/day. Higher doses (1200 mg/day) may be necessary for resistant arrhythmias.

- Renal dysfunction: Dosing interval should be increased to q8h for Cl_{cr} 30 to 40 ml/min; q12h for Cl_{cr} 15 to 30 ml/min; q24h for $Cl_{Cr} < 15$ ml/min. Disopyramide is not dialyzable.
- Congestive heart failure / liver disease: Disopyramide should be avoided.[7]

LIDOCAINE

Lidocaine is widely used and is often the drug of choice for the acute treatment of ventricular arrhythmias. Although oral lidocaine readily penetrates the gastrointestinal membrane, it undergoes extensive first-pass metabolism, limiting its use to intravenous and intramuscular administration.[6]

Clinical Pharmacokinetic/Pharmacodynamic Considerations

- A total loading dose is administered in divided doses.
- The dose is adjusted based on heart and liver function.
- Lidocaine levels are obtained only in special circumstances.
- Central nervous system toxicities may appear before cardiac toxicities.
- Concentration-dependent side effects may include drowsiness, paresthesias, euphoria (3 to 6 µg/ml), fasciculations, visual disturbances, tinnitus (6 to 8 µg/ml), seizures, obtundation (> 8 µg/ml), hypotension, decreased cardiac output (>9 µg/ml).[6,23]
- Lidocaine is bound to α_1-acid glycoprotein, which varies in settings that lidocaine is commonly administered (e.g., cardiac arrest, myocardial infarction).
- Renal dysfunction may predispose to active metabolite accumulation and consequent toxicity with prolonged administration.

Absorption

Intramuscular absorption is 100%, but variable depending on the site of injection (absorption $T_{1/2} = 12$ and 26 min from deltoid and gluteus sites, respectively). For this reason, the intravenous route of administration is preferred.[19]

Distribution

Lidocaine follows a two-compartment model. The volume of distribution (V_c and V_{ss}) varies with concomitant diseases or conditions as shown below. Loading doses, based on V_c, are adjusted depending on concomitant disease because the cardiovascular and central nervous systems behave as though they were located in the central compartment.[23] Following intravenous bolus administration lidocaine concentrations fall rapidly ($T_{1/2\alpha} \cong 8$ min).[19]

Approximately 70% of lidocaine is bound to plasma proteins (70% α_1-acid glycoprotein, 30% albumin). Therefore, conditions that increase α_1-acid glycoprotein (e.g., myocardial infarction, trauma, cardiac arrest) may dramatically alter the unbound fraction. This poses a clinical challenge as lidocaine is often administered

Condition	V_c (l/kg)	V_{ss} (l/kg)
Normal	0.5	1.6
Congestive heart failure	0.3	0.9
Liver disease	0.6	2.3

in cardiac arrest and myocardial infarction where α_1-acid glycoprotein levels are increased.[20,21]

Metabolism

Lidocaine is extensively metabolized (>90%) by cytochrome P-450 primarily to monoethylglycinexylidide (MEGX) and glycinexylidide (GX). MEGX and GX are active metabolites and possess about 90% and 10% of the potency of lidocaine, respectively. Lidocaine extraction ratio is 62 to 81%, so changes in liver blood flow can alter metabolism.[22] For example, maintenance infusions of lidocaine in patients with cirrhosis are empirically decreased by approximately 40% because of decrease in liver blood flow and enzyme activity. The loading dose is not adjusted as V_c is the same or slightly increased in this setting. Conversely, patients with heart failure have a protracted V_c and attenuated liver blood flow and clearance: loading and maintenance infusions are decreased to account for these disease-specific pharmacokinetic changes as shown below.

Condition[a]	Cl (ml/min/kg)	$t_{1/2}$ (h)
Normal	15 (10–18)	2 (1.6–2.5)
CHF (average)	5.5	2–70
NYHA I	10	—
II	5	—
III/IV	2	—
AMI	9.1	5
Liver disease	4–6	5–7

[a] AMI: acute myocardial infarction; CHF: congestive heart failure; NYHA: New York Heart Association classification of heart failure.[20]

Elimination

Renal clearance contributes less than 5% of the total clearance of lidocaine. GX is predominantly renally eliminated (40 to 60%) and may accumulate in renal failure causing toxicity. MEGX concentrations do not appear to accumulate with renal failure.

Drug Interactions

Lidocaine is a CYP3A3/4 substrate (Table 8.5).

TABLE 8.5
Drug Interactions for Lidocaine

Drug	Interaction
Cimetidine	Increase lidocaine serum levels
CYP3A4 inhibitors (e.g., erythromycin, ketoconazole)	Increase lidocaine serum levels
CYP3A4 inducers (e.g., carbamazepine, rifampin)	Decrease lidocaine serum levels
β-Blockers (e.g., propranolol)	May increase lidocaine serum levels

Dosing

Loading dose

Clinically, lidocaine is administered intravenously 1 to 1.5 mg/kg over 1 to 2 min. This dose is decreased empirically by 40 to 50% in patients with congestive heart failure. A loading dose (LD) may also be determined by Equation 8.1:

$$LD \text{ (mg)} = V_c(\text{l/kg}) \times \text{Weight (kg)} \times C \text{ (mg/l)} \tag{8.1}$$

To avoid toxicity, half of the loading dose (calculated or on a mg/kg basis) is usually administered at a rate ≤50 mg/min followed by the remaining dose divided in equal parts at 5-min intervals (see example below). The usual maximum total dose is 300 mg.[24]

Maintenance dose

Clinically, lidocaine is infused 20 to 50 µg/kg/min (1 to 4 mg/min). Using appropriate pharmacokinetic data for heart and liver disease an infusion rate may also be calculated using Equation 8.2:

$$\text{Infusion rate (mg/min)} = \frac{C_{ss}(\mu g/ml) \times Cl(\text{ml/min/kg}) \times IBW(\text{kg})}{1000 \ \mu g/mg} \tag{8.2}$$

Lidocaine levels are not routinely measured except in special circumstances (e.g., observed toxicity at normal doses, arrhythmia not terminated despite maximum doses, in patients that are at high risk for toxicity).

Example

A 64-year-old male patient (5 ft 8 in., 85 kg) is admitted with a diagnosis of cirrhosis of the liver. The patient develops ventricular tachycardia while in the hospital. An appropriate loading and maintenance dose of lidocaine may be calculated as follows:

$IBW = 50 \text{ kg} + (2.2 \times 8) = 67.6 \text{ kg} \approx 68 \text{ kg}$
$LD = V \text{ (l/kg)} \times \text{Weight} \times Cp \text{ (mg/l)} \qquad V_c = 0.6 \text{ l/kg (cirrhosis)}$

LD = 0.6 l/kg × 85 kg × 4 mg/l = 204 mg
LD = 204 mg

Administer loading dose (LD) as 100 mg IV over 2 min (50 mg/min), wait 5 min, then administer 50 mg IV over 1 min, wait 5 min, then another 50 mg over 1 min (total dose = 200 mg). A maintenance dose (MD) is calculated below for this example:

$$MD \ (mg/min) = \frac{Cpss(\mu g/ml) \times Cl \ (ml/min/kg) \times LBW \ (kg)}{1000 \ \mu g/mg}$$

$$MD \ (mg/min) = \frac{4 \ \mu g/ml \times 6 \ (ml/min/kg) \times 68 \ kg}{1000 \ \mu g/ml}$$

$$MD = 1.6 \ mg/min \ or \ 96 \ mg/h$$

PROCAINAMIDE

Procainamide is a Type Ia antiarrhythmic agent used to treat supraventricular and ventricular arrhythmias. This drug is also used in the management of cardiac arrest.[25]

Clinical Pharmacokinetic/Pharmacodynamic Considerations

- Blood pressure and electrocardiogram are monitored during dosing. An increase in QRS or QT intervals by >25% or hypotension necessitates discontinuation of a loading infusion.
- Procainamide and active metabolite N-acetylprocainamide (NAPA) are renally eliminated.
- Procainamide and NAPA levels may be measured for toxicity, particularly in patients with renal dysfunction.
- Assess for proarrhythmia (e.g., torsades de pointes).
- Use an immediate-release formulation in the acute setting.
- Switch from immediate-release to sustained-release formulation for chronic therapy to improve compliance.

Absorption

Procainamide absorption is a first-order process, $T_{1/2\alpha}$ = 20 min (range 8 to 140 min). Procainamide bioavailability is similar for immediate- and sustained-release formulations (F = 0.85). Following a single dose, peak concentrations occur in 1 h. Procainamide HCl is available for oral and intravenous administration (S = 0.85)

Distribution

Procainamide distribution follows a two-compartment model ($T_{1/2\alpha} \sim 5$ min; $V_c \sim$ 0.7 l/kg; $V_{ss} \sim 2.0$ l/kg).

Metabolism

Approximately 50% of procainamide is metabolized in the liver. The major metabolite NAPA possesses activity, primarily class III antiarrhythmic properties. The fraction as NAPA averages 0.33 and 0.20 in fast and slow acetylators, respectively.[26] It is unknown and impossible to predict the contribution of NAPA to arrhythmia suppression. The effective concentration range for procainamide and NAPA are 4 to 10 µg/ml and 7 to 15 µg/ml, respectively.[26]

Elimination

Approximately 50% of procainamide is excreted unchanged in the urine by glomerular filtration with proximal tubular secretion. NAPA is renally eliminated.[27]

	Procainamide $t_{1/2}$, h	Cl (ml/min/kg)	NAPA $t_{1/2}$, h
Normal	2.3–5.5	9	5–8
CHF	2.6–6	Decreased	—
Renal failure	11	Decreased	Prolonged
Anephric	5–14	—	41

Drug Interactions

See Table 8.6.

Dosing

Loading dose

Clinically, an intravenous loading dose of 15 to 18 mg/kg is administered over 30 to 60 min (maximum infusion rate 25 to 50 mg/min). In the presence of severe renal insufficiency or decreased cardiac output the dose is empirically decreased to 12 mg/kg. A loading dose may also be determined by Equation 8.3:

$$LD = \frac{V_{ss}(l/kg) \times IBW\ (kg) \times Cp\ (mg/l)}{S \times F} \tag{8.3}$$

TABLE 8.6
Drug Interactions for Procainamide

Drug	Interaction
Amiodarone	Increases procainamide and NAPA serum levels
Cimetidine	Increases procainamide and NAPA serum levels
Drugs that prolong QT interval (e.g., erythromycin, haloperidol, tricyclic antidepressants, sotolol, bepridil)	Increase the risk of torsades de pointes (proarrhythmia)
Ranitidine	Increases procainamide serum levels
Trimethoprim	Increases procainamide and NAPA serum levels

Prolongation of the QRS or QT intervals on electrocardiogram by greater than 25% or the development of hypotension necessitates discontinuation of the loading infusion.

Maintenance dose
Intravenous: The usual maintenance infusion is 3 mg/kg/h in normal individuals, 2 mg/kg/h in patients with moderate renal failure, and 1 mg/kg/h in patients with moderate congestive heart failure. Long-term therapy is usually avoided in patients with severe renal failure and/or severe congestive heart failure.

Convert IV to oral therapy by Equation 8.4:

$$\text{Total daily oral dose (mg)} = \frac{\text{mg/h infusion} \times (C_{ss_{po}} / C_{ss_{IV}}) \times 24}{F} \qquad (8.4)$$

Oral
Based on patient weight, the usual empiric oral maintenance dose is 50 mg/kg/day. This dose is decreased by 25% in patients with congestive heart failure and in those with liver failure. In renal failure the maintenance dose is empirically decreased by 50%. An empiric daily dose can also be determined by Equation 8.5:

$$\text{Daily dose} = \frac{(C_{ss})_{ave} \times Cl \times 1440}{S \times F} \qquad (8.5)$$

where $(C_{ss})_{ave}$ is 4 to 10 µg/ml, F is 0.85, and S is 0.85. The constant 1440 is the number of minutes in a day. Normal clearance is 9 ml/min/kg. After a maintenance dose is calculated, the dose is empirically reduced depending on patient factors (e.g., renal failure, heart failure).

Example

$$\text{MD (µg/day)} = \frac{Cl \times Cp_{ss} \times 1440}{S \times F} \qquad Cl = 9 \text{ ml/min/kg}$$

$$\text{MD} = \frac{9 \text{ ml/min/kg} \times 68 \text{ kg} \times 6 \text{ µg/ml} \times 1440 \text{ min/day}}{0.85 \times 0.85}$$

MD = 7,318,588 µg or approximately 7000 mg a day in divided doses

QUINIDINE

Oral quinidine is used to treat supraventricular and ventricular arrhythmias. A loading dose is usually not required as C_{ss} is reached in approximately 24 h ($t_{1/2} \sim 6$ h). An oral loading may be associated with severe gastrointestinal side effects. Intravenous quinidine is usually avoided because of hypotension.[6]

Clinical Pharmacokinetic/Pharmacodynamic Considerations

- A dosing regimen should be individualized because of variation in bioavailability, distribution, and elimination between patients.
- Therapeutic index is narrow (2 to 5 µg/ml).
- Oral loading is not necessary.
- IV administration may cause severe hypotension.
- Active metabolites exist (contribution to efficacy/toxicity not clear).
- Assess for potential drug–drug interactions (e.g., digoxin).
- Assess for proarrhythmia (e.g., torsades de pointes).[4,28,29]
- Use immediate-release formulation in the acute setting.
- Switch from immediate-release to sustained-release formulation for chronic therapy to improve compliance.
- Account for differences in salt formulation when switching between products.

Absorption

Quinidine is absorbed from the small intestine ($F \sim 0.80$; range 0.7 to 0.9) and is available in three oral salt dosage forms (see table below). Quinidine gluconate is available for intravenous administration; however, intramuscular administration should be avoided because of erratic absorption and pain at the injection site.[30]

Quinidine salt	S	t_{max} (h)
Sulfate	0.83	1–3[a]
Gluconate (sustained release)	0.62	4–6
Polygalacturonate	0.60	1–3

[a] T_{max} longer for sustained release products.

Distribution

Quinidine follows a two-compartment model ($T_{1/2\alpha} \sim 7$ min, $V_c \sim 0.7$ l/kg). The volume of distribution at steady state (V_{ss}) is dependent on concomitant disease states (see table below).[31]

Population	V_{ss} (l/kg)
Normal	2.5–3.0
Elderly	2.2
Congestive heart failure	1.8
Cirrhosis	3.8

Quinidine is 70 to 90% protein bound predominantly to albumin and α_1-acid glycoprotein. Conditions that increase α_1-acid glycoprotein (e.g., myocardial infarction, trauma, cardiac arrest) or decrease albumin may alter the unbound fraction of quinidine.[31]

Metabolism

Quinidine is primarily eliminated by biotransformation in the liver. About 80% of a dose is metabolized to active metabolites of varying activity. The contribution of these metabolites to efficacy/toxicity is not fully known.[32]

Metabolites	Activity	Unbound Fraction
3-hydroxyquinidine	+	0.26
2'-Quinidinone	+	0.54
Quinidine-*N*-oxide	+	0.05
Quinidine 10,11-dihyrodiol		—
O-Desmethylquinidine	+	<0.10

Elimination

Quinidine clearance decreases with age and congestive heart failure (see table below). Approximately 10 to 20% of the administered dose is excreted unchanged in the urine; however, metabolites are also eliminated through renal pathways. Renal excretion is dependent upon GFR and urine pH. Quinidine is a weak base (pK_a = 8.3), so an increased urine pH increases reabsorption.[31]

Condition	Cl (ml/min/kg)	$t_{1/2}$ (h)
Normal	5	6–7
Elderly (>60 yr)	2.5	up to 9
Congestive heart failure	3.3	7–8
Cirrhosis	No change	9
Renal failure	No change	—

Drug Interactions

Quinidine is a CYP3A4 substrate/inhibitor and CYP2D6 inhibitor. The concomitant use of drugs that prolong the QT interval should be avoided (Table 8.7).

Dosing

Dose regimens should be individualized. Careful consideration of concomitant diseases, potential for proarrhythmia, and drug interactions should be considered when determining acute and chronic therapy.[28]

Empiric dosing
- Oral (sulfate) dose: Initial dosage is 200 mg every 4 to 6 h. Dose is titrated every 3 to 4 days to a usual daily dose of 800 to 2400 mg/day. Maximum single dose = 600 mg. A daily dose may be calculated using pharmacokinetic data (see below).
- Intravenous/intramuscular (not recommended): Quinidine gluconate 5 to 8 mg/kg diluted in D_5W over 20 to 25 min (infusion rate <10 mg/min).

TABLE 8.7
Drug Interactions for Quinidine

Drug	Interaction
Cimetidine	Increases quinidine serum levels
CYP3A4 inhibitors (e.g., erythromycin, ketoconazole)	Increase quinidine serum levels
CYP3A4 inducers (e.g., carbamazepine, rifampin)	Decrease quinidine serum levels
Digoxin	Quinidine increases digoxin levels; decrease digoxin dose by 50%
Drugs that prolong QT interval (e.g., erythromycin, haloperidol, tricyclic antidepressants, sotolol, bepridil)	Increase the risk of torsades de pointes (proarrhythmia)
Urinary alkalinizers (antacids, sodium bicarbonate, acetazolamide)	Increase quinidine serum levels

Pharmacokinetic dosing

An empiric daily dose can be determined by Equation 8.6:

$$\text{Daily dose} = \frac{(C_{ss})_{ave} \times Cl \times 1440}{S \times F} \tag{8.6}$$

where $(C_{ss})_{ave}$ is 2 to 5 µg/ml, F is 0.7, and S is 0.83, 0.62, and 0.6 for the sulfate, gluconate, and polygalacturonate formulations, respectively. The constant 1440 is the number of minutes in a day. Cl (ml/min/kg) should be based on the patient's age and concomitant diseases (see table on p. 157 under Elimination).

Example

The daily quinidine sulfate dose to achieve a serum concentration of 2 µg/ml for a 75-year-old, 67-kg patient can be calculated by Equation 8.7:

$$\text{Daily dose} = \frac{2 \text{ µg/ml (2.5 ml/min/kg)}(1440 \text{ min/day})}{(0.83)(0.7)} \tag{8.7}$$

$$= 12{,}392 \text{ µg/kg/day or } 12.4 \text{ mg/kg/day quinidine sulfate}$$

For this elderly, 67-kg patient, the daily dose would be 67 kg × (12.4 mg/kg/day) or 831 mg/day quinidine sulfate. This daily dose, given in divided doses, should achieve a quinidine serum concentration of 2 µg/mlL (i.e., quinidine sulfate 200 mg q6h). A similar calculation using a desired serum concentration of 5 µg/ml, the upper limit of the therapeutic range, would give a daily quinidine sulfate dose of approximately 2000 mg/day.

DIGOXIN

Digoxin is used to control the ventricular response rate in tachyarrhythmias such as atrial fibrillation and atrial flutter. In the setting of systolic heart failure and normal sinus rhythm, digoxin improves morbidity and decreases the frequency of hospitalizations for exacerbation of heart failure.[33] The therapeutic range for digoxin is 0.8 to 2 ng/ml.[6]

CLINICAL PHARMACOKINETIC/PHARMACODYNAMIC CONSIDERATIONS

A summary of the pharmacokinetic properties of dioxin based on age is shown in Table 8.8. Pharmacodynamic considerations for tachyarrhythmias and systolic heart failure are as follows:

Tachyarrhythmias
- Narrow therapeutic range
- Used to control the ventricular response rate, particularly in patients with congestive heart failure
- Ventricular rate control usually achieved over 24 h
- A loading dose given in divided doses because of a long distribution half-life
- Possible to use higher doses to control rate in the acute setting (>2 ng/ml)

Systolic Heart Failure
- Decreases frequency of hospitalization for exacerbation of heart failure
- No improvement in cardiovascular mortality
- Serum concentration achieved in mortality trial at the lower end of therapeutic range
- Necessary to maintain serum concentrations in the mid- to low therapeutic range (≤1.5 ng/ml)
- Edema and cardiac output changes with severity of heart failure may alter pharmacokinetic parameters

ABSORPTION

Digoxin is nearly completely absorbed from the gut. In some patients, absorption may be decreased due to digoxin inactivation by gut bacteria (i.e., *Eubacterium lentum*).[36] Digoxin absorption is reduced by concomitant administration of antacids, cholestyramine, kaolin/pectin, tetracycline, and neomycin. Radiation malabsorption and gastrointestinal motility drugs such as metoclopramide can also reduce absorption.[35]

METABOLISM/DISTRIBUTION

Digoxin is not extensively metabolized. The volume of distribution for digoxin is 6 to 7 l/kg.[37] Digoxin distributes into lean body tissue but not appreciably into adipose or fatty tissues. For this reason ideal body weight should be used to dose digoxin.[35] Digoxin is 20 to 30% bound to albumin. The distributive phase of digoxin is 6 to 8

TABLE 8.8
Summary of the Pharmacokinetic Properties of Digoxin

	Renal Function	V_d (l/kg)	$t_{1/2}$ (h)	Cl (ml/min/1.73m²)	t_{max} (h)	Onset of Action	F
Adults	Normal	6–7	38 (range 18–72 h)	180 (range 100–300)[a]	1	Oral 1–2 / IV 5–30 min	Tablets: 0.7–0.8 / Capsules: 0.9–1.0 / Elixer: 0.75–0.85
	Impairment hemodialysis	5–6	Variable	—	—	—	—
	Anephric	4–5	106	—	1	—	—
Neonates	Normal	7.5–10	35–45	—	—	—	IM: 0.82
Infants	Normal	—	18–25	—	—	—	IV: 1.0
Children	Normal	16	35	—	—	—	—

[a] Influenced by various disease states and drugs.
— Means information is not available.

Source: Adapted from References 34 through 38.

h.[35] This has two important clinical implications. First, loading doses of digoxin need to be administered in divided doses 6 h apart. Second, serum digoxin levels should be determined at least 8 h after administration.

ELIMINATION

Digoxin is excreted largely unchanged in the urine with renal clearance approximating creatinine clearance. A small proportion of digoxin is cleared by nonrenal routes, biliary excretion and intestinal clearance. The volume of distribution decreases with decreasing renal function.[35] This may be due to a decreased volume of extracellular fluid and a decreased binding to lean tissues.[38] The clearance of digoxin is also reduced with renal impairment.[38] Half-life is extended in patients with renal impairment and can range from 38 h for normal renal function to 106 h for anephric patients.[38] Adjustment of the dose or dosing interval may be necessary when renal function changes.[39]

DISEASE STATE CONSIDERATIONS

Various patient disease characteristics can also play a role in digoxin kinetics and influence the dosing calculations required to attain appropriate steady-state serum concentrations and a desired clinical effect (Table 8.9).[35,40-42]

DRUG INTERACTIONS

Digoxin drug interactions occur predominantly by alterations in absorption or renal clearance. Some of these interactions necessitate a 50% reduction in the maintenance dosage of digoxin as shown in Table 8.10.

DOSING

The Jeliffe method is a simple and straightforward method for calculating loading and maintenance digoxin doses.[43] This equation is designed to produce dosing regimens that will provide a serum digoxin level of approximately 2.0 ng/ml (therapeutic range 0.8 to 2 ng/ml). This method for calculating digoxin loading and maintenance doses does not account for changes in heart function or drug interactions. It is important to adjust the dose empirically if significant drug–disease or drug–drug interactions are present.

Using the Jeliffe method, a loading dose or total body stores (TBS) is calculated by Equation 8.8:

$$\text{LD (or TBS)} = \frac{(10 \ \mu g/kg) \times \text{IBW (kg)}}{F} \tag{8.8}$$

Clinically, half this loading dose is administered intravenously or orally. The remaining dose is divided equally and administered at 6-h intervals (e.g., 1/2 dose, 1/4 dose, 1/4 dose at 6-h intervals). This is necessary to account for the slow distribution half-life of the drug.

TABLE 8.9
Disease Characteristics That Affect Digoxin Pharmacokinetics

Disorder	Effect upon Kinetics	Management
Congestive heart failure	Reduced cardiac output; can decrease V_d Edema may increase V_d	Optimize heart failure therapy; adjust digoxin dose appropriately
Hyperthyroidism	Increases V_d and increases resistance to digoxin	Correct thyroid disorder and follow digoxin levels
Hypothyroidism	Decreases V_d	Correct thyroid disorder; may require lower digoxin doses
Hypokalemia	Low serum potassium levels increase cardiac effects of digoxin	Correct hypokalemia, measure clinical effects (ECG, etc.)
Hyperkalemia	High serum potassium levels decrease cardiac effects	Correct hyperkalemia, measure clinical effects (ECG, etc.)
Hypomagnesemia	Associated with potentiated digoxin toxicity (possibly due to low intracellular potassium levels caused by hypomagnesemia)	Correct hypomagnesemia and any hypokalemia; monitor clinical effects (ECG, nausea, etc.)
Hypercalcemia	Very high levels of calcium have been associated with digoxin toxicity	

ECG = Electrocardiogram.

TABLE 8.10
Drug Interactions for Digoxin

Drug	Interaction
Aluminum/magnesium antacids	Binds digoxin in gastrointestinal tract and reduces absorption by 25%
Amiodarone	Increases digoxin serum levels; decrease digoxin dose by 50%
Antibiotics	Eradicate gut bacteria (*Eubacterium lentum*) responsible for inactivation of digoxin in select patients (increased absorption)
Cholestyramine/dietary fiber	Binds digoxin in gastrointestinal tract and reduces absorption by 20 to 35%.
Kaolin/pectin	Binds digoxin in gastrointestinal tract and reduces absorption by 20 to 60%
Gastrointestinal motility drugs (metoclopramide, neomycin)	Decrease gastronintestinal transit time, decreasing digoxin absorption
Quinidine	Increases serum digoxin levels; decrease digoxin dose by 50%

Maintenance dose

The maintenance dose can then be calculated based on the amount of the loading dose (or total body stores) that is eliminated each day.[43] The amount of digoxin eliminated daily can be calculated by Equation 8.9.

$$\% \text{ Total Body Stores lost daily } = 14 + \frac{Cl_{cr}(\text{ml/min})}{5} \tag{8.9}$$

Therefore, the maintenance dose is as follows:

$$MD = \text{Total Body Stores } (\%TBS \text{ lost each day}) \tag{8.10}$$

Example

JL is a 68-year-old male who is admitted to the hospital for atrial fibrillation. During his hospitalization he is converted to normal sinus rhythm by quinidine. His past medical history includes congestive heart failure. His thyroid function is normal. His weight is 84 kg (IBW = 73 kg) and his estimated creatinine clearance is 68 ml/min. What would be an appropriate oral loading and maintenance dose?

$$LD = \frac{10 \ \mu g/kg \times IBW}{F} = TBS$$

$$LD = \frac{10 \ \mu g/kg \times 73 \ kg}{0.7} = TBS = 1043 \ \mu g \ (1.04 \ mg)$$

Administer 0.5 mg orally, then 0.25 mg in 6 h, then 0.25 mg in 6 h.

$$MD = TBS \ (\%TBS \text{ lost each day})$$

$$\%TBS \text{ lost daily } = 14 + \frac{Cl_{cr}(\text{ml/min})}{5} = 14 + \frac{68 \ \text{ml/min}}{5} = 27.6\%$$

$$MD = 1.04 \ mg \ (27.6\%) = 0.29 \ mg$$

Because JL is also on quinidine the MD would be decreased by 50%.

$$MD = 0.15 \text{ or } 0.125 \text{ mg/day}$$

OTHER CARDIOVASCULAR DRUGS

Angiotensin-converting enzyme (ACE) inhibitors, angiotensin-II receptor blockers (ARBs), β-blockers, and calcium channel blockers are not dosed in the clinical setting using pharmacokinetic equations. The clinical use and selection of a drug within a class of medications is determined, in part, on differences in pharmacokinetic parameters. For example, metoprolol is commonly used in patients with renal disease because it is metabolized in the liver. Atenolol is avoided because it is renally eliminated. Select pharmacokinetic parameters for ACE inhibitors, ARBs, β-blockers, and calcium channel blockers are listed in Tables 8.11 through 8.13.

TABLE 8.11
Pharmacokinetic Parameters of ACE Inhibitors and ARBs*

	Bioavailability (%)	Half-Life (h)	V_d (l/kg)	Protein Binding %	Metabolism/Excretion	Onset of Action (h)	t_{max} (h)
ACE Inhibitors							
Benazepril	37	*10–11	8.71	~96	Metabolism to active metabolite benazeprilat, parent drug entirely eliminated via hepatic route/11–12% eliminated in bile	1	1–1.5
Captopril	70–75	*<2	0.7	25–30	50% metabolized/95% excreted in urine in 24 h	1–1.5	1–2
Enalapril	60	2 (3.4–5.8 with CHF)	—	50–60	Metabolism to enalaprilat/eliminated 60–80% in urine	1	0.5–1.5
Enalaprilat	—	35–38	—	50–60	Eliminated 60–80% in urine	0.25	3–4.5
Fosinopril	30–36	12 (fosinoprilat)	10	99.4	Metabolized to (active) fosinoprilat by intestinal wall and hepatic enzymes/eliminated in the urine and bile as fosinoprilat and conjugates	1	3
Lisinopril	25	11–12	124	0	Excreted almost entirely in urine as unchanged drug	1	ND
Moexipril	13–22	2–9	183	50 (70–90 for moexiprilat)	Metabolized to (active) moexiprilat in liver and intestines/13% excreted in urine, 53% in feces	1.5	1.5
Quinapril	50	0.8 (2 h for quinaprilat)	—	97	Metabolized to (active) quinaprilat/50–60% excreted in urine as quinaprilat	1	1 (2 h for quinaprilat)

Ramipril	60	13–17 (50 h for ramiprilat)		73 (56 for ramiprilat)	Metabolized to (active) ramiprilat in liver/parent drug and metabolites excreted 60% renally, 40% in feces	1–2	1
Trandolapril	10	5 (10 h for trandolaprilat)	18	80	Metabolized to (active) trandolaprilat in liver/metaboiltes excreted in urine	4	6
ARBs							
Candesartan	15	9	0.13	>99	Prodrug metabolized to active drug by intestinal wall	2–4	3–4
Eprosartan	13	5–9	308	98	Glucuronidated (not by P-450)/90% excreted in feces via biliary route, 7% in urine	—	4
Irbesartan	60–80	11–15	53–93	90	<20% metabolized/20% recovered in urine, 80% in feces	2	1.5–2
Losartan	33	2 (6–9 h for active metabolite)	34	99	14% in liver-active metabolite (E3174)	—	1 (2–4 for metabolite)
Telmisartan	42–58	24	500	>99.5	97% in feces	3	0.5–1
Valsartan	25	6	17	95	20% metabolized/13% excreted in urine, 83% in feces	2	2–4

CHF = congestive heart failure; ND = no data.

TABLE 8.12
Select Pharmacokinetic Parameters of β-Blockers (BB)

Medication	Bioavailability (%)	Half-Life (h)	Protein Binding %	Metabolism/Excretion	Onset of Action (h)	T_{max} (h)
Acebutolol	20–60	3–4	15–26	Hepatic metabolism / 30% renal excretion	1.5–3	2–4
Atenolol	50–60	6–9	5–16	~50% excreted in feces unchanged	3	2–4
Betaxolol	84–94	14–24	50–55	Liver / 80 % excreted in urine	1–1.5	2
Bisoprolol	80	9–12	~30	50% excreted in urine unchanged	2–4	1.7–3
Carteolol	80–85	6	20–30	50–70% excreted in urine unchanged	1–1.5	1–3
Carvedilol	25–35	7–10	95–98	Hepatic metabolism-active metabolites	—	1–1.5
Esmolol	N/A	0.15	55	Metabolized in RBC	2–10 min	5 min
Labetalol	18–40	5.5–8	50	55–60% excreted in urine as conjugates or unchanged	Oral: 20 min to 2 h / IV: 2–5 min	oral: 1–2
Metoprolol	50	3–7	10–12	Hepatic metabolism	1	1.5–2
Metoprolol (long acting)	77	3–7	10–12	Hepatic metabolism	—	3.3
Nadolol	30–50	20–24	25–30	Excreted in urine unchanged	3–4	2–4
Penbutolol	~100	5	80–98	Hepatic metabolism / renal excretion of conjugates	oral: 1–3	0.84–3
Pindolol	90–100	3–4	40–57	Urinary excretion of metabolites (60–65%) and active drug 35–40%	—	1–2
Propranolol	30	3–5	90	Extensive metabolism	1–2	2
Propranolol (long acting)	9–18	8–11	90	Extensive metabolism	—	6–10
Sotalol	90–100	12	0	Not metabolized – excreted unchanged in urine	1–2	oral: 2–3
Timolol	75	4	10	55–60% excreted in urine	15–45 min	1.55

N/A = not applicable-IV only; RBC = red blood cells; ND = no data.

TABLE 8.13
Pharmacokinetic Parameters of Calcium Channel Blockers (CCB)

CCB	Bioavailability (%)	Half-Life (h)	V_d (l/kg)	Protein Binding %	Metabolism/ Excretion	Onset of Action (h)	T_{max} (h)
				Dihydropyridines			
Amlodipine	52–90	30–50	21	93–97	Liver — inactive metabolites/bile, gut wall	30–50 min	6–12
Felodipine	10–25	10–36	10.3	>99	Liver — inactive metabolites/70% urine 10% feces	3–5	2.5–5
Isradipine	15–24	8	2.9	95–97	Liver — inactive metabolites/90% urine 10% feces	2	1.5
Nicardipine	35	2–4	0.88	>95	Liver/60% urine, 35% feces	20 min	20–120 min
Nifedipine IR	45–70	2–5	1.4–2.2	92–98	Liver — inactive metabolites/60–80% urine, feces, bile	20 min	0.5
Nifedipine (extended release)	86	2–5	ND	92–98	—	—	6
Nimodipine	13	1–2	0.43	>95	Liver — inactive metabolites/urine	ND	<1
Nisoldipine	4–8	7–12	4–5	>95	Liver — 1 active metabolite/70–75% kidney	ND	6–12
				Non-Dihydropyridines			
Diltiazem	40–67	3.5–6	5.3	70–80	CYP3A4 — active metabolites/2–4% excreted in urine	1/2–1	2–3
Diltiazem SR	40–67	5–7	5.3	70–80	excreted in urine	1/2–1	6–11
Bepridil	59	24	8	>99	Active metabolites	1	2–3
Verapamil	20–35	3–7	4.5–7	83–92	Active metabolites/3–4% excreted in urine	30	1–2.2

SUMMARY

For select cardiovascular drugs it is important to consider patient characteristics, the condition being treated, and the potential for drug and disease interactions before and during therapy. Careful consideration of pharmacokinetic and pharmacodynamic parameters in drug dosing and patient monitoring in each patient will optimize the benefits and minimize the risks of treatment.

REFERENCES

1. Vaughn Williams, E. M., A classification of antiarrhythmic actions reassessed after a decade of new drugs, *Clin. J. Pharmacol.*, 24:129–147, 1984.

2. Echt, D. S. et al., Mortality and morbidity in patients receiving encainide, flecainide, or placebo, The Cardiac Arrhythmia Suppression Trial, *N. Engl. J. Med.*, 149:1524–1527, 1989.

3. Flaker, G. C. et al., Antiarrhythmic drug therapy and cardiac mortality in atrial fibrillation, *J. Am. Coll. Cardiol.*, 20:527–532, 1992.

4. Coplen, S. E. et al., Efficacy and safety of quinidine therapy for maintenance of sinus rhythm after cardioversion: a meta-analysis of randomized control trials, *Circulation*, 82:1106–1116, 1990.

5. Morganroth, J., Bigger, J. T., and Anderson, J. L., Treatment of ventricular arrhythmias by United States cardiologists: a survey before the Cardiac Arrythmia Suppression Trial results were available, *Am. J. Cardiol.*, 65:40–48, 1990.

6. Bauman, J. L., Schoen, M. D., and Hoon, T. J., Practical optimization of antiarrhythmic drug therapy using pharmacokinetic principles, *Clin. Pharmacokinet.*, 20:151–166, 1991.

7. Woosley, R. L., Pharmacokinetics and pharmacodynamics of antiarrhythmic agents in patients with congestive heart failure, *Am. Heart J.*, 114:1280–1291, 1987.

8. Andreason, F., Agerback, H., and Byerregarrd, P., Pharmacokinetics of amiodarone after intravenous and oral administration, *Eur. J. Clin. Pharmacol.*, 19:293–299, 1981.

9. Riva, E. et al., Pharmacokinetics of amiodarone in man, *J. Cardiovasc. Pharmacol.*, 4:270–275, 1982.

10. Freedman, M. D. and Somberg, J. C., Pharmacology and pharmacokinetics of amiodarone, *J. Clin. Pharmacol.*, 31:1061–1069, 1991.

11. Nolan, P. E. et al., Single dose pharmacokinetics of amiodarone, *Drug Intell. Clin. Pharm.*, 19:463–466, 1985.

12. Holt, D. W. et al., Pharmacokinetics of amiodarone in man, *J. Cardiovasc. Pharmacol.*, 4:264–269, 1982.

13. Zipes, D. P., Amiodarone: electrophysiologic action, pharmacokinetics and clinical effects, *J. Am. Coll. Cardiol.*, 3:1059–1065, 1984.

14. LeRoy, G. et al., Torsades de pointes during loading with amiodarone, *Eur. Heart J.*, 8:541–542, 1987.

15. Tham, T. C. K. et al., Pharmacodynamics and pharmacokinetics of the class III antiarrhythmic agent dofetilide (UK-68,798) in humans, *J. Cardiovasc. Pharmacol.*, 21:507–512, 1993.

16. Sedgwick, M. et al., Pharmacokinetic and pharmacodynamic effects of UK-68,798, a new potential class III antiarrhythmic drug. *Br. J. Clin. Pharmacol.*, 31:515–519, 1991.

17. Anderson, J. L. et al., Antiarrhythmic drugs: clinical pharmacology and therapeutic uses, *Drugs*, 15:271–309, 1978.
18. Woosley, R. L. and Shand, D. G., Pharmacokinetics of antiarrhythmic drugs, *Am. J. Cardiol.*, 41:986–994, 1978.
19. Scott, D. B. et al., Plasma lignocaine levels after intravenous and intramuscular injection, *Lancet*, 1:41, 1970.
20. Boyles, R. N. et al., Pharmacokinetics of lidocaine in man, *Clin. Pharmacol. Ther.*, 12:105–116, 1971.
21. LeLorier, J. et al., Pharmacokinetics of lidocaine after prolonged intravenous infusions in uncomplicated myocardial infarction, *Ann. Intern. Med.*, 87:700–702, 1977.
22. Huet, P. M. et al., Bioavailability of lidocaine in normal volunteers and cirrhotic patients, *Gastroenterology*, 75:969–974, 1978.
23. Collinsworth, K. A., Kalman, S. M., and Harrison, D. C., Clinical pharmacology of lidocaine as an antiarrhythmic drug, *Circulation*, 50:1217–1230, 1974.
24. Stargel, W. W. et al., Clinical comparison of rapid infusion and multiple injection methods of lidocaine loading, *Am. Heart J.*, 102:872–876, 1981.
25. Hoffman, B. F., Rosen, M. R., and Wit A. L., Electrophysiology and pharmacology of cardiac arrhythmias. VII. Cardiac effects of quinidine and procainamide, *Am. Heart J.*, 90:117–122, 1975.
26. Connolly, S. J. and Kates, R. E., Clinical pharmacokinetics of N-acetylprocainamide, *Clin. Pharmacokinet.*, 7:206–220, 1982.
27. Gibson, T. P. et al., Kinetics of procainamide and N-acetylprocainamide in renal failure, *Kidney Int.*, 12:422–429, 1977.
28. Bauman, J. L. et al., Torsades de pointes due to quinidine: observations in 31 patients, *Am. Heart J.*, 107:425–430, 1984.
29. Morganroth, J. and Goin, J. E., Quinidine-related mortality in the short-to-medium-term treatment of ventricular arrhythmias: a meta-analysis, *Circulation*, 84:1977–1983, 1991.
30. Greenblatt, D. L. et al., Pharmacokinetics of quinidine in humans after intravenous, intramuscular, and oral administration, *J. Pharmacol. Exp. Ther.*, 202:365–378, 1977.
31. Ochs, H. R., Greenblatt, D. L., and Woo, E., Clinical pharmacokinetics of quinidine, *Clin. Pharmacokinet.*, 5:150–168, 1980.
32. Crevasse, L., Quinidine: an update on therapeutics, pharmacokinetics, and serum concentration monitoring, *Am. J. Cardiol.*, 62:221–231, 1983.
33. The Digitalis Investigation Group, The effect of digoxin on mortality and morbidity in patients with heart failure, *N. Engl. J. Med.*, 336:525–533, 1997.
34. Ohnhaus, E. E., Vozeh, S., and Nuesch, E., Absolute bioavailability of digoxin in chronic renal failure, *Clin. Nephrol.*, 11:302–306, 1979.
35. Ochs, H. R. et al., Disease-related alterations in cardiac glycoside disposition, *Clin. Pharmacokinet.*, 7:424–451, 1982.
36. Doherty, J. E. et al., A multicenter evaluation of the absolute bioavailability of digoxin dosage forms, *Curr. Ther. Res.*, 35:301–306, 1984.
37. Reuning, R. H., Sams, R. A., and Notari, R. E., Role of pharmacokinetics in drug dosage adjustment. I. Pharmacologic effect, kinetics and apparent volume of distribution of digoxin, *J. Clin. Pharmacol.*, 13:127–141, 1973.
38. Dettli, L., Spring, P., and Ryler, S., Multiple dose kinetics and drug dosage in patients with kidney disease, *Acta Pharmacol.*, 29:211–222, 1971.
39. Koup, J. R. et al., Digoxin pharmacokinetics: role of renal failure in dosage regimen design, *Clin. Pharmacol. Ther.*, 18:9–21, 1975.

40. Sellar, R. H. et al., Digitalis toxicity and hypomagnesemia, *Am. Heart J.*, 79:57–68, 1970.

41. Sonnenblick, M. et al., Correlation between manifestations of digoxin toxicity and serum digoxin, calcium, potassium, and magnesium concentrations and arterial pH, *Br. Med. J.*, 286:1089–1091, 1983.

42. Sampson, J. J. et al., The effect on man of potassium administration in relation to digitalis glycoside, with special reference to blood serum potassium, the electrocardiogram, and ectopic beats, *Am. Heart J.*, 26:164–179, 1943.

43. Jelliffe, R. W., An improved method of digoxin therapy, *Ann. Intern. Med.*, 69:703–717, 1968.

9 Psychotropic Agents

Paul J. Perry

CONTENTS

1-56676-973-6/02/$0.00+$1.50
© 2002 by CRC Press LLC

INTRODUCTION

Therapeutic drug monitoring of psychotropic drugs has clinical utility for the following reasons:

- Interindividual pharmacokinetic variation usually makes it unrealistic to dose patients on the basis of body weight.
- Excessively high drug levels may be associated with clinical deterioration of the patient because of drug toxicity.
- Plasma level monitoring may help assure compliance and may reduce medication defaulting because of adverse drug reactions (ADR).
- Blood levels clearly correlate with therapeutic response for a minority of the psychotropic drugs, including the antipsychotics haloperidol, clozapine, and olanzapine; the tricyclic antidepressants; and the mood stabilizer lithium.

TYPICAL ANTIPSYCHOTICS

Available kinetic data for the commonly prescribed typical antipsychotics are presented in Table 9.1. The noted references are recommended for reviews of antipsychotic pharmacokinetics and therapeutic response.[1-5]

CHLORPROMAZINE

Absorption

- Chlorpromazine is only partially absorbed by the gastronintestinal (GI) tract.
- Differences in formulations, the presence of food in the GI tract, and concomitant therapy with other drugs, such as antacids, may significantly alter the absorption of chlorpromazine.
- Plasma levels are four to ten times higher after IM injection than those resulting from an equal dose given orally.

Metabolism

- Chlorpromazine is extensively metabolized by the liver and intestinal wall.
- Many metabolites have been described in humans.
- After 2 to 3 weeks of chronic treatment, a 40% decline in AUC was observed as a result of inhibition of its own absorption due to decreased GI motility, or because chlorpromazine may accelerate its own metabolism through enzyme induction.

Therapeutic Levels

- Therapeutic levels have not been established.

TABLE 9.1
Pharmacokinetics of Typical Antipsychotics

Drug	Oral Bioavailability (%)	Plasma Protein Binding (%)	Clearance (ml/min/kg)	Volume of Distribution (l/kg)	$t_{1/2}$ (h)
Chlorpromazine	32 ± 19	95–98	8.6 ± 2.9	21 ± 9	30 ± 7
Fluphenazine decanoate	n/a	99	—	220	5–12
Fluphenazine enanthate	n/a	99	—	220	3–4
Fluphenazine HCl	< 50	99	—	220	13–58
Haloperidol decanoate	21 days	—	—	—	21 days
Haloperidol HCl and lactate	60 ± 18	92 ± 2	11.8 ± 2.9	18 ± 7	18 ± 5
Loxapine	—	—	—	—	3–4
Molindone	—	—	—	—	1.5–6
Perphenazine	60–80	—	—	10–35	8–12
Thioridazine	10–60	96	18	2–4	26–36
Thiothixene	—	—	—	—	12–36
Trifluoperazine	—	> 90	—	—	7–18

Source: Adapted from References 6 through 8.

FLUPHENAZINE

Absorption

- Bioavailability is <50%.
- Decanoate produces peak serum concentrations after a single injection, which typically occur within 24 h but a second peak has been noted 8 to 12 days after a single injection.
- The half-life of fluphenazine decanoate after a single injection is approximately 8.6 ± 0.2 days.

Therapeutic Levels

- Therapeutic levels have not been established.
- A curvilinear relationship between plasma level and response (therapeutic window) may exist for fluphenazine between 0.1 to 2.8 ng/ml; however, other studies report continued improvement up to 4.5 ng/ml while other studies suggest no relationship.

HALOPERIDOL

Absorption

- Bioavailability is 60%, such that oral doses should be 1.4 to 1.7 times as large as IM or IV doses.
- IV injections of 10 mg produced serum levels of ~30 ng/ml.

Metabolism

- A hydroxylated metabolite is inactive, but may be converted back to haloperidol.

Therapeutic Levels (Acute Schizophrenia)

- Haloperidol serum concentrations after a single decanoate injection typically remain relatively stable over a 4-week interval.
- A meta-analysis of seven haloperidol studies concluded that the therapeutic range was 5 to 18 ng/ml.
- Studies need to be performed using levels <5 ng/ml and also changing a patient's serum level from one range to another to determine improvement or exacerbation of symptoms.
- Multiple linear regression analysis showed a significant interaction between the variables of smoking. The haloperidol plasma concentration-to-dose relationship was best described by Equation 9.1 for nonsmokers and Equation 9.2 for smokers.

$$\text{haloperidol (ng/ml)} = e^{[0.467 \, * \, \ln(\text{dose}) + 3.397]} \qquad (9.1)$$

$$\text{haloperidol (ng/ml)} = e^{[1.088 \, * \, \ln(\text{dose}) + 3.716]} \qquad (9.2)$$

For example, to achieve a steady-state blood level of 10 ng/ml, a smoking patient would require a dose of approximately 0.28 mg/kg/day, whereas a nonsmoking patient would require a dose of only 0.1 mg/kg/day.

Therapeutic Levels (Maintenance Therapy)

- The annualized relapse rate for patients with chronic schizophrenia treated with haloperidol decanoate 50 and 100 mg per month is ~75%.
  - 50 to 100 mg/month doses of haloperidol decanoate produce steady-state levels in the range of ~1.8 to 3.6 ng/ml.
  - These levels are well below the therapeutic range of 5 to 18 ng/ml recommended in acutely ill patients with schizophrenia and are not usually associated with extrapyramidal side effects.
- Generally, the acute typical antipsychotic dose that a patient with schizophrenia responded to initially can be decreased by 20% every 6 months. This generaliztion may not be transferable to atypical antipsychotics.

The conversion of patients from oral haloperidol to the depot form may lead to problems if the differences in half-life of the two formulations are not considered. Steady-state serum concentrations of the oral and depot forms will be achieved after approximately 5 days and 3 months of continuous dosing, respectively. Table 9.2 presents the probable steady-state plasma haloperidol concentrations resulting from differing doses of haloperidol decanoate after the first dose and after three doses are given on a 28-day schedule. A patient's

TABLE 9.2
Projected Plasma Steady-State Haloperidol Plasma Concentrations Expected from Haloperidol Decanoate

Haloperidol Decanoate (mg/4 weeks)	Haloperidol Concentration (ng/ml)	
	Month 1	Month 3
50	1.1	1.8
100	2.2	3.6
150	3.3	5.5
200	4.5	7.5
250	5.7	9.4
300	6.9	11.4
350	8.1	13.4
400	9.3	15.4

Source: Adapted from References 9 and 10.

steady-state haloperidol serum level would be determined on an oral drug. By using Table 9.2, a loading dose of haloperidol decanoate could be chosen based on the month 1 (see Table 9.2) serum level. The corresponding maintenance dose could be determined by finding the matching haloperidol serum level in the "month 3" column.

PERPHENAZINE

Absorption

- Perphenazine is 60 to 80% bioavailable.

Therapeutic Levels

- A level of 0.8 to 2.4 ng/ml has been recommended to improve the clinical response and decrease extrapyramidal side effects.

THIORIDAZINE

Absorption

- Thioridazine is 10 to 60% bioavailable.
- Thioridazine impairs its own absorption at higher plasma levels possibly because of its strong anticholinergic effect, which could influence gastric emptying.

Metabolism

- Thioridazine undergoes extensive side-chain degradation to active metabolites, such as mesoridazine, and inactive metabolites.

Therapeutic Levels

- Therapeutic levels have not been established.

THIOTHIXENE

Absorption

- Peak plasma levels are reached within 1 to 3 h of oral administration.

Therapeutic Levels

- Therapeutic levels have not established.

Therapeutic Levels

- A level of 1 to 2.3 ng/ml indicated, but based on only a single study.

ATYPICAL ANTIPSYCHOTICS

Available kinetic data for the prescribed atypical antipsychotics are presented in Table 9.3. These antipsychotics are termed atypical because they have a lower rate of extrapyramidal side effects in contrast to the typical antipsychotics.

CLOZAPINE

Absorption

- Peak plasma concentrations occur within 1 to 6 h.
- Bioavailability of clozapine is not affected by food.

Metabolism

- Extensively metabolized by the liver and generates three primary metabolites, of which desmethylclozapine is slightly active.

Therapeutic Levels

- Plasma concentrations in males were 69% of females' levels adjusted for weight.
- The average plasma concentration for male and female smokers was 82% of nonsmokers' concentrations; thus, smoking induces the CYP1A2 hepatic metabolism.

TABLE 9.3
Pharmacokinetics of Atypical Antipsychotics

Drug	Oral Bioavailability (%)	Plasma Protein Binding (%)	Clearance (ml/min/kg)	Volume of Distribution (l/kg)	$t_{1/2}$ (h)
Clozapine	55 ± 12	>95	6.1 ± 1.6	5.4 ± 3.5	12 ± 4
Risperidone	66 ± 28	89	5.4 ± 1.4	1.1 ± 0.2	3.2 ± 0.8
Olanzapine	—	93	—	10–20	21–54
Quetiapine	~9	83	—	10	2.7–9.3
Ziprasidone	—	>99	7.5	1.5	6.6

Source: Adapted from References 6 through 8 and 11.

- Concentrations are approximately doubled from the age cohort of 18 to 26 years to 45 to 54 years.
- The threshold blood level for a therapeutic concentration is estimated to be between 350 and 500 ng/ml.
- A dosing model that optimally predicts steady-state clozapine plasma concentrations includes the variables dose (mg/day), smoking (yes = 0 and no − 1), and gender.

$$\text{clozapine (ng/ml)} = 111 \text{ (smoke)} + 0.464 \text{ (dose)} + 145 \qquad \text{(males) (9.3)}$$

$$\text{clozapine (ng/ml)} = 111 \text{ (smoke)} + 1.590 \text{ (dose)} - 149 \qquad \text{(females) (9.4)}$$

RISPERIDONE

Absorption

- Risperidone is rapidly absorbed after oral administration, reaching a peak plasma level within 2 h.
- Administering the drug with or without food does not affect the extent of absorption.

Metabolism

- The active metabolite hydroxy-risperidone has a half-life of 24 h.
- Half-life is longer in elderly people and in patients with renal insufficiency.
- Hepatic insufficiency does not lengthen the half-life of risperidone; a more pronounced effect may occur due to more unbound risperidone.

Therapeutic Levels

- Therapeutic levels have not been established.

OLANZAPINE

Absorption

- Peak plasma levels were seen an average of 4.9 h after administration.

Metabolism

- The major metabolite is olanzapine-N-glucuronide; other metabolites are olanzapine-N-oxide and N-desmethyl olanzapine.
- Clearance of olanzapine is approximately 30% higher in men than in women, 40% higher in those who smoke than in those who do not smoke, and 30% higher in the young than in elderly people. If one compares changes in doses reflecting this increase in clearance then the difference

between male smokers and nonsmokers would be an increase of 6 mg/day or 16 mg/day, respectively (10 mg/day being the average daily dose), whereas the difference between females would be an increase of 9 mg/day or 19 mg/day (10 mg/day being the average daily dose), respectively.

Therapeutic Levels

- The threshold blood level for a therapeutic concentration is 23 ng/ml, drawn 12 h after the last dose.

QUETIAPINE

Absorption

- Peak plasma levels occur in 1 to 1.8 h.

Metabolism

- Quetiapine exhibits linear pharmacokinetics.
- Twenty metabolites have been reported, but only 7-hydroxy and n-dealkly metabolites are active and their contribution to efficacy is unclear.
- Clearance is decreased in elderly people by 40%.
- Hepatic impairment reduces clearance by ~30%.
- Gender, race, smoking, and renal insufficiency do not affect pharmacokinetics.

Therapeutic Levels

- Therapeutic levels have not been established.

ANTIPSYCHOTIC SUMMARY

The clinical interpretation of the above blood level data suggests a three-step treatment algorithm for the treatment of the acutely ill patient with schizophrenia if the clinician intends to base the dosing on data of blood level to therapeutic response.

- Haloperidol for 6 to 12 weeks: Choose a dose that achieves a blood level of 5 to 18 ng/ml if the patient is haloperidol refractory or cannot tolerate the extrapyramidal adverse effects of haloperidol; then
- Olanzapine for 6 to 12 weeks: Choose a dose that achieves a blood level of >23 ng/ml if the patient is olanzapine nonresponsive; then
- Clozapine for 16 weeks: Choose a dose that achieves a blood level of 504 ng/ml.

The recommended sampling time for patients receiving oral antipsychotics should be approximately 12 h after the last dose. Subsequent blood levels should be obtained at the same sampling time. The patient's dose should be fixed for at least 1 week prior to sampling and for the decanoate 3 months prior to sampling. This recommendation will aid in the interpretation of subsequent antipsychotic levels.

ANTIDEPRESSANTS

Table 9.4 presents the pharmacokinetic parameters of bioavailability, free fraction, volume of distribution, and half-life of all the currently available U.S. heterocyclic antidepressants. At steady state, increases or decreases in the dose will result in similar proportional changes in blood level for all the antipsychotics. Importantly, relevant blood level information cannot be elicited if the patient is receiving more than one antipsychotic concurrently.

TRICYCLIC ANTIDEPRESSANTS

ABSORPTION

- Tricyclic antidepressants (TCAs) are rapidly and completely absorbed from the small intestine following oral administration.
- Peak blood concentrations are usually reached within 2 to 8 h.
- Food in the stomach does not delay absorption, but systemic bioavailability is incomplete because 30 to 70% of the TCAs are metabolized on the first pass through the liver.

DISTRIBUTION

- TCAs are highly lipophilic agents that distribute widely throughout the body, as evidenced by their volumes of distribution reaching 60 l/kg.
- TCAs concentrate to the greatest extent in myocardial and cerebral tissues with <1% of the drug present in the plasma.
- Extensive distribution, coupled with 63 to 98% protein binding of the TCAs, explain why hemodialysis and hemoperfusion are inefficient treatments for TCA overdoses.
- Hemodialysis, despite removing 75 to 95% of the offending TCA from the blood, removes only 1% of the ingested drug from the body.
- TCAs are extensively but reversibly bound to α_1-acid glycoproteins (αAG).
- Changes in the concentrations of αAG can result in clinically significant increases in the αAG-bound TCAs in medically ill, depressed patients.
- αAG increase in renal transplantation, liver disease, myocardial infarction, pregnancy, malignancy, ulcerative colitis, chronic alcoholism, and rheumatoid arthritis.
- These patients may require maximal tolerable TCA doses before responding.

TABLE 9.4
Antidepressant Pharmacokinetic Parameters

Drug	Oral Bioavailability (%)	Plasma Protein Binding (%)	Clearance (ml/min/kg)	Volume of Distribution (l/kg)	$t_{1/2}$ (h)
Amitriptyline	48 ± 11	94.8	11.5 ± 3.4	15 ± 3	21 ± 5
Amoxapine	46–82	—	—	—	8.8–14
Bupropion	>90	84 ± 2	35 ± 9	7.2 ± 1.6	12 ± 4
Citalopram	80	50	5.9	12–16	25–35
Clomipramine	36–62	96	—	17	12–36
Desipramine	38 ± 13	82 ± 2	10 ± 2	20 ± 3	22 ± 5
Doxepin	55 ± 12	> 95	6.1 ± 1.6	5.4 ± 3.5	12 ± 4
Fluoxetine	> 60	94	9.6 ± 6.9	35 ± 21	53 ± 41
Fluvoxamine	> 90	77	—	25	8–28
Imipramine	39 ± 7	90.1 ± 1.4	15 ± 4	18 ± 3	12 ± 5
Maprotiline	79–87	88	—	14–22	36–105
Mirtazepine	50	85	—	10–14	20–40
Nefazodone	20	99	—	0.2–0.9	1 -4
Nortriptyline	51 ± 5	92 ± 2	7.2 ± 1.8	18 ± 4	31 ± 13
Paroxetine	a	95	8.6 ± 3.2	17 ± 10	17 ± 3
Protriptyline	77–93	92 ± 0.6	3.6 ± 0.6	22 ± 1	78 ± 11
Sertraline	80–95	99	23	76	24–36
Trazodone	75 ± 30	93	1.8 ± 0.6	1.0 ± 0.3	6.5 ± 1.8
Trimipramine	—	95	—	17–48	16–39
Venlafaxine	~10	27 ± 2	22 ± 10	7.5 ± 3.7	4.9 ± 2.4

[a] Dose dependent.

Source: Adapted from References 6 through 8, 12, and 13.

METABOLISM

- TCAs are metabolized primarily by the liver.
- Dimethylated TCAs, imipramine, amitriptyline, and doxepin, are demethylated to monomethylated tricyclics by CYP4503A4, which are active metabolites.
- When interpreting a TCA plasma level for a dimethylated TCA, both the dimethylated and monomethylated metabolites are measured and their values summed to estimate the concentration of active drug; e.g., for amitriptyline, the total plasma concentration equals amitriptyline + nortriptyline.
- Hydroxylation of the mono- and dimethylated TCAs by the hepatic cytochrome P4502D6 enzyme also occurs.

- Although active, hydroxylated metabolites are not assayed routinely because there is no current evidence to suggest that their measurement yields any useful information for the clinical management of depression.
- No dose alterations are necessary in patients with decreased renal function.

ELIMINATION

- The elimination half-life of TCAs can vary from 6 to 198 h.
- With the exception of protriptyline, the half-lives are approximately 24 h for the monomethylated TCAs allowing for single daily dosing, usually at bedtime.

SAMPLING

- Sampling should be carried out approximately 12 h after the last dose so that plasma concentrations are estimated during the TCA elimination phase.

THERAPEUTIC LEVELS

- A meta-analysis of the TCA blood level literature using ROC curves as the primary statistical tool determined the recommended therapeutic thresholds or ranges for four TCAs: nortriptyline, desipramine, amitriptyline, and imipramine. A summary of these data are presented in Table 9.5.

PROSPECTIVE DOSING

- A linear relationship exists between a test dose plasma concentration and steady-state concentration for the TCA nortriptyline.

TABLE 9.5
TCA Response Rates In and Out of the Suggested Therapeutic Concentrations

Drug	Therapeutic Range (ng/ml)	Response Rate (In vs. Out)[a]
Nortriptyline	58–148	66% vs. 26%
Desipramine	>116	51% vs. 15%
Imipramine	175–350	67% vs. 39%
Amitriptyline	93–140	50% vs. 30%

[a] Response rate in the therapeutic range or lower limit vs. outside the recommended blood level.

Source: Adapted from References 4 and 15.

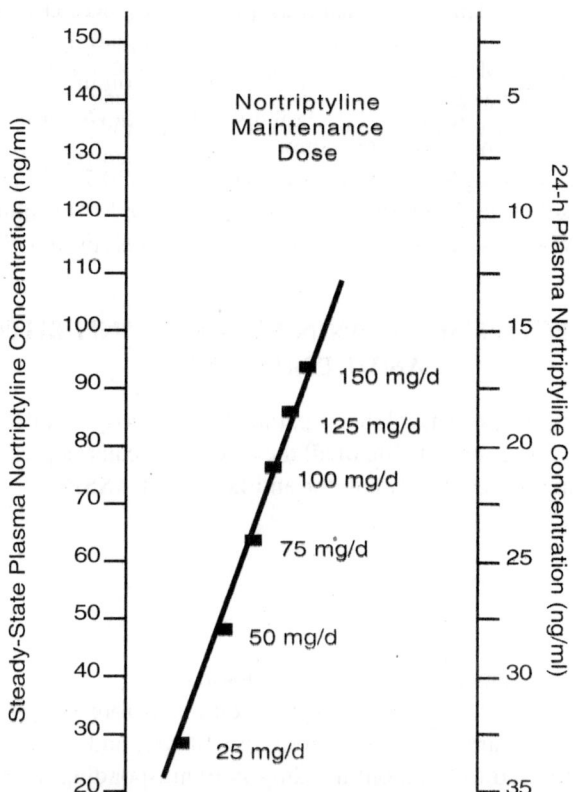

FIGURE 9.1 Nortriptyline dosing nomogram for predicting steady-state plasma concentrations for 25- to 150-mg/day maintenance doses. To use the nomogram, administer the patient a 100-mg nortriptyline test dose and then determine the plasma nortriptyline concentration 24 h following the test dose. To determine the predicted steady-state plasma nortriptyline levels for different maintenance doses, draw a line from the observed 24-h test dose nortriptyline level (right-side vertical axis) through the maintenance dose point of interest (diagonal line) to the steady-state nortriptyline level (left-side vertical axis). The 150-mg/day dose should not be exceeded, because the relationship of nortriptyline plasma concentrations and maintenance doses may not be linear at higher doses. (Adapted from Reference 14.)

- The only information required to predict steady-state doses for maintenance doses of 50 to 150 mg/day using the nomogram shown in Figure 9.1 is a plasma nortriptyline.
- Concentration is measured 24 h following the administration of a 100-mg test dose of nortriptyline.

MONOAMINE OXIDASE INHIBITORS

- Use of monoamine oxidase inhibitors (MAOI) antidepressant blood levels to monitor patients with depression is of no clinical value.

- A positive correlation exists between the degree of platelet MAO inhibition and the clinical response of patients with depression for phenelzine and isocarboxazid, but not necessarily tranylcypromine.
- Measurement of MAOI activity is not clinically practical because of the difficulty in handling platelet samples.
- Phenelzine of 1 mg/kg/day and tranylcypromine of 0.7 mg/kg/day in most patients will usually produce an 80% or greater MAO inhibition. This level of MAOI is necessary to attain an antidepressant response.

SELECTIVE SEROTONIN REUPTAKE INHIBITOR ANTIDEPRESSANTS

Table 9.4 presents the pharmacokinetic parameters of bioavailability, free fraction, volume of distribution, and half-life of all the heterocyclic antidepressants, including the selective seratonin reuptake inhibitor antidepressants (SSRIs).

FLUOXETINE

Metabolism

- Fluoxetine exhibits nonlinear pharmacokinetics.
- A period of 6 to 7 weeks may be required before steady-state serum concentrations are reached due to long half-life, and the same period of time is required to wash out the drug in nonresponding patients.
- Pharmacokinetic parameters of fluoxetine are not altered in patients with decreased renal function.
- The rate of elimination of fluoxetine is reduced in patients with alcohol-induced cirrhosis.
- Patients with liver disease should be treated with lower doses (e.g., 20 mg QOD) to minimize adverse effects.
- Plasma concentrations of fluoxetine and norfluoxetine do not differ between depressed elderly, younger healthy adults, and younger depressed patients after receiving similar doses.

Therapeutic Levels

- Therapeutic levels have not been established.
- There may be a negative correlation between norfluoxetine concentrations and therapeutic response.
- Patients with high ratios of fluoxetine to norfluoxetine are more likely to respond than patients with low ratios.

PAROXETINE

Metabolism

- Paroxetine exhibits nonlinear pharmacokinetics.
- Patients with severe renal impairment ($Cl_{cr} < 30$ mg/min) require dosage reduction, but hepatic dysfunction does not require dosage adjustment.

Therapeutic Levels

- Therapeutic levels have not been established.

SERTRALINE

Absorption

- Ingestion with food results in plasma concentrations that are 32% greater than concentrations observed in the fasting state.

METABOLISM

- Sertraline exhibits linear pharmacokinetics.
- Dose reduction is not necessary in patients with renal failure.
- The less active (five to ten times) demethylated metabolite has a mean half-life of 66 h.

Therapeutic Levels

- Therapeutic levels have not been established.

FLUVOXAMINE

Metabolism

- No active metabolites exist.
- Fluvoxamine exhibits nonlinear pharmacokinetics; drug clearance is decreased by about 50% in elderly patients compared with younger patients.

Therapeutic Levels

- Therapeutic levels have not been established.

CITALOPRAM

Absorption

- Peak levels occur within 2 to 4 h.

Metabolism

- Citalopram exhibits nonlinear pharmacokinetics.
- Clearance and elimination are longer in elderly patients.
- Desmethyl-citalopram is an active metabolite, but is a less potent serotonin reuptake inhibitor.

Therapeutic Levels

- Therapeutic levels have not been established.

THIRD-GENERATION ANTIDEPRESSANTS

AMOXAPINE

Absorption

- Amoxapine is almost completely absorbed following oral administration and achieves peak blood concentrations within 1 to 2 h of ingestion.

Metabolism

- Two active hydroxylated metabolites are formed of which one is an NE agonist (8-OH-amoxapine) and the other a dopamine antagonist (7-OH-amoxapine), with half-lives of 30 and 4 h, respectively.

Therapeutic Levels

- Therapeutic levels have not been established.

BUPROPION

Absorption

- Bupropion is absorbed rapidly after oral administration with serum concentrations peaking 2 h after ingestion.

Metabolism

- Bupropion undergoes extensive first-pass metabolism in the liver; <1% is excreted in the urine unchanged.
- Six urinary metabolites have been identified that are less active and slightly more toxic than the parent compound.
- Bupropion induces its own metabolism as well as that of other drugs.

Therapeutic Levels

- Therapeutic levels have not been established.

MAPROTILINE

Absorption

- Maprotiline is completely absorbed following administration with peak blood concentrations occurring 9 to 16 h after ingestion.

Metabolism

- Maprotiline exhibits linear pharmacokinetics.

Therapeutic Levels

- Therapeutic levels have not been established.

TRAZODONE

Absorption

- Trazodone is well absorbed following oral administration, with concentrations peaking within 2 h.

Metabolism

- Trazodone is highly metabolized in the liver by hydroxylation, pyridine ring splitting, oxidation, and N-oxidation, with <1% of the unchanged drug appearing in the urine and feces.
- It is noteworthy that the primary metabolite of trazodone, m-CPP, is a serotonin agonist that is eliminated from the body at a slower rate than trazodone.

Therapeutic Levels

- Therapeutic levels have not been established.

VENLAFAXINE

Absorption

- Venlafaxine is rapidly absorbed after oral administration.
- Food has no effect on the absorption of this drug.

Metabolism

- Venlafaxine is metabolized in the liver to a demetyhylated active metabolite, ODV, that has a 10-h half-life.
- Linear kinetics are observed between low doses of 25 to 75 mg/day.
- An exponential rather than a linear increase occurs for C_{max} and AUC for the 150-mg dose.
- Venlafaxine and its metabolites are primarily renally cleared.

Therapeutic Levels

- Therapeutic levels have not been established.

NEFAZODONE

Absorption

- Nefazodone is rapidly absorbed after oral administration with an absolute bioavailability of 20%.
- Food delays absorption by 20 min.

Metabolism

- Nefazodone is highly (>99%) protein bound, which may cause clinical drug interactions with other medications that are highly protein bound.
- Nefazodone is extensively metabolized in the liver to three active metabolites: hydroxy-nefazodone (OH-NEF), triazole-dione, and *m*-chlorophenylpiperazine (mCPP).
- Nefazodone exhibits nonlinear kinetics.
- No adjustments in dose are required in patients with renal insufficiency.
- In patients with severe but not moderate hepatic dysfunction, caution is warranted due to the altered kinetics of nefazodone.

MIRTAZAPINE

Absorption

- Mirtazapine exhibits ~50% bioavailability.

Metabolism

- Mirtazapine is metabolized in the liver primarily by demethylation and oxidation with the metabolite demethyl-mirtazapine undergoing conjugation.
- The desmethyl metabolite is three- to fourfold less active than the parent drug, and it is not yet known what the contribution of this metabolite will have on the clinical outcome.
- Caution is indicated in administering mirtazapine in patients with compromised renal or hepatic function.
- Serum concentrations have been reported to be higher in elderly people with an increased incidence of certain side effects such as dry mouth, dizziness, and constipation; specific dosage adjustments have not been suggested.

MOOD STABILIZERS

Table 9.6 presents the pharmacokinetic parameters of bioavailability, free fraction, volume of distribution, and half-life of all the currently available U.S. mood stabilizers.

LITHIUM

Absorption

- Lithium is completely absorbed from the GI tract.
- Absorption takes place over the whole length of the intestine and is not affected by food.

TABLE 9.6
Mood Stabilizer Pharmacokinetic Parameters

Drug	Oral Bioavailability (%)	Plasma Protein Binding (%)	Clearance (ml/min/kg)	Volume of Distribution (l/kg)	$t_{1/2}$ (h)
Lithium	100	0	0.35 ± 0.11	0.66 ± 0.16	22 ± 8
Carbamazepine	>70	74 ± 3	1.3 ± 0.5	1.4 ± 0.4	15 ± 5
Valproic acid	100 ± 10	93 ± 1	0.11 ± 0.02	0.22 ± 0.07	14 ± 3

Source: Adapted from References 6 through 8.

Distribution

- The volume of distribution is slightly less in patients >65 years, leading to smaller daily dose requirements than in younger patients.
- Lithium is not bound to plasma proteins.
- Lithium crosses the placenta and fetal lithium concentrations equal maternal levels.

Elimination

- Lithium exhibits first-order kinetics.
- Clearance varies in individuals from about 10 to 40 ml/min and is 20% lower during the night than during the day.
- Clearance increases with peritonealdialysis and hemodialysis.
- Lithium is not metabolized, but almost completely eliminated from the body by the kidneys.
- Changes in renal function or in sodium and fluid balance will significantly affect serum lithium concentrations.

Half-Life

- In euthymic psychiatric patients it is estimated to range from 15 to 55 h.
- In manic patients with normal renal function, the half-life ranges from 8 to 64 h.

Formulations

- Table 9.5 contrasts the average pharmacokinetic values of various lithium carbonate products derived from a single 900-mg lithium carbonate dose.
- Formulations in the United States include lithium citrate syrup, lithium immediate-release capsules and tablets, and two sustained-release products.
- No clinically significant differences exist between lithium immediate-release capsules and tablets, as they are 95 to 100% absorbed, regardless of the manufacturer.
- Switching between the citrate, capsule, and tablet dosage forms should not result in significantly different 12-h steady-state lithium levels.
- Sustained-release formulations, developed in an attempt to decrease the adverse effects, are associated with peak and rapidly rising serum lithium concentrations.
- If slow-release lithium carbonate preparations cause fewer adverse effects, the difference may be restricted to a minority of patients.

Sampling

- Therapeutic lithium levels are based on the serum sample obtained in the morning 12 h after the last dose.
- Acutely manic patients require and tolerate higher lithium doses due to an increased lithium clearance.
- Higher levels and doses are not tolerated once the manic episode abates.
- Frequent lithium levels, normally drawn twice weekly, are required in the treatment of an acute manic episode because of the possibility of a rapid rise in lithium levels as the episode begins to resolve.

Therapeutic Levels

Acute Mania — 0.9 to 1.4 meq/l
- Patients with concentrations >1.4 meq/l experience no greater improvement in their manic symptoms than patients with lower serum lithium concentrations.
- Patients with serum lithium concentrations <0.9 meq/l generally do not experience complete remissions.
- To achieve this therapeutic range, interindividual variability is quite large such that patients cannot be prospectively dosed simply on a mg/kg/day basis to achieve this therapeutic range.

Prophylaxis
- 0.45 to 0.59 meq/l based on single daily dosing — This level is preferred because fewer adverse effects at this level enhance compliance.
- A level of 0.8 to 1.0 meq/l is indicated for patients >65 years old.
- Clinicians should instruct patients that, because they may be at higher risk of relapse at lower concentrations, they should increase their dose by 1.5 times at the first signs of any manic or depressive symptoms and then slowly taper downward to the lower concentration several months after the symptoms are resolved.

Lithium Prospective Dosing — Pharmacokinetic Method

- A linear relationship exists between concentration in serum or plasma at steady state and a single lithium concentration at some time after a dose of lithium carbonate.
- Equation 9.5 can be utilized to dose lithium prospectively for which C is the steady-state lithium concentration at the 12-h postdosing period for an 1800-mg/day dose, and C_{24h} is the 24-h serum lithium concentration following an initial 1200-mg dose. The dosing nomogram in Figure 9.2 can be utilized in place of equations.

$$C_{ss12h} = 0.13 + 3.3 \, (C \times 24) \qquad\qquad (9.5)$$

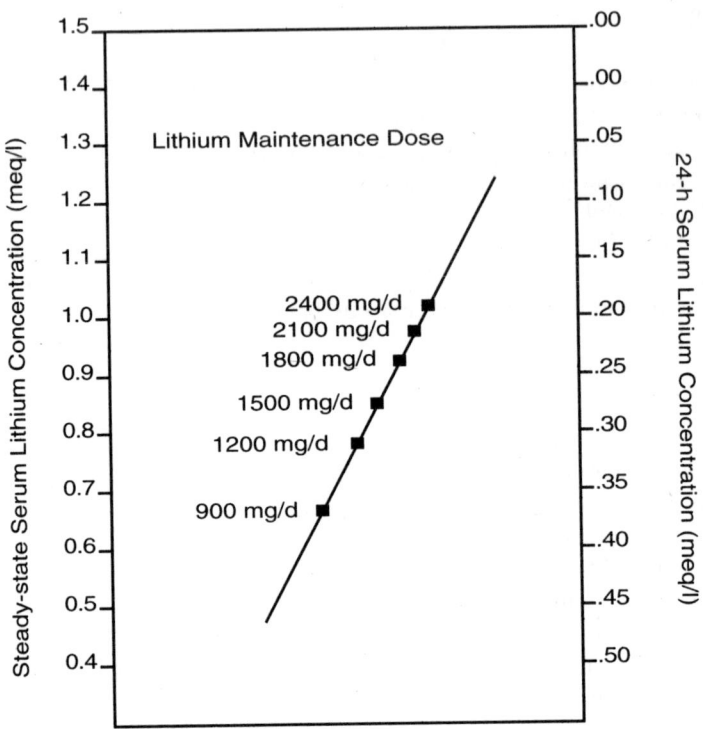

FIGURE 9.2 Lithium dosing nomogram for predicting steady-state serum concentrations for 900 to 2400 mg/day maintenance doses. To use the nomogram, administer the patient a 1200-mg lithium carbonate test dose and then determine the serum lithium concentration 24 h following the test dose. To determine the predicted steady-state lithium levels for different maintenance doses, draw a line from the observed 24-h test dose lithium level (right-side vertical axis) through the maintenance dose point of interest (diagonal line) to the steady-state lithium level (left-side vertical axis). (Adapted from Reference 16.)

Lithium Prospective Dosing — Nonpharmacokinetic Method

- There are two stepwise multiple linear regression analyses that have derived a mathematical formula for determining a lithium dose based on a number of dependent variables.
- The model was constructed incorrectly in that the true dependent variable, the lithium level, was treated as an independent variable, whereas dose should be an independent variable.
- The true amount of variance explained between dose and lithium serum level is low enough to make these models clinically insignificant.

CARBAMAZEPINE

Absorption

- Absorption of carbamazepine (CBZ) is slow and erratic as peak plasma concentrations may vary from 2 to 24 h.
- The bioavailability of tablets is approximately 85 to 90%.

Metabolism

- CBZ is extensively metabolized by the liver to the primary active metabolite, 10,11-epoxide (CBZE).
- CBZE appears in the plasma at concentrations of 10 to 40% of CBZ.

Half-Life

- The single-dose half-life in patients with epilepsy ranges between 30 and 40 h.
- Multiple dose half-lives are reduced to 12 to 20 h due to enzyme induction.
- The half-life for CBZE after autoinduction is estimated to be 6 to 20 h.
- Maximal enzyme induction occurs within 1 week with CBZ daily doses of 100 to 1200 mg/day in patients with bipolar affective disorder.

Formulations

- CBZ is available as a 200-mg immediate-release tablet, 100-mg chewable tablet, and a suspension.
- 100 and 200 mg dosage forms have comparable bioavailability and may be exchanged without a change in the dose or dosing interval.
- The suspension results in a rapid absorption of drug, which may lead to peak-related adverse effects.
- Tablets administered on a twice a day schedule and the suspension given on a three times a day schedule yield similar AUCs as well as similar minimum and maximum serum concentrations.
- Suspensions may be administered as an enema if the oral route is unavailable.
- Patients started on a generic form of CBZ should continue to receive that brand throughout treatment.
- For patients switched from a brand name to a generic product or from generic to generic, the plasma concentrations should be monitored weekly for 2 weeks to determine the effect, if any, of the substitution.

Therapeutic Levels

Acute Treatment — Mania and Depression

- Therapeutic levels have not been established, but it is recommended that a nonresponding patient's CBZ dose be increased until initial adverse effects occur.
- Adverse effects usually occur with levels of 10 to 12 mg/ml.

Maintenance Treatment — Mania and Depression

- Therapeutic levels have not been established.
- Levels of 4 to 7 mg/ml (mean 5.7 mg/ml) are effective in the majority of patients, but the investigation was a naturalistic study.

Valproic Acid

Absorption

- Valproic acid (VA) is almost completely bioavailable.
- Food will delay the absorption of all oral products except the sprinkle formulation.

Distribution

- VA is rapidly distributed within minutes into the central nervous system.
- As valproate concentrations exceed 100 mg/ml, there is a disproportionate increase in the free fraction of drug.
- The distribution characteristics of VA may explain some dose-related adverse effects (e.g., GI, neurological, hepatic).

Metabolism

- VA is primarily metabolized by oxidation in the liver, producing active and inactive glucuronide metabolites.
- VA exhibits linear kinetics, so proportional increases or decreases in dose should result in like changes in the serum level.

Formulations

- VA is available as a syrup, a soft gelatin capsule, an enteric-coated delayed-release tablet, and as coated pellets in a capsule to be "sprinkled" on food.
- Syrup and soft capsules result in peak valproate concentrations within 0.5 to 1 h.
- Both the coated tablet and the sprinkle peak in 4 to 6 h.
- Enteric-coated tablets and sprinkle products are divalproex sodium, which is a 50/50 combination of sodium valproate and valproate.

- Thc above products reduce the GI adverse effects of valproate by reducing mucosal contact, slowing the rate of absorption, and lowering the peak valproic concentration.
- The absorption of an enteric-coated tablet may be so delayed that distribution may still be occurring at 12 h after a dose.
- A true trough may not occur until 14 to 16 h after the last dose.
- The pharmacokinetics of the different dosage forms are similar.
- Dosage adjustments when switching from one product to another typically are not necessary.

Therapeutic Levels

- Therapeutic levels have not been established.
- Uncontrolled reports recommend levels between 50 and 120 µg/ml.

BENZODIAZEPINES

Table 9.7 presents the pharmacokinetic parameters of bioavailability, free fraction, volume of distribution, and half-life of all the currently available U.S. benzodiazepines (BZD) and buspirone.

TABLE 9.7
Benzodiazepine and Buspirone Pharmacokinetic Parameters

Drug	Oral Bioavailability (%)	Plasma Protein Binding (%)	Clearance (ml/min/kg)	Volume of Distribution (l/kg)	$t_{1/2}$ (h)
Alprazolam	88 ± 16	71 ± 3	0.74 ± 0.14	0.72 ± 0.12	12 ± 2
Buspirone	95	—	—	?	1.3 – 6.6
Chlordiazepoxide	100	96.5 ± 1.8	0.54 ± 0.49	0.30 ± 0.03	10.0 ± 3.4
Clonazepam	98 ± 31	86 ± 0.5	1.55 ± 0.28	3.2 ± 1.1	23 ± 5
Clorazepate	N/A	N/A	1.8 ± 0.2	0.33 ± 0.17	2.0 ± 0.9
Diazepam	100 ± 14	98.7 ± 0.2	0.38 ± 0.06	1.1 ± 0.3	43 ± 13
Estazolam	—	93	—	?	10–24
Flurazepam	—	96.6	4.5 ± 2.3	22 ± 7	74 ± 24
Halazepam	—	?	—	1.0	14–16
Lorazepam	93 ± 10	91 ± 2	1.1 ± 0.4	1.3 ± 0.2	14 ± 5
Midazolam	44 ± 17	95 ± 2	6.6 ± 1.8	1.1 ± 0.6	1.9 ± 0.6
Oxazepam	97 ± 11	98.8 ± 1.8	1.05 ± 0.36	0.60 ± 0.20	8.0 ± 2.4
Temazepam	91	97.6	1.0 ± 0.3	0.95 ± 0.34	11 ± 6
Triazolam	44	90.1 ± 1.5	5.6 ± 2.0	1.1 ± 0.4	2.9 ± 1.0

N/A = Not applicable.

Source: Adapted from References 6 through 8.

ABSORPTION

- After oral administration, chlordiazepoxide, diazepam, lorazepam, halazepam, and alprazolam are rapidly and completely absorbed, with peak serum levels occurring at about 1 to 2 h.
- Clorazepate, itself inactive, is rapidly converted to the active metabolite desmethyldiazepam, which reaches peak levels in 1 to 2 h.
- Absorption of oxazepam may require up to 8 h, although the average peak is 2.7 h.
- Prazepam is slowly absorbed, as peak levels of the major metabolite desmethyldiazepam are reached 6 h after administration.
- Sustained-release diazepam peaks about 3 h after a dose.

DISTRIBUTION

- All BZDs are highly protein bound.
- Differences in lipid and water solubility are used to explain the duration of action of available BZDs after a single dose of the drug.
- Changes in distribution of the drugs as a result of aging or liver disease may increase or decrease the therapeutic or ADRs of the drug depending upon the direction of the change (Table 9.8).

METABOLISM AND ELIMINATION

- BZDs exhibit first-order (linear) kinetics.
- BZDs fall into two classes depending upon their biotransformation pathways:
  1. One group includes those that are transformed primarily by oxidative (phase 1) pathways (N-dealkylation, N-demethylation, or aliphatic hydroxylation) to active metabolites and then conjugated to inactive metabolites.
  2. The second group includes those that are metabolized only by conjugation to water-soluble glucuronides (subsequently excreted in the urine), which are pharmacologically inactive (e.g., lorazepam and oxazepam).

THERAPEUTIC LEVELS

- Therapeutic levels have not been established.
- The Cross-National Collaborative Panic Study found that alprazolam plasma concentrations between 20 and 40 ng/ml produced significantly greater decreases in panic attacks than lower concentrations, as well as optimum improvement in symptoms of anxiety and dysphoria.

TABLE 9.8
Cirrhosis and Age Effects on BZD Pharmacokinetics

Drug	Volume of Distribution		Elimination Half-Life		Plasma Clearance	
	Cirrhosis	Age	Cirrhosis	Age	Cirrhosis	Age
Alprazolam	NI	↓	NI	↑	NI	↓
Chlordiazepoxide	↑	↑	↑	↑	↓	↓
Clorazepate (as DD)	NI	↑	NI	↑ M	NI	↓ M
Diazepam	↑	↑	↑	↑	↓	↔
Lorazepam	↔	↓	↑	↔ or ↑	↔	↓
Oxazepam	↔	↔	↑	↔	↔	↔
Prazepam (as DD)	NI	↑	NI	↑ M and F	NI	↓ M and F

DD = desmethyldiazepam, M = male, F = female, ↓ = decrease, ↑ = increase, ↔ = no difference, NI = no information.

REFERENCES

1. Dahl, S. G., Plasma level monitoring of antipsychotic drugs: clinical utility, *Clin. Pharmacokinet.*, 11:36–61, 1986.
2. Sramek, J. J., Potkin, S. T., and Hahn, R., Neuroleptic plasma concentrations and clinical response: in search of a therapeutic window, *Drug Intell. Clin. Pharm.*, 22:373–380, 1988.
3. Garver, D. L., Neuroleptic drug levels and antipsychotic effects: a difficult correlation; potential advantage of free (or derivative) versus total plasma levels, *J. Clin. Psychopharmacol.*, 9:277–281, 1989.
4. Balant-Gorgia, A. E., Blanat, L. P., and Andreoli, A., Pharmacokinetic optimization of the treatment of psychosis, *Clin. Pharmacokinet.*, 25:217–236, 1993.
5. Preskorn, S. H. and Burke, M. J., Fast GA, therapeutic drug monitoring, *Psychiatr. Clin. North Am.*, 16:611–645, 1993.
6. Goodman, L. S., Limbird L. E., and Milinoff, P. B., Eds., *Goodman & Gilman's The Pharmacological Basis of Therapeutics*, 9th ed., McGraw-Hill, New York, 1996.
7. Perry, P. J., Zeilmann, C., and Arndt, S., Tricyclic antidepressant concentrations in plasma: an estimate of their sensitivity and specificity as a predictor of response, *J. Clin. Psychopharmacol.*, 14:230–240, 1994.
8. Baselt, R. C., *Disposition of Toxic Drugs and Chemicals in Man*, 5th ed., Chemical Toxicology Institute, Foster City, CA, 2000.
9. Perry, P. J. and Alexander, B., Switching haloperidol from oral to im depot formulation, DICP, *Ann. Pharmacother.*, 25:1270–1271, 1991.
10. Reyntgens, A. J. M. et al., Pharmacokinetics of haloperidol decanoate, *Int. Pharmacopsychiatr.*, 17:238–246, 1982.
11. Goren, J. L. and Levin, G. M., Quetiapine, an atypical antipsychotic, *Pharmacotherapy*, 18:1183–1194, 1998.

12. DeVane, C. L. and Jarecke, C. R., Cyclic antidepressants, in *Applied Pharmacokinetics, Principles of Therapeutic Drug Monitoring*, Evans, W. E., Schentag, J. J., and Jusko, W. J., Eds., Applied Therapeutics, San Francisco, 1992, 1–34.
13. DeVane, C. L., Pharmacokinetics of the newer antidepressants, *Am. J. Med.*, 97 (Suppl. 6A):6A–13A, 1994.
14. Perry, P. J. et al., Two prospective dosing methods for nortriptyline, *Clin. Pharmacokinet.*, 9:555–563, 1991.
15. Perry, P. J., Alexander, B. and Liskow, B. I., *Psychotropic Drug Handbook*, American Psychiatric Press, Washington, D.C., 1997.
16. Perry, P. J. and Alexander, B., Dosage and serum levels, in *Modern Lithium Therapy*, Johnson, F. N., Ed., MTP Press, Lancaster, U.K., 1987, 70.

10 Theophylline

Bernd Meibohm

CONTENTS

INTRODUCTION

Theophylline is one of the classical drugs, for which a pharmacokinetically guided dosage optimization has been widely used in clinical settings in the last three decades. This adjustment for dosage requirements to the individual needs of the patient allowed for safe and effective therapy despite a narrow therapeutic range in

1-56676-973-6/02/$0.00+$1.50
© 2002 by CRC Press LLC

the presence of high interindividual pharmacokinetic variability. The continued benefit of therapeutic drug monitoring for theophylline substantially contributed to its historical place in the management of inflammatory respiratory diseases as well as its current role in clinical practice alongside with other new, powerful agents.

Although other xanthine derivatives similar to theophylline have been used in therapy in the United States and other countries in the past, they have no or only minor roles in contemporary pharmacotherapy and will not be considered.

PHARMACOLOGICAL ACTIVITY

Theophylline was initially classified as a bronchodilator; nonspecific inhibition of phosphodiesterase isoenzymes was thought to be its major mechanism of action. It can, for example, completely block exercise-induced bronchospasms at concentrations in the upper therapeutic range. Because phosphodiesterase inhibition by theophylline is only about 20% with clinical concentrations,[1] however, involvement of other mechanisms of action is highly likely. The only other known molecular mechanism of theophylline at clinically relevant concentrations is a nonselective antagonism of specific cell-surface receptors for adenosine.

More importantly, the effects of theophylline are not limited to bronchodilation, but also include immunomodulatory, anti-inflammatory, and bronchoprotective activity that substantially contribute to its usefulness as a prophylactic drug in asthma and other respiratory diseases. Additional effects include an increase in mucociliary clearance, a decrease of microvascular leakage into the airways, and an improvement of respiratory muscle fatigue, especially that of the diaphragm. Theophylline furthermore acts centrally, blocking the decrease in ventilation that occurs with sustained hypoxia. While some of these effects are the rationale for its use in asthma, others form the basis for its effectiveness in chronic obstructive pulmonary disease (COPD) or in the treatment of apnea in premature newborns.

Although theophylline has extensive pharmacological actions in the respiratory system, it has also central nervous system, cardiovascular, hepatic, digestive, and metabolic activity.[2] Some of these activities contribute to the profile of adverse effects that might be encountered during its use in clinical pharmacotherapy (Table 10.1).

THERAPEUTIC USE

Theophylline is used in three major areas of clinical pharmacotherapy.

ASTHMA

In chronic asthma, the main use of theophylline, is as adjuvant therapy according to the "Guidelines for the Diagnosis and Management of Asthma" of the National Heart, Lung and Blood Institute.[3] It is recommended to be used in combination with anti-inflammatory therapy with inhaled corticosteroids as well as inhaled or oral β_2-agonist therapy, leukotriene modifiers, and cromolyn sodium or nedocromil, particularly for controlling nocturnal asthma symptoms.[3] It is especially useful in

TABLE 10.1
Pharmacologic Activity of Theophylline with Potential for Adverse Effects

Organ	Adverse Effect
Central nervous system	
Stimulation of respiratory center	Tachypnea
Stimulation of vasomotor center	Tachycardia
Stimulation of spinal reflexes	Convulsion
Cardiovascular system	
Increase in cardiac output	Tachycardia
Increase of pulse	Palpitation
Stimulation of β_1-adrenoceptor	Arrhythmia
Vasodilation	Hypotension
Renal system	
Increase of renal blood flow	Polyuria
Increase of glomerular filtration rate	Polyuria
Increase of the excretion of sodium and chloride	Polyuria
Digestive system	
Increase of gastric secretion	Nausea, vomiting, peptic ulcer
Metabolic system	
Glycogenolysis	Hyperglycemia
Gluconeogenesis	Hyperglycemia

Modified from Kawai and Kato.[2]

those patients with asthma who are not adequately controlled with conventional doses of inhaled steroids. In addition, it might be used as primary therapy in cases where the administration of an inhaled corticosteroid is cumbersome or difficult, e.g., in young children, or in cases where the patient is more likely to adhere to oral rather than inhaled pharmacotherapy.[4] Furthermore, low-dose theophylline has been shown to reduce requirements for inhaled corticosteroid therapy in patients with asthma, and this steroid-sparing effect may reduce overall treatment costs.[5]

In the management of acute asthma, theophylline appears superfluous for routine treatment of acute exacerbations in patients who are receiving optimal therapy with β_2-agonists and corticosteroids.[6] It may be indicated, however, for patients with severe acute symptoms who do not respond rapidly to these other measures.[4]

CHRONIC OBSTRUCTIVE PULMONARY DISEASE

In COPD, theophylline is recommended by the "Standards for the Diagnosis and Care of Patients with Chronic Obstructive Pulmonary Disease" of the American Thoracic Society as secondary therapy in addition to inhaled or oral β_2-agonists and anticholinergic agents.[7] Similar to the therapy of asthma, theophylline is of particular value in those patients with COPD who are less compliant or who are less capable of using inhaled drug therapy because they can readily take long-acting oral theophylline once or twice daily.

APNEA

Theophylline has also been used for the treatment of preterm neonatal apnea, but recent evaluations of the available data suggest that caffeine is therapeutically advantageous compared with theophylline in this indication.[8]

PHARMACODYNAMICS

Efficacy and toxicity of theophylline are closely related to its serum drug concentrations. The degree of bronchodilation[9] and the decrease in airway responsiveness to exercise[10] change in parallel with serum theophylline concentration. A similar dependency on serum drug levels has been described for the frequency and severity of asthma exacerbations and lung function parameters, as depicted, for example, in Figure 10.1.[11] Pharmacokinetic/pharmacodynamic modeling of the concentration–effect relationship for the increase in peak expiratory flow rate (PEFR) by theophylline in adult patients has provided mean estimates for E_{max}, the maximum drug-induced PEFR increase, and EC_{50}, the concentration producing 50% of E_{max}, of 344 l/min and 11 µg/ml, respectively.[12,13]

Some clinical benefits of theophylline, including anti-inflammatory, immunemodulatory, and bronchodilatory effects, already occur at serum concentrations above 5 µg/ml. Thus, these concentrations may be adequate for some patients. In patients receiving theophylline as monotherapy for chronic asthma, however, doses resulting in serum concentrations beyond 10 µg/ml have clearly been shown as most likely to prevent asthma symptoms and decrease the occurrence and severity of exacerbations.

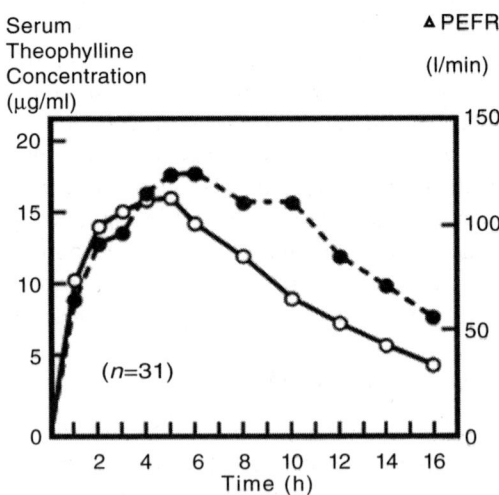

FIGURE 10.1 Relationship between serum theophylline concentration and improvement in pulmonary function, measured as increase in PEFR, among 31 adult subjects with asthma given 7.5 mg/kg of a rapidly absorbed theophylline formulation. (From Richer, C. et al., *Clin. Pharmacol. Ther.*, 31(5):579–586, 1982. With permission.)

Similar to the clinical efficacy of theophylline, side effects are also closely related to serum concentrations. No critical side effects have been observed at serum theophylline levels up to 20 µg/ml. However, transient comparatively slight side effects similar to those of caffeine, including nausea, vomiting, headache, diarrhea, and insomnia, occur in some patients at theophylline serum concentrations below 20 µg/ml. These side effects may be present in up to 50% of patients when serum concentrations of 10 to 20 µg/ml are rapidly achieved, but they are usually much less frequent when the initial dose is low and the final drug concentration is gradually achieved by increasing doses in intervals of at least 3 days.

Nausea, vomiting, headache, insomnia, and tachycardia become more frequent when serum concentrations exceed 20 µg/ml. Above 40 µg/ml, a frequent occurrence of cardiac arrhythmias and seizures was reported.[2] Other adverse events at high serum concentrations, especially after acute overdose, include hyperglycemia, hypokaliemia, hypotension, seizures, toxic encephalopathy, hyperthermia, brain damage, and death (Table 10.2). The risk of serious theophylline-induced toxicity requiring hospitalization has been described as rare with <1 per 1000 patient-years, but was reported five times greater among elderly patients and those taking cimetidine.[14]

Although a single overdose in a patient not previously taking theophylline is unlikely to cause a seizure or encephalopathy unless the peak serum concentration exceeds 100 µg/ml, these might occur at substantially lower concentrations in patients regularly taking theophylline or after repeated excessive doses. These adverse effects, however, have rarely if ever been reported at concentrations below 30 µg/ml.[4] Thus, as illustrated by Figure 10.2,[15] side effects are observed at concentrations higher than 20 µg/ml, but generally only become serious at substantially higher concentrations. Although typically present when looked for, minor symptoms of toxicity such as nausea and vomiting do not necessarily precede severe toxicity and cannot be relied upon as dosing end point. Only theophylline serum concentration measurements can reliably forewarn the clinician of impeding life-threatening toxicity.

TABLE 10.2
Relationship between Serum Concentration and Effects of Theophylline

Serum Theophylline Concentration (µg/ml)	Effects
0–5	Not effective
5–10	Anti-inflammatory effect
8–15	Effective for many patients
15–20	Effective for most patients
	Side effects appear in some patients
20–25	Side effects (abdominal symptoms)
25–40	Side effects for many patients (tachycardia, tachypnea, convulsion)
>40	Toxic dose (seizure, arrhythmias, etc.)

Modified from Kawai and Kato.[2]

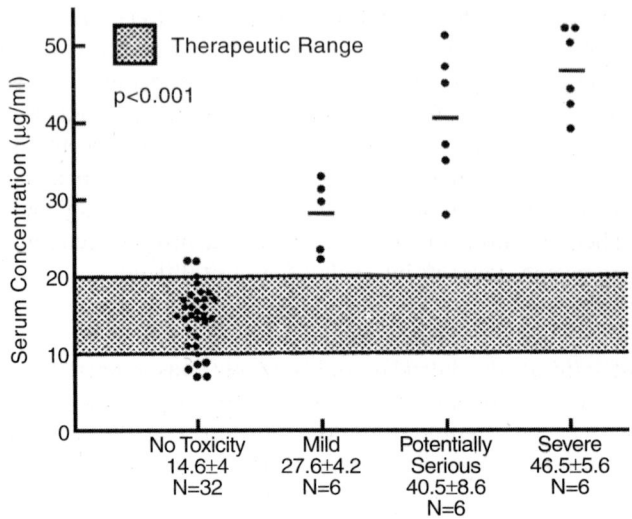

FIGURE 10.2 Toxicity related to theophylline serum concentration in 48 patients receiving intravenous aminophylline. (From Hendeles, L. et al., *Drug Intell. Clin. Pharm.*, 11:12–18, 1977. With permission.)

THERAPEUTIC RANGE

Based on the outlined relationships between serum concentration and therapeutic effects and toxicity, theophylline is considered a drug with narrow therapeutic index. Although generally well tolerated when appropriate serum concentrations are maintained, theophylline has the greatest potential of all antiasthmatics for serious toxicity at high serum concentrations. Individual patients might already experience efficacy at serum concentrations above 5 µg/ml, but below 10 µg/ml. The highest likelihood of optimal efficacy, however, is achieved at concentrations beyond 10 µg/ml. Because an increased frequency of toxicity was observed beyond 20 µg/ml, serum concentrations of 10 to 20 µg/ml are generally considered the therapeutic range for theophylline in asthma therapy.[4,16] This range, however, has recently come under discussion[16,17] as the "Guidelines for the Diagnosis and Management of Asthma" of the National Heart, Lung and Blood Institute recommend a targeted theophylline steady-state serum concentration of 5 to 15 µg/ml.[3] This approach accounts for the fact that some therapeutic activity of theophylline already occurs at concentrations lower than 10 µg/ml and maintaining these concentrations might be sufficient in some patients. In addition, as each dose is not absorbed in an identical manner, peak serum concentrations near the upper end of the therapeutic range may be associated with an increased risk of adverse effects if absorption of some doses results in drug concentrations in excess of 20 µg/ml. Furthermore, a target serum theophylline concentration in routine clinical use not exceeding 15 µg/ml would provide an additional buffer zone to minimize the risk of adverse effects from processes that might slow theophylline elimination during therapy,[18] For the management of COPD, the American Thoracic Society recommends targeting a similar theophylline serum concentration range of 8 to 12 µg/ml.[7]

For the treatment of preterm neonatal apnea, theophylline serum concentrations of 5 to 10 µg/ml are being considered as effective in decreasing the number of apnea episodes. The pharmacodynamics of theophylline in this situation, however, have not well been studied, and theophylline in neonates is converted to caffeine to varying degrees, which also is effective in suppressing apnea and thus confounds the concentration–effect relationship.

REQUIREMENT FOR DOSAGE INDIVIDUALIZATION

Theophylline is not only characterized by a narrow therapeutic index and distinct relationships between serum concentration and therapeutic and toxic effects, but also by a high interindividual pharmacokinetic variability. This variability is predominantly based on a high patient-to-patient variability in the metabolic clearance of theophylline that is confounded by numerous additional physiological, pathophysiological, and environmental factors. Intravenous theophylline (given as aminophylline), for example, has been shown to result in considerable variations in serum concentrations among patients despite the same dose,[19] and theophylline dose requirements to maintain serum concentration in the range of 10 to 20 µg/ml varied from 400 to 3200 mg/day.[20]

Reports like these form the basis for the currently practiced approach to use therapeutic drug monitoring based on serum drug concentration measurements for optimizing theophylline dosing according to the needs of the individual patient. This approach allows for individualized drug therapy by accounting for the interindividual pharmacokinetic variability and thus assures that drug serum concentrations can reliably be maintained within the therapeutic range thereby providing a safe as well as effective therapy.

PRODUCT FORMULATIONS

Theophylline is most frequently used in intravenous or oral dosage forms. Several so-called salts of theophylline have been used as an attempt to improve water solubility without changing the pharmacological activity of theophylline. One of them, aminophylline, a mixture of theophylline and the base ethylenediamine, is still frequently used, especially as an intravenous dosage form. To assure comparability between different dosage forms with regard to drug content, all dosage and labeling should be in terms of anhydrous theophylline content. Theophylline monohydrate, for example, contains 90.9% anhydrous theophylline; aminophylline dihydrate contains 78.9% anhydrous theophylline.

Oral dosage forms include solutions as well as immediate-release and slow-release tablets and capsules. Since theophylline is cleared from the body relatively rapidly, particularly in children, slow-release formulations have been developed that minimize fluctuations in steady-state serum concentrations during an 8- to 12-h dosing interval. These slow-release products are designed to release theophylline at an approximately constant zero-order rate over the dosing interval. This concept, however, is rarely achieved clinically, and some formulations are hampered by erratic drug delivery in certain individuals.

Slow-release formulations differ among themselves in the rate of release, and most products require 8-h dosing intervals to prevent serum concentration fluctuations greater than 100%. For formulations with very slow release, 12-h dosing intervals are often possible.[21] Products for once-daily dosing have to provide the entire 24-h dose at one time. This entails the risk of toxicity that might occur after erratic drug release from such products. Some 24-h dosage forms have shown erratic absorption after fasting, and some have also exhibited dose dumping after food intake leading to increases in the rate and extent of absorption.[22,23]

PHARMACOKINETICS

ABSORPTION

Absorption after oral administration is rapid, consistent, and nearly complete when given in solution or as immediate-release dosage form. Theophylline can be detected in blood within 10 to 15 min after ingestion of a liquid formulation, and food does not have any major influence when administered with these dosage forms.[24] Thus, oral bioavailability can be assumed as 100% in these cases.[25] For oral immediate-release dosage forms, peak serum concentrations are reached 1 to 2 h after administration.[26]

For most sustained-release formulations, the extent of absorption is rather complete, but some show incomplete absorption. Furthermore, the rate of absorption of many sustained release formulations is affected by the administration of food and other factors delaying gastric emptying, but the extent of absorption remains relatively unchanged except for some 24-h dosage forms as described in the previous section. For the sake of switchability between dosage forms and consistency in drug exposure, only those dosage forms with complete absorption should be used clinically.[4]

Intrasubject variability in the extent of absorption has also been described. The net result is large fluctuations in serum concentrations in some, but not all dosing intervals, which may cause major problems in terms of interpretation of measured serum concentrations.[27,28] Similarly, diurnal variations in theophylline absorption with slower rates at night have to be taken into account.[29]

DISTRIBUTION

After absorption, theophylline quickly reaches equilibrium with peripheral tissues, except for fat tissues. Serum concentrations follow an open mamillary two-compartment pharmacokinetic model after intravenous administration. Because the early distribution phase is complete within 35 to 45 min, however, theophylline serum concentration profiles are adequately characterized by a one-compartment model when given orally.

Theophylline is predominantly distributed throughout the extracellular water. It freely crosses the placenta and passes into breast milk, although only minor adverse effects have been reported in infants receiving the drug in this manner. Theophylline also distributes into saliva, cerebrospinal fluid, and presumably other body fluids.

In plasma, a moderate fraction of approximately 40% of theophylline is reversibly bound to plasma proteins.[30] In patients with hepatic cirrhosis and reduced serum albumin concentration, binding is reduced to 30 to 35%.[31] In full-term neonates, approximately one third of theophylline is protein bound.[32] Microdialysis investigations revealed theophylline tissue exposures of 55% of serum concentrations in muscle and subcutaneous tissue indicating equilibrium between unbound theophylline concentrations in serum and tissues.[33]

The volume of distribution V_d averages 0.5 l/kg in healthy children and adults,[34] and this value is not changed by age, gender, smoking habit, or disease. It is slightly increased in premature newborns (0.7 l/kg)[32] and adults with hepatic cirrhosis,[31] and decreased in obese patients.[35]

ELIMINATION

Theophylline elimination is predominantly mediated by hepatic metabolism. The liver metabolizes about 85 to 90% of the absorbed dose of theophylline; the remainder is excreted unchanged by the kidneys.[36,37] Therefore, dosage adjustments for renal failure are almost unnecessary.[38] Moreover, the hepatic extraction ratio of theophylline is only 10%, indicating a negligible first-pass inactivation after oral administration.[34]

Hepatic metabolism is predominantly mediated via the cytochrome P-450 (CYP) isoenzymes CYP2E1 and CYP3A4 (8-hydroxylation to 1,3-dimethyluric acid, 40 to 50%), and CYP1A2 (N-demethylation to 3-methylxanthine, 10 to 15%, and 1-methylxanthine). 1–Methylxanthine is further oxidized, by xanthine oxidase, to 1-methyluric acid (15 to 20%).[39–42] Caffeine and 3-methylxanthine are the only theophylline metabolites with pharmacological activity, but only 3-methlyxanthine may accumulate in end-stage renal disease to have relevant pharmacological effects.

The pharmacokinetics of theophylline, and in particular its metabolic clearance, are modulated by numerous aspects including physiological, pathophysiological, and environmental factors as well as interactions with concurrently administered drugs. Despite the resulting large interindividual variability in theophylline elimination among patients, however, variability in individual patients is usually relatively small in the absence of confounding factors.

The average clearance Cl and resulting half-life values $t_{1/2}$ for theophylline in various subpopulations and disease states are summarized in Table 10.3. Although the metabolism of theophylline appears to proceed as a first-order process, single metabolism pathways may already reach saturation within therapeutic concentrations, resulting in nonlinear elimination. Thus, increases in theophylline dose may result in disproportionally larger increases in serum concentrations.[36,43,44]

TABLE 10.3
Mean and Range of Total Body Clearance and Half-Life of Theophylline Related to Selected Physiological and Pathophysiological States[a]

Population Characteristics	Total Body Clearance[b] Mean (range)[c] (ml/kg/min)	Half-Life Mean (range)[c] (h)
Age		
Premature neonates		
Postnatal age 3–15 days	0.29 (0.09–0.49)	30 (17–43)
Postnatal age 25–57 days	0.64 (0.04–1.2)	20 (9.4–30.6)
Term infants		
Postnatal age 1–2 days	NR[d]	25.7 (25–26.5)
Postnatal age 3–30 weeks	NR	11 (6–29)
Children		
1–4 years	1.7 (0.5–2.9)	3.4 (1.2–5.6)
4–12 years	1.6 (0.8–2.4)	NR
13–15 years	0.9 (0.48–1.3)	NR
6–17 years	1.4 (0.2–2.6)	3.7 (1.5–5.9)

Population Characteristics	Total Body Clearance[b] Mean (range)[c] (ml/kg/min)	Half-Life Mean (range)[c] (h)
Physiological or Pathophysiological State		
Acute pulmonary edema	0.33[e] (0.07–2.45)	19[e] (3.1–82)
COPD >60 years, stable nonsmoker >1 year	0.54 (0.44–0.64)	11 (9.4–12.6)
COPD with cor pulmonale	0.48 (0.08–0.88)	NR
Cystic fibrosis (14–28 years)	1.25 (0.31–2.2)	6.0 (1.8–10.2)
Fever associated with acute viral respiratory illness (children 9–15 years)	NR	7.0 (1.0–13)
Liver disease		
Cirrhosis	0.31[e] (0.1–0.7)	32[e] (10–56)
Acute hepatitis	0.35 (0.25–0.45)	19.2 (16.6–21.8)
Cholestatis	0.65 (0.25–1.45)	14.4 (5.7–31.8)

Adults (17–60 years) otherwise healthy non-smoking asthmatics	0.65 (0.27–1.03)	8.7 (6.1–12.8)
Elderly (>60 years) — nonsmokers with normal cardiac, liver, and renal function	0.41 (0.21–0.61)	9.8 (1.6–18)
Pregnancy		
1st trimester	NR	8.5 (3.1–13.9)
2nd trimester	NR	8.8 (3.8–13.8)
3rd trimester	NR	13.0 (8.4–17.6)
Sepsis with multiorgan failure	0.47 (0.19–1.9)	18.8 (6.3–24.1)
Thyroid disease		
Hypothyroid	0.38 (0.13–0.57)	11.6 (8.2–25)
Hyperthyroid	0.8 (0.68–0.97)	4.5 (3.7–5.6)

[a] For various North American patient populations from literature reports. Different rates of elimination and consequent dosage requirements have been observed among other peoples.

[b] Clearance values listed are generally determined at serum concentrations <20 µg/ml; clearance may decrease and half-life may increase at higher concentrations due to nonlinear pharmacokinetics.

[c] Reported range or estimated range (mean ± 2 SD) where actual range not reported.

[d] NR = not reported or not reported in comparable format.

[e] Median.

Modified from Food and Drug Administration.[82]

FACTORS AFFECTING THEOPHYLLINE PHARMACOKINETICS

Physiological Perturbations

Age. Theophylline undergoes age-dependent metabolism. In premature infants and newborns, the N-demethylation pathway is not developed because of immaturity of the metabolizing enzyme systems, and 50% of the theophylline dose is excreted unchanged instead of 10% after 1 year of age.[45,46] Since renal clearance of theophylline is dependent on urinary flow rate, dosage adjustments are necessary if urine output in neonates is <2 ml/kg/h. At all ages, a small amount of theophylline is N-methylated to caffeine. This minor conversion becomes clinically relevant only in premature infants, in whom caffeine has an extremely long half-life (96 h), which results in caffeine accumulation and pharmacological action.[2] During the first year of life, drug-metabolizing enzyme systems gradually mature and clearance rates normalized for body weight approach maximum values by 8 to 12 months of age. This is followed by a gradual decline to the adult values during childhood and adolescence. In parallel, elimination half-life gradually deceases in the first year of life from 24 h in neonates to 4 h (range 2 to 10 h) in 1- to 9-year-old children, and then progressively increases to 8 h (range 3 to 16 h) in adults.[47] As a result, children beyond infancy require weight-adjusted doses of theophylline almost twice those of adults, and the range of doses needed by both children and adults varies at least fourfold. In elderly patients (>60 years), theophylline clearance may be decreased by an average of 30% compared to healthy young persons.[48]

Pathophysiological Perturbations

Obesity. The effect of obesity on theophylline clearance appears to be negligible.[35] The loading dose administered to obese patients, however, should be based on the volume of distribution for an ideal weight, since theophylline is only to a very limited extent distributed into fat tissue.[49–51]

    Hepatic Disease. Hepatic dysfunction may result in major changes in the metabolic clearance of theophylline, especially in cases of decompensated cirrhosis, acute hepatitis, and cholestasis.[52] Theophylline clearance may be remarkably reduced through impaired hepatic blood flow and injured hepatocytes to 50% or less of normal. In addition, plasma protein binding is decreased, resulting in an increase in volume of distribution and thus a prolongation of the elimination half-life up to 26 h.[31,53]

    Cardiac Disease. Theophylline clearance is reduced in congestive heart failure (up to 50% of normal), especially in the decompensated state and if liver congestion and associated damage develop secondary to congestive heart failure.[54]

    Pulmonary Disease. Theophylline clearance is decreased prominently (up to 50%) in patients with acute pulmonary edema induced by left ventricular failure, but these changes differ among patients.[55] Similarly, theophylline clearance is decreased (80% of normal) and the elimination half-life prolonged in patients with severe COPD with cor pulmonale.[56]

    Thyroid Gland Disorders. The elimination half-life of theophylline is markedly prolonged in hypothyroidism.[57] Conversely, clearance is increased in hyperthyroidism,[58] but reduces toward normal values after initiation of treatment for this disorder.[59]

Infectious Diseases. Prolonged fever may also slow theophylline elimination (up to 50% reduced Cl).[60] Although the level and duration of fever necessary for this effect is not known, fever must be sustained to result in an appreciable theophylline accumulation. Although this is unlikely for fever lasting less than 24 h, dosage should be reduced in patients previously maintained within the therapeutic range, if fever is high and sustained, e.g., >102°F for more than 24 h.

Cystic Fibrosis. Theophylline is more rapidly eliminated in patients with cystic fibrosis than in healthy individuals. The observed theophylline clearance reaches values twice that in healthy individuals.[61,62] Although the mechanism for this increase is unknown, clinicians should be aware of the increased dose requirements in this patient population and serum concentration monitoring should be used as guiding tool. In addition, the volume of distribution is also increased to approximately 0.6 l/kg.

Environmental Perturbations

Smoking. Cigarette or marijuana smoking, even if passive, significantly increases the clearance of theophylline through induction of drug-metabolizing enzymes.[63,64] Theophylline clearance is increased 50 to 80% compared to nonsmokers, with a similar increase in old and young smokers and half-lives of 4 to 6 h.[65] The magnitude of increase correlates with the degree of smoking.[66] It is not clear how long this enzyme induction persists after cessation of smoking. Although no change in clearance was observed after 3 months since smoking cessation,[66] a decrease compared to active smokers was noted in patients 2 years after smoking cessation, but clearance levels were still increased compared to nonsmokers.[67]

Diet. Metabolic theophylline clearance is also increased by ingestion of a high-protein, low-carbohydrate diet, presumably by increased liver enzyme activity.[68] Theophylline half-life was consequently reduced from 8.1 to 5.2 h, but changed again to 7.2 h when a high carbohydrate diet was initiated. Charbroiled beef was also described as shortening theophylline half-life by approximately 22%, presumably via a similar mechanism as smoking through enzyme induction by polycyclic hydrocarbons formed during the roasting process over charcoal fire.[69] Although this effect is unlikely to be clinically significant, it visualizes the wide variety of potential factors affecting theophylline clearance. Dietary intake of other methylxanthines, especially caffeine, prevents theophylline metabolism by competitively interacting with drug-metabolizing enzymes. Elimination of caffeine from the diet of patients consuming large quantities of coffee might thus result in reduced theophylline serum concentrations.[70]

Drug–Drug Interactions

Numerous concomitantly administered drugs alter the pharmacokinetics of theophylline, in particular its metabolic clearance. Table 10.4 summarizes the most relevant of these drug–drug interactions that may result in altered theophylline serum concentrations; Figure 10.3 shows one representative example. It should be noted that, although Table 10.4 is not intended to be conclusive and drugs might interact with theophylline without being listed, most commonly used drugs do not interact adversely with theophylline.

TABLE 10.4
Pharmacokinetic Drug–Drug Interactions Resulting in Changes in Theophylline Serum Concentration (percentage values represent average changes)

Drug	Effect	Clinical Relevance
Alcohol	A single large dose of alcohol (3 ml/kg of whiskey) decreases theophylline Cl by 30% for up to 24 h.[84]	Decrease theophylline dose accordingly.
Allopurinol	Modest decrease in theophylline Cl by 25%, probably due to unspecific CYP inhibition.[85]	Effect on theophylline Cl may be dose-dependent and was modest at 600 mg/day; monitoring of serum concentrations and appropriate dose reduction if indicated.
Aminoglutethimide	Increase in theophylline Cl by 18–46% through CYP induction.[86]	Increase theophylline dose by 25% and measure theophylline serum concentration after 1 week of concurrent therapy.
Antacids	Increase or decrease in rate but not extent of absorption, probably by altering gastric emptying time.[87]	Effects seem insignificant, except for formulations with pH-dependent dissolution, where more rapid absorption was reported.
Antiepileptics	Induction of CYP enzyme activity results in increases in theophylline Cl: 30% for phenobarbital[88,89] 40–75% for phenytoin[90,91] 59% for carbamazepine[92]	Decrease in theophylline serum concentration requires according adjustment of individual theophylline dosing.
β-Receptor agonists	Oral or inhaled orciprenaline has no effect on theophylline Cl,[93] i.v. isoproterenol increases theophylline Cl by 21%.[94] Oral terbutaline and inhaled albuterol may slightly increase theophylline Cl.[95,96]	Potential decrease in theophylline serum concentrations should be monitored by concentration measurements and dosing adjusted accordingly.
β-Receptor blocker	Propranolol reduces theophylline Cl by 30–50%.[97,98] Other β-blockers like metoprolol and atenolol have only little if any effect.[99,100]	Since asthmatic patients are generally not treated with β-blockers, this interaction is only of limited clinical relevance.

Calcium channel blockers	Theophylline Cl was reduced by 18 and 12% after concomitant therapy with verapamil and diltiazem, respectively; Nifedipine had no relevant effect on theophylline Cl.[101]	The modest reduction in theophylline Cl is not likely to produce clinically significant theophylline serum concentration increases in most patients.
Disulfiram	Disulfiram dose dependently decreases theophylline metabolism by inhibiting hydroxylation and demethylation.[102]	Increase in theophylline serum concentration may require dose reduction by up to 50% dependent on the disulfiram dose.
Fluvoxamine	Fluvoxamine is a potent CYP1A2 inhibitor, thereby reducing theophylline Cl.[103–105]	Serum concentration monitoring and appropriate dose reduction should be performed.
H$_2$-blockers	Cimetidine reduces theophylline Cl by CYP1A2 inhibition dose-dependently up to 1200 mg/day.[106–109] Theophylline metabolism is hardly influenced by ranitidine at therapeutic doses, including high doses up to 4200 mg/day.[109,110] Similarly, famotidine and nizatidine do not alter theophylline Cl.[111,112]	Theophylline serum concentrations may increase 30–40% within 1 day after initiation of cimetidine therapy, with a maximum effect after 48–72 h. Theophylline serum concentrations recover within 2 days of cimetidine cessation.[113] If an H$_2$-blocker is required, a noninteracting form should be used. If, however, cimetidine is used, the dose of theophylline should be reduced by 40% and patients should be carefully monitored within the first 72 h of therapy. According dose reductions might be performed but the clinical significance remains questionable.
α-Interferon	Concomitant administration of recombinant human α-interferon decreased theophylline clearance by 15%.[114]	
Macrolides	Most macrolides are potent inhibitors of CYP3A4. Erythromycin and its salts as well as troleandomycin reduce theophylline Cl concentration-dependently: 20–40% reduction after 5–7 days of therapy[115–117] with erythromycin, 50% after 10 days with troleandomycin.[118] Results for clarithromycin are similar to those for erythromycin.[119] Azithromycin does not seem to influence theophylline metabolism, although reported results are contradictory.[120,121]	Increase in theophylline serum concentrations is generally not more than 25% after erythromycin. Dose reduction by 25% might be appropriate. Use of other antibiotics should be considered, especially if troleandomycin is used.

(continued)

TABLE 10.4 (CONTINUED)
Pharmacokinetic Drug–Drug Interactions Resulting in Changes in Theophylline Serum Concentration (percentage values represent average changes)

Drug	Effect	Clinical Relevance
Methotrexate	Low-dose methotrexate therapy decreases theophylline Cl by 19%, probably due to CYP isoenzyme inhibition.[122] Higher methotrexate doses might have a greater effect.	Serum concentration monitoring and appropriate dose reduction should be performed. The effect might not be clinically significant for the majority of patients.
Mexiletine	Decrease in theophylline Cl by 40–60% through inhibition of demethylation of theophylline (Figure 10.3).[123,124]	Decrease theophylline dose by half and monitor serum concentrations.
Moricizine	Concomitant moricizine administration increases theophylline Cl by 32–36%, probably via enzyme induction.[125]	Increase theophylline dose by 25–30% and monitor serum concentrations.
Oral contraceptives	Oral contraceptives during chronic administration decrease theophylline Cl by approximately 30% via inhibition of CYP isoenzymes.[126–128] The effect may be less with low-estrogen preparations.	Theophylline dose should be reduced by 30% and serum concentrations measured after 5 days of concurrent therapy. Low-estrogen-containing preparations should be preferred if clinically justifiable.
Pentoxifylline	Pentoxifylline decreases theophylline elimination by 30%.[129]	Serum concentration monitoring and appropriate dose reductions should be performed.
Propafenone	Propafenone decreases theophylline Cl by approximately 40–60% via inhibition of CYP3A4 and CYP1A2.[130,131]	Decrease theophylline dose and monitor serum concentrations. β_2-Blocking effect may also decrease pharmacological efficacy of theophylline.

Quinolones	Enoxacin decreases theophylline Cl by 40–60%, pefloxacin and ciprofloxacin by 30%.[106,132] No effect was described for ofloxacin and norfloxacin.[132,133]	Appropriate dose reductions and serum concentration monitoring should be performed if theophylline is concomitantly used with interacting quinolones.
Rifampin	Theophylline Cl increased by 25–80% after concurrent rifampin administration by induction of CYP1A2 and CYP3A4.[134–136]	Potential decrease in theophylline serum concentrations should be avoided by concentration measurement 5 days after addition of rifampin and subsequent dose increases accordingly.
Sulfinpyrazone	Increase in theophylline Cl by 22% via enzyme induction.[137]	Increase dose by 20% and perform serum concentration monitoring if potential decrease in efficacy is noticed.
Tacrine	Tacrine is a specific substrate for CYP1A2. Coadministration of tacrine and theophylline inhibits theophylline Cl up to 50%.[138]	Decrease dose accordingly and monitor serum concentrations.
Thiabendazole	Thiabendazole reduces theophylline elimination by more than 50% due to CYP isoenzymes inhibition.[139]	The similar antiparasitic agent mebendazole does not affect theophylline pharmacokinetics and should be therapeutically preferred in patients receiving theophylline.
Ticlopidine	Ticlopidine decreases theophylline Cl by 37%.[82,140]	Serum concentration monitoring and appropriate dose reductions should be performed.

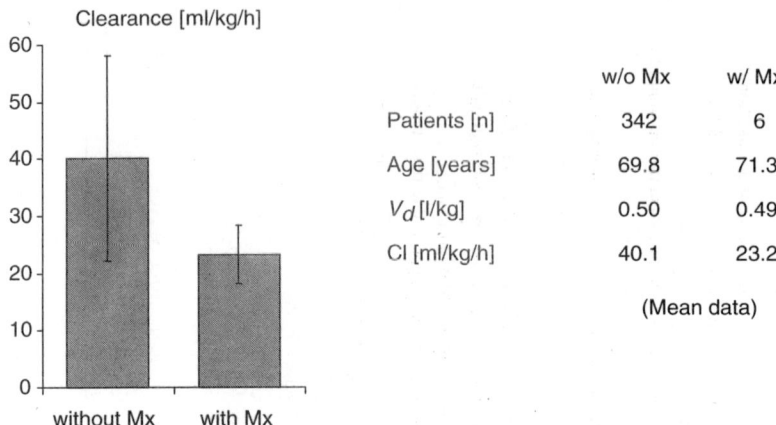

	w/o Mx	w/ Mx
Patients [n]	342	6
Age [years]	69.8	71.3
V_d [l/kg]	0.50	0.49
Cl [ml/kg/h]	40.1	23.2

(Mean data)

FIGURE 10.3 Effect of concomitantly administered mexiletine (Mx; 200 mg Q8h) on the clearance of theophylline (mean ± SD, $p < 0.05$) as determined by a retrospective analysis of theophylline pharmacokinetic data in 348 patients obtained from routinely performed therapeutic drug monitoring. (Modified from Meibohm and Wegener.[124])

Theophylline may also affect the pharmacokinetics of other concomitantly administered drugs. Lithium serum concentrations may be decreased due to an increased renal clearance of lithium induced by theophylline.[71,72] This effect, however, may only be transient and may not be sustained during chronic theophylline therapy. Theophylline furthermore inhibits phenytoin absorption when administered together, resulting in decreased phenytoin serum levels.[73]

In addition to pharmacokinetic drug–drug interactions, pharmacodynamic effects have been reported as well. Halothane increases the susceptibility to ventricular arrhythmias under theophylline therapy as a result of increased sensitivity of the myocardium to endogenous catecholamine release by theophylline.[74] Ketamine lowers the theophylline seizure threshold.[75] Benzodiazepines like midazolam, diazepam, lorazepam, and flurazepam increase the central nervous system concentration of adenosine, a potent central nervous system depressant. As theophylline also blocks adenosine receptors, it counteracts benzodiazepine-induced sedation, resulting in increased dosage requirements for these compounds.[76]

THERAPEUTIC DRUG MONITORING

Treatment with theophylline requires consideration of its interpatient variability in dose requirements, the relationship between efficacy and toxicity vs. serum concentration, and its narrow therapeutic index. Thus, therapeutic drug monitoring is generally applied for dosing individualization, using theophylline serum concentration measurements and simultaneous clinical assessments as guiding tools for dosage adjustments.

Since unspecific routine performance of serum level measurements in all theophylline-treated patients is cost-prohibitive, drug level monitoring should only be performed where clearly indicated. Theophylline serum concentration measurements are recommended during the initial phase of therapy to guide final dosage adjustment after titration. This will also provide a reference for future dose adjustments in the particular patient. In addition, theophylline serum concentrations should be determined any time a patient on chronic theophylline seeks medical attention with an exacerbation of the disease. This will provide a baseline level for any subsequent therapeutic interventions during this exacerbation phase and evaluate whether theophylline concentrations are subtherapeutic. Furthermore, theophylline drug level monitoring is indicated in cases of suspected theophylline-induced toxicity and in cases of suspected noncompliance. Although dosage requirements, once established, generally remain stable for extended periods, additional individual situations might make serum concentration measurements beneficial as well, for example, intercurrent illness or drug–drug interactions.

SAMPLING TIME

Because the $t_{1/2}$ for theophylline is approximately 8 h in nonsmoking adults, meaningful steady-state serum concentrations cannot be obtained until approximately 24 to 48 h after initiation or change of therapy; i.e., approximately 3.3 to 5 times the assumed $t_{1/2}$ in the treated patient is needed for achieving 90% and >95% of steady-state concentrations, respectively.[77] Corresponding adjustments to this time period have to be made if $t_{1/2}$ is known to be increased or decreased in other patient subpopulations.

In patients receiving multiple-dose regimens of immediate-release dosage forms, blood sampling should be performed at the expected peak serum concentration 1 to 2 h after a dose. A trough concentration measurement at the end of the dosing interval may lead to an inappropriate dose increase since peak concentrations can be two or more times higher than trough concentration with an immediate-release formulation. In patients receiving the clinically more frequently used multiple-dose regimens of sustained-release formulations, timing is less critical as concentrations are presumably relatively constant. Although trough concentrations are recommended for these regimens, samples obtained at the midpoint of the dosing interval may also be acceptable.[78]

DOSAGE INDIVIDUALIZATION

Oral Dosing for Chronic Therapy

The steady-state peak serum theophylline concentration is a function of dose and dosing interval that are controlled by the respective health-care professional, and theophylline absorption rate and clearance determined by the dosage form used and the individual patient treated. Because rate and completeness of absorption vary for different formulations, only products with complete absorption should be preferred, especially if switching between formulations. The formulation selected should be capable of maintaining stable serum concentrations when taken no more often than twice daily.

An oral theophylline dosage regimen should start with a low dose allowing tolerance to develop to the minor caffeine-like side effects frequently associated with the initiation of therapy. This dose should be increased stepwise until slowly approaching the recommended average dose for the population treated.[4,79] The incremental dosage changes should be made cautiously and in small increments because of potentially dose-dependent pharmacokinetics. Furthermore, after each dosage change, a waiting period should be followed allowing serum concentrations to reach steady state before any additional dosage adjustments are performed. Table 10.5 illustrates the latest published revision of dosing recommendations for children older than 6 months and adults who have no risk factors for altered theophylline clearance.[4] If factors confounding the theophylline elimination are present in individual patients, the recommended doses have to be adjusted accordingly.

TABLE 10.5
Dosage Guideline for Theophylline (based on anhydrous theophylline) in Children Older Than 6 Months and Adults Who Have No Risk Factors for Decreased Theophylline Clearance[a]

Variable	Weight-Adjusted Maximal Dose	Comments/Interventions
Initial dose	~10 mg/kg of body weight/day; maximum 300 mg/day	If initial dose is tolerated, increase the dose no sooner than 3 days later to the first increment
First increment	~13 mg/kg/day; maximum 450 mg/day	If the first incremental increase is tolerated, increase the dose no sooner than 3 days later to the second increment
Second increment	~16 mg/kg/day; maximum 600 mg/day	Measure an estimate of the peak serum concentration after at least 3 days at the highest tolerated dose
Theophylline serum concentration interpretation, μg/ml		
< 10		Increase dose approximately 25%
10–15		Maintain dose if tolerated
15.1–19.9		Consider a dose reduction of approximately 10%[b]
20–25		Withhold next dose, then resume treatment with next lower dose increment
>25		Withhold next two doses, then resume treatment with initial dose or lower dose

[a] For infants 6 weeks to 6 months of age, the initial daily dose is calculated according to the following regression equation: dose (in milligrams per kilogram per day) = (0.2)(age in weeks) + 5.0. Subsequent increases in the dose in this age group should be based on peak serum concentrations measured no sooner than 3 days after the start of therapy.

[b] This decreases the likelihood of side effects due to fluctuations in the absorption or elimination rate that may result in serum concentrations above 20 μg/ml and is especially important for patients who require doses higher than those used in the second increment.

Source: Modified from References 4 and 81.

Maximum doses should never be maintained or exceeded unless a serum level has been reliably obtained at steady state and confirms this dosage requirement. Steady-state conditions can only be assumed if no doses have been missed in the previous 48 h, no extra doses were taken, approximately equal dosing intervals were used, and no other confounding factors have altered theophylline clearance. Although patients with rapid theophylline metabolism may have low serum concentrations, noncompliance, inappropriate sampling times, or erratic absorption of some slow release formulations must be considered as alternative causes. With the use of the guidelines specified in Table 10.5, one or two measurements of serum theophylline are usually sufficient to determine the dose requirement, with annual checks thereafter unless clinical indications suggest the need for more frequent assessments. In addition to this approach, Bayesian-based pharmacokinetic parameter estimation has successfully been applied to support dosage individualization.[80]

Recent investigations on mean dosage requirements for different age groups revealed a consistent decrease of approximately 25% for all of them over the last 12 to 18 years.[81] The reasons for the related decrease in mean theophylline clearance remains so far unknown, but were speculated to be related to reduced environmental exposure to cigarette smoke due to social and societal changes. These decreased population dosage requirements are reflected in lower recommended doses in the guideline listed in Table 10.5,[4,81] whereas previous guidelines[79] including the FDA labeling guideline for theophylline[82] are still based on the historically higher dose requirements.

Intravenous Dosing for Acute Therapy

Although theophylline is not any longer considered a first-line therapy in the acute management of asthma or COPD,[3,7] it might be indicated if a patient is unresponsive to other pharmacotherapeutic interventions. To achieve therapeutic serum concentrations rapidly in these cases, an intravenous loading dose (LD, in mg/kg) may be given that is determined by the target concentration C_{target} (in µg/ml) to be reached, the measured predose serum concentration $C_{measured}$ (in µg/ml) resulting from preceding theophylline dosing, and the volume of distribution V_d (in l/kg):

$$\text{LD} = \frac{C_{target} - C_{measured}}{V_d} \qquad (10.1)$$

Assuming a V_d of 0.5 l/kg, each mg/kg theophylline results in an increase in serum concentration of 2 µg/ml. Thus, for a patient who has not received theophylline within the last 24 h, i.e., $C_{measured}$ is negligible, a loading dose of 5 mg/kg theophylline given as short-term infusion over 15 to 30 min (equivalent to 6 mg/kg aminophylline) will result in a serum concentration of approximately 10 µg/ml. A loading dose should not be given before obtaining serum theophylline concentrations if the patient received any theophylline in the previous 24 h. If measured serum concentrations are unavailable, alternative agents should be used for managing the patient's condition.

A serum concentration measurement obtained 30 min after the LD, when distribution is complete, can be used to assess the need for and size of subsequent loading doses and for guidance of the continuation therapy. Once a serum concentration of 10 to 15 µg/ml has been achieved by loading doses, a constant intravenous infusion is started. The required infusion rate R_0 (in mg/kg/h) can then be determined based on the target steady-state serum concentration $C_{\text{ss,target}}$ (in µg/ml) and the estimated theophylline population clearance Cl_{pop} (in l/kg/h), taking any known confounding factors into account.

$$R_0 = C_{\text{ss, target}} \cdot \text{Cl}_{\text{pop}} \tag{10.2}$$

For a healthy nonsmoker, an intravenous theophylline infusion of 0.4 mg/kg/h (0.5 mg/kg/h as aminophylline) will result in a steady-state concentration of approximately 10 µg/ml.

Theophylline serum concentrations should be monitored frequently and clinical response carefully until steady-state concentrations have been reached after approximately 24 to 48 h. Subsequent dosage adjustments can be performed based on serum concentration measurements at this time ($C_{\text{ss,measured}}$) through assessing the patient's individual clearance, Cl,

$$\text{Cl} = \frac{R_0}{C_{\text{ss, measured}}} \tag{10.3}$$

and readjusting the infusion rate R_0 via Cl and Equation 10.2. To account for potentially nonlinear pharmacokinetics of theophylline, increases in infusion rate should not be higher than increments of 25%, regardless of the calculations.

Alternatively, the patient's individual theophylline clearance Cl can already be assessed from two non-steady-state serum concentration measurements during continuous infusion by the method of Chiou et al.:[83]

$$\text{Cl} = \frac{2 \cdot R_0}{C_1 + C_2} + \frac{2 \cdot V_d \cdot (C_1 - C_2)}{(C_1 + C_2) \cdot (t_2 - t_1)} \tag{10.4}$$

where,   Cl = Total body clearance for theophylline, in l/h
              R_0 = Theophylline infusion rate, in mg/h
              V_d = Volume of distribution (0.5 l/kg)
              C_1 = First theophylline serum concentration, in µg/ml obtained at time t_1
              C_2 = Second theophylline serum concentration, in µg/ml obtained at time t_2, at least one estimated half-life from t_1

The Chiou equation only provides accurate clearance estimates if the infusion rate was constant, serum concentration sample drawings were separated by an interval of at least one half-life, and measurements were performed with a reliable and precise assay technique. Lack of adherence to these stringent conditions can result in substantial errors of clearance estimates and thus potentially serious theophylline overdosing.

When the patient has been stabilized on an effective intravenous regimen, conversion to oral therapy may be performed by dividing the 24-h intravenous theophylline dosage into two or three equal doses of a sustained release formulation.

SUMMARY

Theophylline is a pharmacological agent with a narrow therapeutic index and high interindividual pharmacokinetic variability. Pharmacokinetically guided dosage individualization based on serum concentration measurements has been shown to contribute substantially in ensuring a safe and effective pharmacotherapy with theophylline despite these, for clinical practice, unfavorable properties. Thus, theophylline is frequently considered a role model for successfully applied therapeutic drug monitoring with relevant clinical benefit.

REFERENCES

1. Polson, J. B. et al., Inhibition of human pulmonary phosphodiesterase activity by therapeutic levels of theophylline, *Clin. Exp. Pharmacol. Physiol.*, 5(5):535–539, 1978.
2. Kawai, M. and Kato, M., Theophylline for the treatment of bronchial asthma: present status, *Methods Find. Exp. Clin. Pharmacol.*, 22(5):309–320, 2000.
3. National Heart, Lung and Blood Institute, Expert Panel Report 2: Guidelines for the Diagnosis and Management of Asthma, Pub. 97–4051, DHHS National Institutes of Health, Bethesda, MD, 1997.
4. Weinberger, M. and Hendeles, L., Theophylline in asthma, *N. Engl. J. Med.*, 334(21):1380–1388, 1996.
5. Markham, A. and Faulds, D., Theophylline. A review of its potential steroid sparing effects in asthma, *Drugs*, 56(6):1081–1091, 1998.
6. Carter, E. et al., Efficacy of intravenously administered theophylline in children hospitalized with severe asthma, *J. Pediatr.*, 122(3):476, 1993.
7. Standards for the diagnosis and care of patients with chronic obstructive pulmonary disease, American Thoracic Society, *Am. J. Respir. Crit. Care Med.*, 152(5 Pt. 2):S77–121, 1995.
8. Steer, P. A. and Henderson-Smart, D. J., Caffeine versus theophylline for apnea in preterm infants, *Cochrane Database Syst. Rev.*, 2, 2000.
9. Simons, F. E., Luciuk, G. H., and Simons, K. J., Sustained-release theophylline for treatment of asthma in preschool children, *Am. J. Dis. Child*, 136(9):790–793, 1982.
10. Pollock, J. et al., Relationship of serum theophylline concentration to inhibition of exercise-induced bronchospasm and comparison with cromolyn, *Pediatrics*, 60(6):840–844, 1977.
11. Richer, C. et al., Theophylline kinetics and ventilatory flow in bronchial asthma and chronic airflow obstruction: influence of erythromycin, *Clin. Pharmacol. Ther.*, 31(5):586, 1982.

12. Holford, N. et al., Theophylline target concentration in severe airways obstruction — 10 or 20 mg/L? A randomised concentration-controlled trial, *Clin. Pharmacokinet.*, 25(6):495–505, 1993.

13. Holford, N., Hashimoto, Y., and Sheiner, L. B., Time and theophylline concentration help explain the recovery of peak flow following acute airways obstruction. Population analysis of a randomised concentration controlled trial, *Clin. Pharmacokinet.*, 25(6):506–515, 1993.

14. Derby, L. E. et al., Hospital admission for xanthine toxicity, *Pharmacotherapy*, 10(2):112–114, 1990.

15. Hendeles, L. et al., Frequent toxicity from IV aminophylline infusions in critically ill patients, *Drug Intell. Clin. Pharm.*, 11:12–18, 1977.

16. Weinberger, M. M. and Hendeles, L., Reassessing the therapeutic range for theophylline: another perspective, *Pharmacotherapy*, 13(6):598–601, 1993.

17. Self, T. H. et al., Reassessing the therapeutic range for theophylline on laboratory report forms: the importance of 5–15 micrograms/ml, *Pharmacotherapy*, 13(6):590–594, 1993.

18. Hendeles, L. et al., Safety and efficacy of theophylline in children with asthma, *J. Pediatr.*, 120(2 Pt. 1):177–183, 1992.

19. Hendeles, L. et al., Unpredictability of theophylline saliva measurements in chronic obstructive pulmonary disease, *J. Allergy Clin. Immunol.*, 60(6):335–338, 1977.

20. Jenne, J. W. et al., Pharmacokinetics of theophylline. Application to adjustment of the clinical dose of aminophylline, *Clin. Pharmacol. Ther.*,13(3):349–360, 1972.

21. Hendeles, L., Hochhaus, G., and Kazerounian S., Generic and alternative brand-name pharmaceutical equivalents: select with caution, *Am. J. Hosp. Pharm.*, 50(2):323–329, 1993.

22. Hendeles, L. et al., Food-induced "dose-dumping" from a once-a-day theophylline product as a cause of theophylline toxicity, *Chest*, 87(6):758–765, 1985.

23. Karim, A. et al., Food-induced changes in theophylline absorption from controlled-release formulations. Part I. Substantial increased and decreased absorption with Uniphyl tablets and Theo-Dur Sprinkle, *Clin. Pharmacol. Ther.*, 38(1):77–83, 1985.

24. Welling, P. G. et al., Influence of diet and fluid on bioavailability of theophylline, *Clin. Pharmacol. Ther.*, 17(4):475–480, 1975.

25. Hendeles, L., Weinberger, M., and Bighley, L., Absolute bioavailability of oral theophylline, *Am. J. Hosp. Pharm.*, 34(5):525–527, 1977.

26. Hendeles, L., Iafrate, R. P., and Weinberger, M., A clinical and pharmacokinetic basis for the selection and use of slow release theophylline products, *Clin. Pharmacokinet.*, 9(2):95–135, 1984.

27. Rogers, R. J. et al., Inconsistent absorption from a sustained-release theophylline preparation during continuous therapy in asthmatic children, *J. Pediatr.*, 106(3):496–501, 1985.

28. Szefler, S. J., Inter- and intra-subject variability in theophylline pharmacokinetics, *Br. J. Clin. Pract. Suppl.*, 35:10–16, 1984.

29. Scott, P. H. et al., Sustained-release theophylline for childhood asthma: evidence for circadian variation of theophylline pharmacokinetics, *J. Pediatr.*, 99(3):476–479, 1981.

30. Shaw, L. M., Fields, L., and Mayock, R., Factors influencing theophylline serum protein binding, *Clin. Pharmacol. Ther.*, 32(4):490–496, 1982.

31. Mangione, A. et al., Pharmacokinetics of theophylline in hepatic disease, *Chest*, 73(5):616–622, 1978.

32. Aranda, J. V. et al., Pharmacokinetic aspects of theophylline in premature newborns, *N. Engl. J. Med.*, 295(8):413–416, 1976.

33. Muller, M. et al., Theophylline kinetics in peripheral tissues *in vivo* in humans, *Naunyn Schmiedebergs Arch. Pharmacol.*, 352(4):441, 1995.

34. Ogilvie, R. I., Clinical pharmacokinetics of theophylline, *Clin. Pharmacokinet.*, 3(4):267–293, 1978.

35. Rohrbaugh, T. M. et al., The effect of obesity on apparent volume of distribution of theophylline, *Pediatr. Pharmacol*, 2(1):75–83, 1982.

36. Gundert-Remy, U. et al., Non-linear elimination processes of theophylline, *Eur. J. Clin. Pharmacol.*, 24(1):71–78, 1983.

37. Jonkman, J. H. et al., Measurement of excretion characteristics of theophylline and its major metabolites, *Eur. J. Clin. Pharmacol.*, 20(6):435–441, 1981.

38. Bauer, L. A., Bauer, S. P., and Blouin, R. A., The effect of acute and chronic renal failure on theophylline clearance, *J. Clin. Pharmacol.*, 22(1):65–68, 1982.

39. Rasmussen, B. B. and Brosen, K., Theophylline has no advantages over caffeine as a putative model drug for assessing CYPIA2 activity in humans, *Br. J. Clin. Pharmacol.*, 43(3):253–258, 1997.

40. Zhang, Z.Y. and Kaminsky, L. S., Characterization of human cytochromes P450 involved in theophylline 8-hydroxylation, *Biochem. Pharmacol.*, 50(2):205–211, 1995.

41. Sarkar, M. A. et al., Characterization of human liver cytochromes P-450 involved in theophylline metabolism, *Drug Metab. Dispos.*, 20(1):31–37, 1992.

42. Ha, H. R. et al., Metabolism of theophylline by cDNA-expressed human cytochromes P-450, *Br. J. Clin. Pharmacol.*, 39(3):321–326, 1995.

43. Sarrazin, E. et al., Dose-dependent kinetics for theophylline: observations among ambulatory asthmatic children, *J. Pediatr.*, 97(5):825–828, 1980.

44. Weinberger, M. and Ginchansky, E., Dose-dependent kinetics of theophylline disposition in asthmatic children, *J. Pediatr.*, 91(5):820–824, 1977.

45. Tateishi, T. et al., Developmental changes in urinary elimination of theophylline and its metabolites in pediatric patients, *Pediatr. Res.*, 45(1):66–70, 1999.

46. Grygiel, J. J. and Birkett, D. J., Effect of age on patterns of theophylline metabolism, *Clin. Pharmacol. Ther.*, 28(4):456–462, 1980.

47. Weinberger, M. and Hendeles, L., Theophylline, in *Allergy: Principles and Practice*, 4th ed., Middleton, E., Jr. et al., Eds., Mosby-Year Book, St. Louis, MO, 1995, 816–855.

48. Shin, S. G., Juan, D., and Rammohan, M., Theophylline pharmacokinetics in normal elderly subjects, *Clin. Pharmacol. Ther.*, 44(5):522–530, 1988.

49. Zell, M. et al., Volume of distribution of theophylline in acute exacerbations of reversible airway disease. Effect of body weight, *Chest*, 87(2):212–216, 1985.

50. Jewesson, P. J. and Ensom, R. J., Influence of body fat on the volume of distribution of theophylline, *Ther. Drug Monit.*, 7(2):197–201, 1985.

51. Zahorska-Markiewicz, B. et al., Pharmacokinetics of theophylline in obesity, *Int. J. Clin. Pharmacol. Ther.*, 34(9):393–395, 1996.

52. Staib, A. H. et al., Pharmacokinetics and metabolism of theophylline in patients with liver diseases, *Int. J. Clin. Pharmacol. Ther. Toxicol.*, 18(11):500–502, 1980.

53. Piafsky, K. M. et al., Theophylline disposition in patients with hepatic cirrhosis, *N. Engl. J. Med.*, 296(26):1495–1497, 1977.

54. Powell, J. R. et al., Theophylline disposition in acutely ill hospitalized patients. The effect of smoking, heart failure, severe airway obstruction, and pneumonia, *Am. Rev. Respir. Dis.*, 118(2):229–238, 1978.

55. Piafsky, K. M. et al., Theophylline kinetics in acute pulmonary edema, *Clin. Pharmacol. Ther.*, 21(3):310–316, 1977.

56. Vicuna, N. et al., Impaired theophylline clearance in patients with cor pulmonale, *Br. J. Clin. Pharmacol.*, 7(1):33–37, 1979.

57. Aderka, D., Shavit, G., and Garfinkel, D., Life-threatening theophylline intoxication in a hypothyroid patient, *Respiration*, 44(1):77–80, 1983.

58. Bauman, J. H., Teichman, S., and Wible, D. A., Increased theophylline clearance in a patient with hyperthyroidism, *Ann. Allergy*, 52(2):94–96, 1984.

59. Vozeh, S. et al., Influence of thyroid function on theophylline kinetics, *Clin. Pharmacol. Ther.*, 36(5):634–640, 1984.

60. Chang, K. C. et al., Altered theophylline pharmacokinetics during acute respiratory viral illness, *Lancet*, 1(8074):1132–1133, 1978.

61. Isles, A. et al., Theophylline disposition in cystic fibrosis, *Am. Rev. Respir. Dis.*, 127(4):417–421, 1983.

62. Georgitis, J. W. et al., Oral theophylline disposition in cystic fibrosis, *Ann. Allergy*, 48(3):175–177, 1982.

63. Powell, J. R. et al., The influence of cigarette smoking and sex on theophylline disposition, *Am. Rev. Respir. Dis.*, 116(1):17–23, 1977.

64. Jusko, W. J. et al., Enhanced biotransformation of theophylline in marihuana and tobacco smokers, *Clin. Pharmacol. Ther.*, 24(4):405–410, 1978.

65. Cusack, B. et al., Theophylline kinetics in relation to age: the importance of smoking, *Br. J. Clin. Pharmacol.*, 10(2):109–114, 1980.

66. Hunt, S. N., Jusko, W. J., and Yurchak, A. M., Effect of smoking on theophylline disposition, *Clin. Pharmacol. Ther.*, 19(5 Pt. 1):546–551, 1976.

67. Grygiel, J. J. and Birkett, D. J., Cigarette smoking and theophylline clearance and metabolism, *Clin. Pharmacol. Ther.*, 30(4):491–496, 1981.

68. Kappas, A. et al., Influence of dietary protein and carbohydrate on antipyrine and theophylline metabolism in man, *Clin. Pharmacol. Ther.*, 20(6):643–653, 1976.

69. Kappas, A. et al., Effect of charcoal-broiled beef on antipyrine and theophylline metabolism, *Clin. Pharmacol. Ther.*, 23(4):445–450, 1978.

70. Monks, T. J., Lawrie, C. A., and Caldwell, J., The effect of increased caffeine intake on the metabolism and pharmacokinetics of theophylline in man, *Biopharm. Drug Dispos.*, 2(1):31–37, 1981.

71. Sierles, F. S. and Ossowski, M. G., Concurrent use of theophylline and lithium in a patient with chronic obstructive lung disease and bipolar disorder, *Am. J. Psychiatr.*, 139(1):117–118, 1982.

72. Thomsen, K. and Schou, M., Renal lithium excretion in man, *Am. J. Physiol.*, 215(4):823–827, 1968.

73. Hendeles, L. et al., Decreased oral phenytoin absorption following concurrent theophyllien administration, *J. Allergy Clin. Immunol.*, 63:156, 1979.

74. Roizen, M. F. and Stevens, W. C., Multiform ventricular tachycardia due to the interaction of aminophylline and halothane, *Anesth. Analg.*, 57(6):738–741, 1978.

75. Hirshman, C. A. et al., Ketamine-aminophylline-induced decrease in seizure threshold, *Anesthesiology*, 56(6):464–467, 1982.

76. Bonfiglio, M. F. and Dasta, J. F., Clinical significance of the benzodiazepine–theophylline interaction, *Pharmacotherapy*, 11(1):85–87, 1991.

77. Pesce, A. J., Rashkin, M., Kotagal, U., Standards of laboratory practice: theophylline and caffeine monitoring. National Academy of Clinical Biochemistry, *Clin. Chem.*, 44(5):1124–1128, 1998.

78. Ellis, E. and Hendeles, L., Theophylline, in *A Textbook for the Clinical Application of Therapeutic Drug Monitoring*, Taylor, W. J. and Diers Caviness, M. H., Eds., Abbott Laboratories, Diagnostics Division, Irving, TX, 1986, 185–201.

79. Hendeles, L., Jenkins, J., and Temple, R., Revised FDA labeling guideline for theophylline oral dosage forms, *Pharmacotherapy*, 15(4):409–427, 1995.

80. Erdman, S. M., Rodvold, K. A., and Pryka, R. D., An updated comparison of drug dosing methods. Part II: Theophylline, *Clin. Pharmacokinet.*, 20(4):280–292, 1991.

81. Asmus, M. J. et al., Apparent decrease in population clearance of theophylline: implications for dosage, *Clin. Pharmacol. Ther.*, 62(5):483–489, 1997.

82. Food and Drug Administration, Theophylline Labeling Guideline, DHHS, Rockville, MD, 1995.

83. Chiou, W. L., Gadalla, M. A., and Peng, G. W., Method for the rapid estimation of the total body drug clearance and adjustment of dosage regimens in patients during a constant-rate intravenous infusion, *J. Pharmacokinet. Biopharm.*, 6(2):135–151, 1978.

84. Thompson, P. J., Social nocturnal alcohol consumption and pharmacokinetics of theophylline, *Chest*, 101(1):156–159, 1992.

85. Manfredi, R. L. and Vesell, E. S., Inhibition of theophylline metabolism by long-term allopurinol administration, *Clin. Pharmacol. Ther.*, 29(2):224–229, 1981.

86. Lonning, E, Kvinnsland, S., and Bakke, O. M., Effect of aminoglutethimide on antipyrine, theophylline, and digitoxin disposition in breast cancer, *Clin. Pharmacol. Ther.*, 36(6):796–802, 1984.

87. Gugler, R. and Allgayer, H., Effects of antacids on the clinical pharmacokinetics of drugs. An update. *Clin. Pharmacokinet.*, 18(3):210–219, 1990.

88. Landay, R. A., Gonzalez, M. A., and Taylor, J. C., Effect of phenobarbital on theophylline disposition, *J. Allergy Clin. Immunol.*, 62(1):27–29, 1978.

89. Saccar, C. L. et al., The effect of phenobarbital on theophylline disposition in children with asthma, *J. Allergy Clin. Immunol.*, 75(6):719, 1985.

90. Marquis, J. F. et al., Phenytoin–theophylline interaction, *N. Engl. J. Med.*, 307(19):1189–1190, 1982.

91. Miller, M. et al., Influence of phenytoin on theophylline clearance, *Clin. Pharmacol. Ther.*, 35(5):666–669, 1984.

92. Rosenberry, K. R. et al., Reduced theophylline half-life induced by carbamazepine therapy, *J. Pediatr.*, 102(3):472–474, 1983.

93. Conrad, K. A. and Woodworth, J. R., Orciprenaline does not alter theophylline elimination, *Br. J. Clin. Pharmacol.*, 12(5):756–757, 1981.

94. Hemstreet, M. P., Miles, M. V., and Rutland, R.O., Effect of intravenous isoproterenol on theophylline kinetics, *J. Allergy Clin. Immunol.*, 69(4):360–364, 1982.

95. Danziger, Y. et al., Reduction of serum theophylline levels by terbutaline in children with asthma, *Clin. Pharmacol. Ther.*, 37(4):471, 1985.

96. Joad, J. P. et al., Relative efficacy of maintenance therapy with theophylline, inhaled albuterol, and the combination for chronic asthma, *J. Allergy Clin. Immunol.*, 79(1):78–85,1987.

97. Conrad, K. A. and Nyman, D. W., Effects of metoprolol and propranolol on theophylline elimination, *Clin. Pharmacol. Ther.*, 28(4):463–467, 1980.

98. Miners, J. O. et al., Selectivity and dose-dependency of the inhibitory effect of propranolol on theophylline metabolism in man, *Br. J. Clin. Pharmacol.*, 20(3):219–223, 1985.

99. Cerasa, L. A. et al., Lack of effect of atenolol on the pharmacokinetics of theophylline, *Br. J. Clin. Pharmacol.*, 26(6):800–802, 1988.

100. Corsi, C. M. et al., Lack of effect of atenolol and nadolol on the metabolism of theophylline, *Br. J. Clin. Pharmacol.*, 29(2):265–268, 1990.

101. Sirmans, S. M. et al., Effect of calcium channel blockers on theophylline disposition, *Clin. Pharmacol. Ther.*, 44(1):29–34, 1988.

102. Loi, C. M. et al., Dose-dependent inhibition of theophylline metabolism by disulfiram in recovering alcoholics, *Clin. Pharmacol. Ther.*, 45(5):486, 1989.

103. Sperber, A. D., Toxic interaction between fluvoxamine and sustained release theophylline in an 11-year-old boy, *Drug Saf.*, 6(6):460–462, 1991.

104. Brosen, K. et al., Fluvoxamine is a potent inhibitor of cytochrome P4501A2, *Biochem. Pharmacol.*, 45(6):1211–1214, 1993.

105. van den Brekel, A. M. and Harrington, L., Toxic effects of theophylline caused by fluvoxamine, *CMAJ*, 151(9):1289–1290, 1994.

106. Loi, C. M. et al., Individual and combined effects of cimetidine and ciprofloxacin on theophylline metabolism in male nonsmokers, *Br. J. Clin. Pharmacol.*, 36(3):195–200, 1993.

107. Reitberg, D. P., Bernhard, H., and Schentag, J. J., Alteration of theophylline clearance and half-life by cimetidine in normal volunteers, *Ann. Intern. Med.*, 95(5):582–585, 1981.

108. Cohen, I. A. et al., Cimetidine–theophylline interaction: effects of age and cimetidine dose, *Ther. Drug Monit.*, 7(4):426–434, 1985.

109. Powell, J. R. et al., Inhibition of theophylline clearance by cimetidine but not ranitidine, *Arch. Intern. Med.*, 144(3):484–486, 1984.

110. Kelly, H. W., Powell, J. R., and Donohue, J. F., Ranitidine at very large doses does not inhibit theophylline elimination, *Clin. Pharmacol. Ther.*, 39(5):577–581, 1986.

111. Verdiani, P., Di Carlo, S., and Baronti, A., Famotidine effects on theophylline pharmacokinetics in subjects affected by COPD — comparison with cimetidine and placebo, *Chest*, 94(4):807–810, 1988.

112. Secor, J. W. et al., Lack of effect of nizatidine on hepatic drug metabolism in man, *Br. J. Clin. Pharmacol.*, 20(6):710–713, 1985.

113. Vestal, R. E. et al., Cimetidine inhibits theophylline clearance in patients with chronic obstructive pulmonary disease: a study using stable isotope methodology during multiple oral dose administration, *Br. J. Clin. Pharmacol.*, 15(4):411–418, 1983.

114. Jonkman, J. H. et al., Effects of alpha-interferon on theophylline pharmacokinetics and metabolism, *Br. J. Clin. Pharmacol.*, 27(6):795–802, 1989.

115. Pfeifer, H. J., Greenblatt, D. J., and Friedman, P., Effects of three antibiotics on theophylline kinetics, *Clin. Pharmacol. Ther.*, 26(1):36–40, 1979.

116. Branigan, T. A. et al., The effects of erythromycin on the absorption and disposition of kinetics of theophylline, *Eur. J. Clin. Pharmacol.*, 21(2):115–120, 1981.

117. Maddux, M. S. et al., The effect of erythromycin on theophylline pharmacokinetics at steady state, *Chest*, 81(5):563–565, 1982.

118. Weinberger, M. et al., Inhibition of theophylline clearance by troleandomycin, *J. Allergy Clin. Immunol.*, 59(3):228–231, 1977.

119. Rodvold, K. A., Clinical pharmacokinetics of clarithromycin, *Clin. Pharmacokinet.*, 37(5):385–398, 1999.

120. Pollak, P. T. and Slayter, K. L., Reduced serum theophylline concentrations after discontinuation of azithromycin: evidence for an unusual interaction, *Pharmacotherapy*, 17(4):827–829, 1997.

121. Nahata, M., Drug interactions with azithromycin and the macrolides: an overview, *J. Antimicrob. Chemother.*, 37 (Suppl. C):133–142, 1996.

122. Glynn-Barnhart, A. M. et al., Effect of low-dose methotrexate on the disposition of glucocorticoids and theophylline, *J. Allergy Clin. Immunol.*, 88(2):180–186, 1991.

123. Ueno, K. et al., Mechanism of interaction between theophylline and mexiletine, *Drug Intell. Clin. Pharm.*, 25(7–8):727–730, 1991.

124. Meibohm, B. and Wegener, S., Mexiletine–Theophylline-Interaction, *Krankenhaus-pharmazie*, 13(7):331–333, 1992.

125. Pieniaszek, H. J., Jr., Davidson, A. F., and Benedek, I. H., Effect of moricizine on the pharmacokinetics of single-dose theophylline in healthy subjects, *Ther. Drug Monit.*, 15(3):199–203, 1993.

126. Gardner, M. J. et al., Effects of tobacco smoking and oral contraceptive use on theophylline disposition, *Br. J. Clin. Pharmacol.*, 16(3):271–280, 1983.

127. Tornatore, K. M. et al., Effect of chronic oral contraceptive steroids on theophylline disposition, *Eur. J. Clin. Pharmacol.*, 23(2):134, 1982.

128. Roberts, R. K. et al., Oral contraceptive steroids impair the elimination of theophylline, *J. Lab. Clin. Med.*, 101(6):821–825, 1983.

129. Ellison, M. J. et al., Influence of pentoxifylline on steady-state theophylline serum concentrations from sustained-release formulations, *Pharmacotherapy*, 10(6):383–386, 1990.

130. Lee, B. L. and Dohrmann, M. L., Theophylline toxicity after propafenone treatment: evidence for drug interaction, *Clin. Pharmacol. Ther.*, 51(3):353–355, 1992.

131. Spinler, S. A. et al., Propafenone–theophylline interaction, *Pharmacotherapy*, 13(1):68–71, 1993.

132. Wijnands, W. J., Vree, T. B., and van Herwaarden, C. L., The influence of quinolone derivatives on theophylline clearance, *Br. J. Clin. Pharmacol.*, 22(6):677–683, 1986.

133. Sano, M. et al., Comparative pharmacokinetics of theophylline following two fluoroquinolones co-administration, *Eur. J. Clin. Pharmacol.*, 32(4):431–432, 1987.

134. Straughn, A. B. et al., Effect of rifampin on theophylline disposition, *Ther. Drug. Monit.*, 6(2):153–156, 1984.

135. Robson, R. A. et al., Theophylline–rifampicin interaction: non-selective induction of theophylline metabolic pathways, *Br. J. Clin. Pharmacol.*, 18(3):445–448, 1984.

136. Boyce, E. G. et al., The effect of rifampin on theophylline kinetics, *J. Clin. Pharmacol.*, 26(8):696–699, 1986.

137. Birkett, D. J., Miners, J. O., and Attwood, J., Evidence for a dual action of sulphinpyrazone on drug metabolism in man: theophylline–sulphinpyrazone interaction, *Br. J. Clin. Pharmacol.*, 15(5):567–569, 1983.

138. Spaldin, V. et al., Determination of human hepatic cytochrome P4501A2 activity *in vitro* use of tacrine as an isoenzyme-specific probe, *Drug Metab. Dispos.*, 23(9):929–934, 1995.

139. Schneider, D. et al., Theophylline and antiparasitic drug interactions. A case report and study of the influence of thiabendazole and mebendazole on theophylline pharmacokinetics in adults, *Chest*, 97(1):84–87, 1990.

140. Colli, A. et al., Ticlopidine-theophylline interaction, *Clin. Pharmacol. Ther.*, 41(3):358–362, 1987.

11 Anticonvulsant Agents

Gail D. Anderson

CONTENTS

1-56676-973-6/02/$0.00+$1.50
© 2002 by CRC Press LLC

INTRODUCTION

Approximately 1% of the population has epilepsy and the majority of the patients will be receiving one or more antiepileptic drug (AED). The goal of therapy is 100% seizure control; however, often a balance between adverse effects and seizure control is necessary to improve the patient's quality of life. The drug of choice is based on the type of seizure disorder, patient tolerability of the AED, and other patient factors (for example, age, concurrent disease, other medications). Prior to the 1990s, therapy was largely limited to phenytoin (PHT), phenobarbital (PB), carbamazepine (CBZ), valproate (VPA), and ethosuximide (ETH). Since 1992, eight new antiepileptic drugs have been released; felbamate (FLB), gabapentin (GBP), lamotrigine (LTG), topiramate (TOP), tiagabine (TGB), zonisamide (ZON), levetiracetam (LEV), and oxcarbazepine (OXC). The traditional AEDs have many limitations including lack of efficacy in a subgroup of patients, similar neurotoxicity, the occurrence of idiosyncratic reactions, and complex pharmacokinetics. The new AEDs have overcome some of these limitations; they have improved efficacy and pharmacokinetics and reduced adverse effects. However, their role is primarily as second-line for refractory seizures or for patients intolerant to the first line AEDs.[1]

CARBAMAZEPINE

INDICATION

- Carbamazepine[2-4] (CBZ) is FDA approved for the treatment of monotherapy and adjunctive therapy of partial and generalized seizures. CBZ is very effective in the treatment of generalized tonic-clonic and partial seizures,[5] but is not effective against myoclonic and absence seizures.[6]
- CBZ is the drug of choice for the treatment of pain associated with trigeminal neuralgia.[7]
- Off-label use: Certain psychiatric disorders including bipolar disorders and post-traumatic stress disorder, non-neuritic pain syndromes, restless leg syndrome, management of alcohol, cocaine and benzodiazepine withdrawal, and chorea in children.[8]

ABSORPTION AND DOSAGE FORMS

- $F = 0.75$ to 0.85
- Dosage forms available: oral and chewable tablets, suspension, two controlled-release formulations (Tegretol-XR, Carbatrol)
- Time to peak: 4 to 8 h (tablet, controlled-release), 1 to 2 h (suspension)
- No significant effect of food on absorption
- No parenteral formulation available due to low aqueous solubility
- Controlled Release Formulations:
  - Can convert TID or QID dosing to BID dosing
  - Tegretol XR product loses its sustained-release properties if broken or chewed

- Carbatrol: capsule can be emptied onto food and maintain sustained release
- Humidity can reduce CBZ bioavailability by 50%; CBZ needs to be stored in a dry place, i.e, not in the bathroom cabinets.[9]
- Generic substitution: Breakthrough seizures and/or increased side effects with generic switching have been reported and therefore should be done cautiously.[10]

DISTRIBUTION

- V_d = 0.8 to 2.0 l/kg
- Protein binding: 75% is bound to albumin and α_1-acid glycoprotein (f_p = 0.25)

CLEARANCE

- CBZ is predominately eliminated by hepatic metabolism via CYP3A4 to carbamazepine epoxide (CBZ-epoxide), an active metabolite. CBZ-epoxide is eliminated by hepatic metabolism via epoxide hydrolase to an inactive dihydrodiol metabolite. Minor metabolism via CYP1A2, CYP2C8, and UGT.[11]
- CBZ-epoxide has equal anticonvulsant and potency to CBZ and has also been shown to be effective against trigeminal neuralgia, a pain condition in which CBZ is the drug of choice.[12]
- CBZ-epoxide may be responsible for the neurotoxicity associated with CBZ.
- CBZ is a potent enzyme inducer and induces its own metabolism (auto-induction), which affects how CBZ therapy is initiated.
- Requires 3 to 4 weeks to achieve maximum autoinduction.
  Single dose:

$T_{1/2}$ = 33 h (18 to 55 h) Cl = 0.26 to 0.50 l/kg/h

Chronic dosing:

$T_{1/2}$ = 15 to 25 h Cl = 1.3 to 1.7 l/kg/day (adults, monotherapy)

$T_{1/2}$ = 3 to 15 h Cl = 1.9 to 3.1 l/kg/day(children, monotherapy)

$T_{1/2}$ = 5 to 13 h Cl = 2.3 to 4.0 l/kg/day (adults, polytherapy with inducers)

DOSING

- Loading doses are not recommended due to central nervous system and gastrointestinal toxicity.
- Maintenance dosing: Initiate therapy at 100 to 200 mg BID.

- Titration: Increase 200 mg daily every 3 to 7 days to seizure control or toxicity until usual maximum doses = 800 to 1200 mg/day.
- TID to QID dosing is usually needed except with controlled release formulations.
- Once daily is not tolerated and is associated with large fluctuations in plasma levels.

THERAPEUTIC DRUG MONITORING

- Therapeutic range: 4 to 12 µg/ml
- Remember CBZ-epoxide is not clinically measured but does contribute to efficacy and adverse effects.

DRUG INTERACTIONS[13]

- CBZ increases the clearance (Cl) of other drugs that are eliminated by hepatic metabolism via CYP2C, CYP3A, and conjugation with glucuronic acid resulting in decreased plasma concentrations and possible loss of efficacy if doses are not adjusted.
- CBZ plasma concentrations will be increased with inhibitors of CYP3A4 (Table 11.1).
- CBZ plasma concentrations have also been reported to increase with CYP2C19 inhibitors (Table 11.1). CBZ plasma concentrations will be decreased and CBZ-epoxide concentrations will be increased with CYP inducers (PHT, PB, rifampin).
- CBZ-epoxide plasma concentrations will be increased with inhibitors of epoxide hydrolase (VPA).

SPECIAL POPULATIONS

- Children — The weight-adjusted Cl of CBZ is higher in children than adults, which result in lower CBZ plasma concentrations in children given equivalent mg/kg doses. Children may need approximately 50 to 100% higher mg/kg doses than adults. Children have higher ratios of CBZ-epoxide to CBZ.[14]
- Elderly People — The weight-adjusted Cl and protein binding of CBZ is lower in elderly people than in young adults. With patients older than 65 years, CBZ doses may need to be decreased by approximately 40%. Total CBZ plasma concentrations underestimate unbound CBZ plasma concentrations. Monitor patients clinically for signs of toxicity and lack of efficacy instead of relying on therapeutic drug monitoring (TDM).[15,16]
- Liver Cirrhosis — Decreased hepatic metabolism and decreased protein binding due to hypoalbuminemia with severe liver disease will require decreased doses. Total CBZ plasma concentrations underestimate unbound CBZ plasma concentrations. Patients should be monitored clinically for signs of toxicity and lack of efficacy instead of relying on TDM.

TABLE 11.1

Inhibitors of Cytochrome P-450 (CYP) Isozymes and UDP Glucuronosyltranferase (UGT) Involved in Antiepileptic Drug Metabolism

Enzyme	Antiepileptic Drug	Inhibitors
CYP1A2	Carbamazepine	Fluvoxamine
CYP2C9	Carbamazepine	Sulfaphenazole
	Phenobarbital	Ketoconazole
	Phenytoin	Fluconazole
	Valproate	Amiodarone
		Miconazole
		Propoxyphene
		Valproate
CYP2C19	Diazepam	Felbamate
	Phenobarbital	Fluoxetine
	Phenytoin	Cimetidine
	Valproate	Ticlopidine
CYP3A4	Carbamazepine	Erythromycin
	Diazepam	Clarithromycin
	Ethosuximide	Diltiazem
	Midazolam	Fluconazole
		Grapefruit juice
		Ketoconazole
		Troleandomycin
		Verapamil
UGT	Lamotrigine	Valproate
	Lorazepam	

- Pregnancy — No change in the metabolism of CBZ or CBZ-epoxide occurs during the course of pregnancy; however, decreased protein binding results in decreased total plasma concentrations but not significant changes in unbound CBZ concentrations. Patients are monitored clinically for signs of toxicity and lack of efficacy instead of relying on TDM.[17]
- Lactation — Reported infant plasma concentrations <1 μg/ml (lower limit of assay sensitivity) when mothers are receiving therapeutic doses of CBZ. Breast-feeding should not cause accumulation of CBZ in infants.[18]

ETHOSUXIMIDE

INDICATION

- Ethosuximide[19,20] (ETH) has the narrowest spectrum of activity of any of the AEDs. ETH is only effective in the treatment of absence seizures and has no activity against partial and generalized tonic-clonic seizure disorders.[21]

ABSORPTION AND DOSAGE FORMS

- $F = 1.00$
- Dosage forms available: capsules, syrup
- Time to peak: <3 h

DISTRIBUTION

- $V_d = 0.6$ to 0.9 l/kg
- Protein binding: 0% ($f_p = 1.00$)

CLEARANCE

- 80 to 90% of ETH is eliminated by hepatic metabolized via cytochrome-P-450-dependent hydroxylation. The isozymes involved have not been identified in humans.
- $T_{1/2}$: 40 to 60 h (adults); 25 to 35 h (children)
- Cl: 9.2 ± 1.9 ml/h/kg (monotherapy); 15.3 ± 3.8 ml/h/kg (polytherapy)
- Time to steady state: 6 to 12 days

DOSING

- Loading doses are not recommended.
- Maintenance dosing: Initial dose: 250 mg q d.
- Titrate dose over 1 to 2 weeks to 20 mg/kg/day to minimize gastrointestinal toxicity.
- It is necessary to dose BID or TID to minimize gastrointestinal toxicity in most patients.

THERAPEUTIC DRUG MONITORING

- Therapeutic range: 40 to 100 µg/ml

DRUG INTERACTIONS[13]

- ETH plasma concentrations will be decreased when coadministered with CYP-inducing drugs.
- ETH plasma concentration may be increased with coadministration with VPA.

SPECIAL POPULATIONS

- Children — The weight-adjusted Cl of ETH is higher in children than adults, resulting in ETH plasma concentrations that are lower in children given equivalent mg/kg doses. Therefore, children require higher doses (mg/kg).[22]
- Liver or Renal Disease — Not studied, but a significant effect is not expected.

- Pregnancy — Cl may increase during pregnancy; monitor ETH plasma concentrations and adjust dose as necessary.
- Lactation — Plasma concentrations of ETH in an infant are 30 to 50% of those in the mother receiving ETH. Breast-feeding will result in clinically significant plasma concentrations in the infant. The risk/benefit of breast-feeding needs to be discussed with the mother.[18]

FELBAMATE

INDICATION

- Felbamate[23–25] (FLB) is FDA approved for monotherapy and adjunctive therapy for partial seizures with and without secondary generalizations and in the treatment of Lennox–Gastaut syndrome in children ≥ 2 years.
- FLB use is restricted for treatment of refractory epilepsy that has failed other AEDs, due to a risk of aplastic anemia (estimated risk of 1 in 4000).

ABSORPTION AND DOSAGE FORMS

- $F = 0.9$
- Dosage forms available: tablets, suspension
- Time to peak: 2 to 4 h
- No effect of food on absorption

DISTRIBUTION

- $V_d = 0.7$ to 0.9 l/kg
- Protein binding: 25% bound to albumin ($f_p = 0.75$)

CLEARANCE

- FLB is eliminated by renal excretion of unchanged drug (40 to 60%) and hepatic metabolism, which occurs by hydrolysis and cytochrome P-450 isozymes, CYP3A4 and CYP2E1.[26]
- $T_{1/2}$: 14 to 23 h; Cl: 31 ± 7 ml/h/kg
- Time to steady state: 5 to 7 days

DOSING

- Loading dose is not recommended.
- Maintenance dosing: Initiate therapy at 1200 mg/day given TID or QID.
- Increase doses at 1 to 2 week intervals to minimize toxicity to a dose of 3600 mg/day.

THERAPEUTIC DRUG MONITORING

- There is not good evidence for a concentration–effect relationship with FLB. Therapeutic drug monitoring (TDM) is not routinely recommended.

DRUG INTERACTIONS[26]

- FLB induces the metabolism of drugs hepatically eliminated by CYP3A4 (CBZ).
- FLB inhibits the metabolism of drugs eliminated by CYP2C19 (PHT) and glucuronidation (VPA).
- FLB plasma concentrations are increased by coadministration of VPA and decreased by coadministration of enzyme-inducing drugs.

SPECIAL POPULATIONS

- Children — The weight-adjusted Cl of FLB is approximately 40% higher in children than adults, resulting in FLB plasma concentrations lower in children given equivalent mg/kg doses.[14]
- Pregnancy — No data are available on the pharmacokinetics of FLB in pregnancy. Adjust doses based on clinical need.
- Lactation — The moderate protein-binding profile of FLB suggests that the infant may have measurable plasma concentrations if breast-fed. In addition, safety issues regarding rare aplastic anemia associated with FLB in patients with epilepsy may increase the theoretical risks of breast-feeding.

GABAPENTIN

INDICATION

- Gabapentin[25,27] (GBP) is FDA approved for the adjunctive therapy of partial seizures with and without secondary generalization in adults.
- Off-label use in neuropathic pain accounts for the majority of the GBP use.

ABSORPTION AND DOSAGE FORMS

- $F = 0.6$ at doses < 900 mg/day
- $F = 0.27$ to 0.47 at doses 1200 to 4800 mg/day due to saturation of L-amino transporter[28]
- Dosage forms available: capsules
- Time to peak: 2 to 3 h
- No significant effect of food on absorption

Distribution

- V_d = 0.7 to 0.8 l/kg
- Protein binding: 0% (f_p = 1.00)

Clearance

- GBP is eliminated predominately by renal excretion of unchanged drug with minimal or no hepatic metabolism.
- $T_{1/2}$: 5 to 7 h
- Cl/F is proportional to creatinine clearance (Cl_{cr})[29]

Cl_{cr} (ml/min)	Cl/F (ml/min)	$T_{1/2}$ (h)
>60	190	6.5
<30	20	52

Dosing

- Loading doses are not needed.
- Maintenance dosing: Initial dose: 300 mg qd.
- Titrate to dose of 1200 mg over 1 to 2 days, increase doses prn until usual maintenance dose of 900 to 3600 mg/day TID is reached.
- Decrease GBP doses for Cl_{cr} < 60 ml/min.

Therapeutic Drug Monitoring

- TDM is not used clinically; concentrations in clinical trials were 1 to 8 µg/ml.

Drug Interactions[13]

- No clinically significant drug interactions are apparent.

Special Populations

- Children — No pharmacokinetic data are available. However, the weight-adjusted Cl would not be expected to be significantly different in children compared to adults.
- Elderly People — GBP doses need to be decreased proportional to Cl_{cr}.[15,16]
- Renal Disease — BP doses need to be decreased proportional to Cl_{cr}.[29]
- Pregnancy — No data are available on the effect of pregnancy on GBP plasma concentrations. Adjust doses based on clinical need (change in seizure frequency or adverse effects).
- Lactation — No data are available, but based on the lack of protein binding breast-feeding may result in clinically significant plasma con-

centrations in the infant. The risks/benefits of breast-feeding need to be discussed with the mother.

LAMOTRIGINE

INDICATION

- Lamotrigine[25,30] (LTG) is FDA approved in adults for monotherapy and adjunctive therapy for partial seizures with and without secondary generalization and as adjunctive therapy in children ≥2 years with Lennox–Gastaut syndrome.
- Off-label use has demonstrated that LTG has broad-spectrum antiepileptic activity. In addition, LTG is used for the treatment of bipolar disease and pain syndromes.

ABSORPTION AND DOSAGE FORMS

- $F = 1.0$
- Dosage forms available: oral and dispersible tablets
- Time to peak: 2 to 4 h
- No effect of food on absorption

DISTRIBUTION

- $V_d = 0.9$ to 1.3 l/kg
- Protein binding: 55% protein bound to albumin ($f_p = 0.45$)

CLEARANCE

- LTG is predominately eliminated by hepatic metabolism as glucuronide conjugates, which is a reaction catalyzed by UDP glucuronosyltranferase isozyme, UGT1A4.[31]
- $T_{1/2}$: 25 h; Cl: 0.24 to 1.15 ml/min/kg (monotherapy)
- $T_{1/2}$: 13 h; Cl: 0.66 to 1.82 ml/min/kg (polytherapy with CBZ, PHT, or PB)
- $T_{1/2}$: 70 h; Cl: 0.12 to 0.33 ml/min/kg (polytherapy with VPA)
- Time to steady state: 3 to 5 days (monotherapy); 2 to 3 weeks (polytherapy with VPA)

DOSING

- Loading doses are not recommended
- Maintenance dosing: Initial doses of 25 to 50 mg/day are recommended if a patient is not receiving VPA concurrently or 25 mg q o d if a patient is receiving VPA to decrease the risk of a rash.
- Titrate q 7 days until 300 to 500 mg given BID if not receiving VPA concurrently or 100 to 150 given BID if currently receiving VPA

THERAPEUTIC DRUG MONITORING

- There is not good evidence for a concentration–effect relationship with LTG. However, TDM is not routinely recommended. LTG plasma concentrations usually range from 1 to 10 µg/ml.

DRUG INTERACTIONS[13]

- LTG plasma concentrations are decreased by coadministration of enzyme-inducing drugs.
- LTG plasma concentrations are increased by coadministration of VPA. Higher incidence of rash occurs with the combination of LTG + VPA, and the risk is significantly lower with low starting doses and a slow rate of escalation of LTG.

SPECIAL POPULATIONS

- Children — Weight-normalized Cl is one- to threefold higher in infants and children than in adolescents and adults, suggesting that children need a higher mg/kg dose.[14]
- Elderly People — The Cl of LTG is not significantly different in elderly patients. No dosage adjustment is necessary based on pharmacokinetic changes.[32]
- Renal Disease — Renal disease has no effect on LTG pharmacokinetics.
- Liver Disease — The oral Cl of LTG is 1/2 to 1/4 the Cl of inpatients with moderate to severe hepatic cirrhosis (Childs–Pugh classification of B or C, respectively) compared to healthy control subjects.
- Pregnancy — There is little known regarding the effect of pregnancy on LTG plasma concentrations. One case report found a decrease in total lamotrigine plasma concentrations with pregnancy; however, no data were reported regarding the unbound lamotrigine plasma concentration.[33] Doses should be adjusted based on clinical need (change in seizure frequency or adverse effects).
- Lactation — Based on an LTG f_p of 55%, infants would receive sufficient LTG to obtain clinically significant plasma concentrations. One case report documents an infant with plasma concentrations of LTG 25% of the concentrations in the mother.[33] The risk/benefit of breast-feeding needs to be discussed with the mother.

LEVETIRACETAM

INDICATION

- Levetiracetam[34] (LEV) is FDA approved for the treatment of adjunctive therapy in adults in partial seizures.
- Off-label use has demonstrated activity in generalized seizure types.

ABSORPTION AND DOSAGE FORMS

- $F = 1.0$
- Dosage forms available: tablets
- Time to peak: 1 h
- No significant effect of food on absorption

DISTRIBUTION

- $V_d = 0.5$ to 0.7 l/kg
- Protein binding: <10% bound ($f_p > 0.9$)

CLEARANCE

- LEV is eliminated by renal excretion of unchanged drug (66%) and enzymatic hydrolysis, a noncytochrome-P-450- or UGT-dependent pathway.
- $T_{1/2}$: 7.2 ± 1.1 h
- Cl = 0.96 ml/min/kg
- Time to steady state: 2 to 3 days

DOSING

- No loading dose is needed.
- Maintenance dosing: Initial doses = 500 mg BID.
- If needed, increase dose weekly by 500 mg BID to a maximum recommended 3000 mg/day.

THERAPEUTIC DRUG MONITORING

- There is not good evidence for a concentration–effect relationship with LEV. TDM is not routinely recommended.

DRUG INTERACTIONS[35,36]

- No drug interactions are reported.

SPECIAL POPULATIONS

- Children — Weight-normalized Cl is 40% higher in children than in adults suggesting that children need a higher mg/kg dose.
- Elderly People — Weight-normalized Cl is 38% less in elderly people compared to young adults suggesting that elderly people need a lower mg/kg dose.

- Renal Impairment — Cl_{cr} of LEV is correlated with creatinine clearance. Cl was reduced by 40% in patients with Cl_{cr} of 50 to 80 ml/min, 50% in patients with Cl_{cr} of 30 to 50 ml/min, and 60% with $Cl_{cr} < 30$ ml/min. Dosage of LEV should be reduced in renal impairment.
- Hepatic Disease — LEV has no effect on Cl.
- Pregnancy and Lactation — No data are available on the pharmacokinetics of LEV in pregnancy. However, the low-protein-binding profile of LEV suggests that an infant will have measurable plasma concentrations if breast-fed. The risk/benefit of breast-feeding needs to be discussed with the mother.

OXCARBAZEPINE

INDICATION

- Oxcarbazepine[25,37] (OXC) is FDA approved for monotherapy and adjunctive therapy of partial seizures in adults and adjunctive therapy for partial seizures in children ≥4 years old.
- OXC is closely related to carbamazepine, but is a prodrug that is rapidly and almost completely converted to 10,11-dihydro,10-hydroxycarbazepine (MHD), the pharmacologically active element. Therefore, the pharmacokinetics of MHD are of more clinical importance than OXC.
- Off-label use has demonstrated that OXC has activity in generalized seizure types.

ABSORPTION AND DOSAGE FORMS

- $F = 0.96$ to 1.0
- Dosage forms available: Oral tablets
- Time to peak: 1 h for OXC and 7 h for MHD
- No significant effect of food on OXC absorption

DISTRIBUTION

- $V_d = 0.7$ to 0.8 l/kg
- Protein binding: OXC is 60% protein bound ($f_p = 0.40$) and MHD is 40% protein bound ($f_p = 0.60$)

CLEARANCE

- OXC is almost completed converted to MHD through a noncytochrome-P-450-dependent reduction reaction in the liver. MHD is excreted in the urine unchanged (50%) or conjugated with glucuronic acid (50%). Approximately 9% of OXC is also metabolized to a glucuronide conjugate with less than 1% excreted unchanged in the urine.
- OXC does not form carbamazepine epoxide, the active metabolite of CBZ.

- $T_{1/2}$: MHD = 8 to 12 h
- Time to steady state: 2 to 3 days

Dosing

- Loading doses are not used.
- Maintenance dosing:
  - Adults: 300 mg BID, increased q 3 to 5 days up to 1200 mg/day
  - Children: 8 to 10 mg/kg/day given BID with a total starting dose not to exceed 600 mg/day; increased to target maintenance doses of 15 to 60 mg/kg/day
- Replacement of CBZ with OXC: Multiply the CBZ dose by 1.5 to approximate the equivalent maintenance dose of OXC. It is not necessary to cross-taper the drugs, due to similar pharmacokinetic and toxicity profile.
- If replacing CBZ with OXC in patients on other medications, increases in the plasma concentrations of other CYP or UGT drugs will occur due to de-induction (except for drugs metabolized by CYP3A4).

Therapeutic Drug Monitoring

- There is not a well-defined relationship between MHD plasma concentrations and effect. Routine TDM is not recommended. Clinically, MHD plasma concentrations are 12 to 32 µg/ml, approximately ninefold higher than OXC plasma concentrations at steady state.

Drug Interactions[35]

- OXC does not autoinduce its own metabolism.
- OXC does induce the metabolism of CYP3A4 substrates, but not other CYP isozymes.
- OXC and MHD are not significantly affected by other drug interactions.

Special Populations

- Children — The weight-adjusted Cl of MHD is higher in children aged 2 to 6 years. than in children age 6 to 12 years and adults; however, there was large intersubject variability. Younger children may need higher mg/kg doses of OXC.[14]
- Elderly People — The Cl of MHD is reduced in elderly people compared to a young adult population; however, only in patients with Cl_{cr} < 30 ml/min will OXC doses need to be significantly lowered to half the target dose. However, the dose in all elderly patients should be titrated slowly to the desired effect and not a target dose.[38]

- Renal Disease — The Cl of MHC is reduced in patients with renal disease; however, dosage reduction is necessary only when Cl_{cr} is less than 30 ml/min.[39]
- Pregnancy — No information is available on the effects of pregnancy on the pharmacokinetics of OXC or MHD.
- Lactation — No information is available on the effects of lactation. However, given the protein-binding profile of OXC and MHD, the infant may have measurable plasma concentrations of MHD if breast-fed. The risk/benefit of breast-feeding needs to be discussed with the mother.

PHENOBARBITAL

INDICATION

- Phenobarbital[40–42] (PB) is FDA approved for the treatment of partial and generalized seizure disorders. PB is also frequently used for treatment of neonatal seizures not due to proven superior efficacy, but to familiarity of use.
- Due to sedative side effects and paradoxical hyperactivity in children and elderly people, PB is used as third-line therapy.

ABSORPTION AND DOSAGE FORMS

- $F = 1.0$
- Dosage forms available: capsules, tablets, elixir, injection
- Injection, which contains propylene glycol and intravenous administration, not to exceed 100 mg/min
- Time to peak: 1 to 4 h
- No significant effect of food on absorption

DISTRIBUTION

- $V_d = 0.5$ to 0.6 l/kg (adults), 0.15 to 0.97 l/kg (neonates)
- Protein binding: PB is 50% protein bound to albumin ($f_p = 0.50$)

CLEARANCE

- PB is eliminated by a combination of excretion of unchanged drug (25%) and by hepatic metabolism to two main metabolites, p-OH-PB and PB-N-glucoside. CYP2C9 and CYP2C19 have been identified as the cytochrome P-450 isozymes involved in PB metabolism.[43]
- $T_{1/2}$: 72 to 144 h
- Time to steady state: 14 to 21 days
- Cl: Adults: 3.7 ± 0.7 ml/h/kg

DOSING

- Loading dose is 10 to 20 mg/kg (adults and children); 15 to 20 mg/kg (neonates).
- If a loading dose is given by the oral route, the dose is divided by three and given 1/3 q 24 h.
- Initial maintenance dose: 90 mg is administered every day or 1 to 3 mg/kg/day adults, 3 to 4 mg/kg/day for neonates and children.
- Usual dose: 180 to 300 mg/day.
- Take doses at bedtime to avoid excessive sedation.

THERAPEUTIC DRUG MONITORING

- Therapeutic range: 10 to 40 µg/ml

DRUG INTERACTIONS[13]

- PB is a strong inducer of hepatic enzymes and will decrease the plasma concentrations of drugs metabolized by CYP2C, CYP3A, and UGT family of enzymes.
- PB plasma concentrations are significantly increased by VPA.

SPECIAL POPULATIONS

- Children — Weight-normalized Cl (based on ratio of PB concentration to dose) increases with increasing age, suggesting that children need a higher mg/kg maintenance dose.[14]
- Neonates — Weight-normalized Cl is lower in neonates than in adults, suggesting that neonates need a lower maintenance mg/kg dose. Weight-normalized V is higher in neonates than in adults, suggesting that neonates need a higher mg/kg loading dose.[22]
- Elderly People — There is some evidence that the Cl of PB is decreased in elderly people compared with young adults, suggesting that elderly people need a lower mg/kg dose.[15,16]
- Hepatic or Renal Disease — No change in dose is needed.
- Pregnancy — There is not a significant effect of pregnancy on the pharmacokinetics of PB.[17] The dose is adjusted during pregnancy based on clinical need and not TDM.
- Lactation — Excess sedation and poor suckling occurs in infants whose mothers are receiving treatment with PB. Reported infant plasma concentrations are 25 to 50% of the plasma concentration of new mothers. The risk/benefit of breast-feeding needs to be discussed with the mother.[18,44]

PHENYTOIN

INDICATION

- Phenytoin[45,46] (PHT) is FDA approved for the treatment of partial and generalized seizure disorders.

ABSORPTION AND DOSAGE FORMS

- $F - 0.9$ to 1.0
- Dosage forms available: capsules, oral and chewable tablets, suspension, injection
- Capsules and injection: Phenytoin sodium (92% phenytoin acid)
- Tablets and suspensions contain phenytoin acid (100%).
- Dilantin Kapseals are "extended" with a 6 to 8 h absorption phase allowing once-a-day dosing.
- Tablets and suspension have more rapid absorption and should be administered BID to avoid high peak concentrations with associated toxicity.
- Beware of suspensions: Shake well to avoid inaccurate dosing.
- Nasogastric feeding impairs absorption to a highly variable degree.
- Saturable absorption: Time to reach maximum PHT plasma concentrations after a single oral dose increases with the dose with doses >400 mg.[47]
- Take with food to decrease gastrointestinal adverse effects.
- Phenytoin sodium injection: The injection can be safely mixed with 0.9% NaCl, but will precipitate in dextrose. A maximum infusion rate of 50 mg/min is recommended.
- Fosphenytoin is a phosphate ester of phenytoin and labeled as 100% phenytoin acid equivalents (PE) and can be safely mixed with all diluents. A maximum infusion rate of 150 mg/min is recommended.
- Intramuscular use: Phenytoin sodium: erratic and painful, $T_{max} = 3$ to 5 days. Fosphenytoin: safe and effective with T_{max} 1 to 2 h.

DISTRIBUTION

- $V_d = 0.6$ to 0.7 l/kg
- Protein binding: 90% protein bound to albumin, $f_p = 0.1$
- Unbound phenytoin plasma concentrations correlate with effect and toxicity.
- A decrease in protein binding occurs due to hypoalbuminemia or protein binding displacement interactions.
- Causes of hypoalbumiemia include age (neonates, elderly people), pregnancy, hepatic cirrhosis, burns, trauma, renal failure, and nephrotic syndrome.
- Displacement interactions can occur with valproate, sulfonamides, salicylic acid, and hyperbilirubinemia.

CLEARANCE

- PHT is predominately eliminated by hepatic metabolism via cytochrome P-450 isozymes, CYP2C (major), and CYP2C19 (minor).[48]
- Nonlinear (Michaelis–Menten) metabolism is observed with the PHT doses that are used clinically.
- Clearance (Cl) decreases with an increase in PHT plasma concentrations according to Equation 11.1:

$$Cl = \frac{V_{max}}{K_m + C_{ss}} \qquad (11.1)$$

where V_{max} is the maximum metabolic capacity, and K_m is an affinity constant that is mathematically equal to the substrate concentration at which the rate of reaction (metabolism) is one-half V_{max}.
- For adults, the average V_{max} and K_m values are 6 to 7 mg/kg/day and 4 to 6 µg/ml.
- For children, the average V_{max} and K_m are 8 to 19 mg/kg/day and 2 to 8 µg/ml.
- Due to the saturable metabolism, the time to eliminate 50% of drug ($T_{50\%}$) increases with increasing dose with an average $T_{50\%}$ of 22 h (7 to 42 h).
- Time to steady state therefore increases with increasing dose (7 to 28 days)

DOSING

- Loading doses:
  - If the PHT concentration is zero: A loading dose of 15 to 20 mg/kg is recommended according to Equation 11.2:

$$LD = \frac{V_d \cdot (C_{desired} - C_{measured})}{S \cdot F}$$

F = 1 for IV and 0.9 for an oral dose

S = 0.92 for sodium phenytoin and 1.0 for fosphenytoin

$\qquad (11.2)$

  - If subtherapeutic PHT concentrations are present ($C_{measured}$), use Equation 11.2 to calculate the loading dose needed to produce a desired PHT concentration ($C_{desired}$). If given by the oral route, divide the 400 mg doses so that each is given q 3 h to avoid saturable and prolonged absorption.
- Maintenance doses: 3 to 5 mg/kg/day
- If a steady-state plasma concentration is reached, the dose is increased to improve seizure control by one of the following two methods:
  1. Method 1 — Graves method:[49]

$$\text{New Dose} = \text{First Dose} \cdot (C_{ss}^{desired})^{0.2} \cdot (C_{ss}^{First})^{-0.2} \tag{11.3}$$

2. Method 2 — Privitera method:[50] The following rule produces less than 10% PHT concentrations greater than 25 µg/ml:

If C_{ss} is:	Then:
<7 µg/ml	Increase dose by 100 mg/day
7–11.9 µg/ml	Increase dose by 50 mg/day
≥12 µg/ml	Increase dose by 30 mg/day

If steady states (i.e., dose–concentration pairs) are reached by two subsequent doses and desired target concentration has not been reached, it is possible to solve the Michaelis–Menten equation algebraically by using simultaneous equations or by use of a graphical method to determine the patient's K_m and V_{max}.

- Ludden method:[51] A plot of dose (mg/day) vs. Cl or dose/C_{ss} (l/day) is constructed using both dose–concentration pairs. Based on Equation 11.4 and shown in Figure 11.1, the slope of the line obtained is $-K_m$ and the Y-intercept is V_{max}.

$$D/\tau = V_{max} - K_M \cdot \frac{\text{Dose}}{C_{ss}} \tag{11.4}$$

- The new dose can then be calculated for the desired C_{ss} using Equation 11.5:

$$D/\tau = \frac{V_{max} \cdot C_{ss}}{K_m + C_{ss}} \tag{11.5}$$

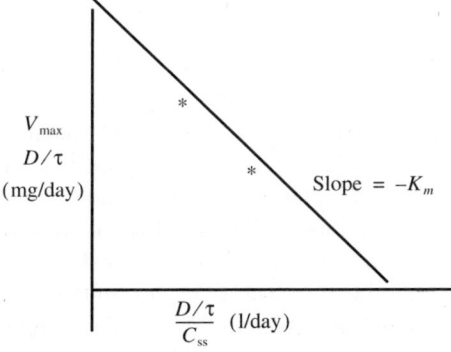

FIGURE 11.1 Graphical illustration of the use of the Ludden method[51] for which the slope and intercept can be calculated yielding K_m and V_{max}, respectively. Modified from Ludden, T. H. et al., *Clin. Pharmacol. Ther.*, 21:287, 1977.

THERAPEUTIC DRUG MONITORING[52]

- Therapeutic range: Total Phenytoin = 10 to 20 µg/ml; unbound phenytoin = 1 to 2 µg/ml. Estimation of normalized PHT plasma concentration in situations with hypoalbuminemia can be done using Equation 11.6 when unbound PHT plasma concentrations are not obtainable.[53]

$$[PHT]_{normalized} = \frac{[PHT]_{measured}}{0.25 + [ALB] + 0.1} \tag{11.6}$$

where $[PHT]_{measured}$ is in µg/ml; [PHT] is an estimate of the measured total PHT under situations of normal albumin concentrations; and [ALB] is serum albumin (mg/dl).

DRUG INTERACTIONS[13]

- PHT induces the metabolism of other drugs that are eliminated by hepatic metabolism via CYP2C, CYP3A, and/or conjugation with glucuronic acid (UGT) resulting in decreased plasma concentrations and possible loss of efficacy if doses are not adjusted.
- PHT plasma concentrations are increased with inhibitors of CYP2C9 or CYP2C19 (see Table 11.1). PHT plasma concentrations will be decreased with the addition of CYP inducers including carbamazepine, phenobarbital, and rifampin.
- PHT + Valproate (VPA) interaction: VPA displaces PHT from albumin and inhibits PHT metabolism. Therefore, unbound PHT is measured TDM or Equation 11.7 is used to estimate unbound PHT concentrations given a measured PHT total and VPA plasma concentrations.[54]

$$[PHT]_{unbound} = (0.095 + 0.01 \cdot [VPA]) + [PHT]_{total} \tag{11.7}$$

SPECIAL POPULATIONS

- Neonates and Elderly People — Decreased metabolism ($\downarrow V_{max}$) and decreased protein binding result in a decreased weight-normalized, unbound Cl. Total PHT plasma concentrations will underestimate the unbound PHT concentrations. Unbound PHT is measured or normalized PHT is estimated using Equation 11.6 for TDM.[55]
- Children — Increased metabolism ($\uparrow V_{max}$) results in a higher weight-normalized Cl in children than adults. Children will need higher mg/kg doses.[14]
- Pregnancy — Increased metabolism ($\uparrow V_{max}$) and decreased protein binding will result in decreased total and unbound PHT concentrations. Total PHT plasma concentrations will underestimate the unbound PHT concentrations. Unbound PHT is measured or normalized PHT is estimated using Equation 11.6 for TDM.[17]

- Lactation — Because of the high protein binding of PHT, an infant does not have measurable PHT plasma concentrations if the mother is taking therapeutic doses of PHT.[18]
- Liver Cirrhosis — Decreased metabolism ($\downarrow V_{max}$) and decreased protein binding result in a decreased unbound Cl. Total PHT plasma concentrations will underestimate the unbound PHT concentrations. Unbound PHT is measured or normalized PHT is estimated using Equation 11.6 for TDM.
- Renal Disease — Decreased protein binding is associated with changes in affinity and decreased albumin. Total PHT plasma concentrations will underestimate the unbound PHT concentrations. Estimating normalized PHT is not useful with diseases associated with changes in affinity; consequently, it is not necessary to measure unbound PHT for TDM.

PRIMIDONE

INDICATION

- Primidone[56] (PRM) is FDA approved for the treatment of partial and generalized seizure disorders either as monotherapy or adjunctive therapy with other AEDs.
- Unlabeled indication: Benign familial tremor
- Along with its two metabolites, phenobarbital and phenylethylmalonamide (PEME), primidone is an active anticonvulsant.

ABSORPTION AND DOSAGE FORMS

- $F = 1.0$
- Dosage forms available: tablets and suspension
- Time to peak: 1 to 3 h
- No effect of food on absorption

DISTRIBUTION

- $V_d = 0.4$ to 0.8 l/kg
- Protein binding: PRM is 20% protein bound ($f_p = 0.8$)

CLEARANCE

- PRM is excreted unchanged in the urine (40%) and metabolized via cytochrome P-450 to its two active metabolites, phenobarbital and PEMA.
- $T_{1/2} =$ 5 to 18 h for PRM
  72 to 144 h for PB
  10 to 25 h for PEMA
- Time to maximum effect is dependent on time to steady state of PB = 14 to 21 days.

DOSING

- Initial doses: 125 to 250 mg BID
- Maintenance doses: Adults: 750 to 1500 mg/day; children: 3.5 to 5.0 mg/kg day given TID

THERAPEUTIC DRUG MONITORING

- Therapeutic range: 5 to 15 PRM and 10 to 40 PB

DRUG INTERACTIONS[13]

- The clinically significant drug interactions with PRM are due to the effects of and on PB (see PB in the section above).
- PB is a strong inducer of hepatic enzymes and will decrease the plasma concentrations of drugs metabolized by CYP2C, CYP3A, and the UGT family of enzymes.
- PB plasma concentrations are significantly increased by VPA, which inhibits the metabolism of PB.

SPECIAL POPULATIONS

- Children — Weight-normalized Cl is higher in children than in adults, suggesting that children need a higher mg/kg dose.[22]
- Neonates — PB is not recommended in neonates due to a lack of information of the ability of the neonate to metabolize PRM to PB.[22]
- Elderly People — Weight-normalized Cl is decreased in elderly compared to young adults, suggesting that elderly people need a lower mg/kg dose.[57]
- Renal Impairment and Hepatic Disease — No change in dose is needed.
- Pregnancy — There is not a significant effect of pregnancy on the pharmacokinetics of RM.[17]
- Lactation — Excess sedation and poor suckling occur in infants whose mothers are receiving treatment with PRM probably due to the accumulation of PB. Lactation is potentially unsafe. The risk/benefit of breast-feeding needs to be discussed with the mother.[18]

TIAGABINE

INDICATION

- Tiagabine[58,59] (TGB) is FDA approved for adjunctive treatment of adults and children ≥12 years old with partial seizures.

ABSORPTION AND DOSAGE FORMS

- $F = 0.90$
- Dosage forms available: Oral tablets

- Time to peak concentrations: <1 h
- No effect of food on absorption

DISTRIBUTION

- V_d = 0.8 to 1.2 l/kg
- Protein binding: 96% of plasma concentrations of drug are protein-bound to albumin and α_1-acid-glycoprotein (f_p = 0.04)

CLEARANCE

- TGB is extensively metabolized by the liver via CYP3A4 and UGT.
- $T_{1/2}$ = 7 to 9 h monotherapy; 4 to 7 h polytherapy with inducing AEDs

DOSING

- Loading doses are not recommended.
- Initial doses: 4 mg/day is recommended and increased by 4 mg/day q 7 days until a usual maintenance dose of 32 to 56 mg/day; given BID to QID if the drug is administered as add-on therapy with enzyme-inducing drugs. Initiate move slowly if non-enzyme-inducing agents are used concurrently.

THERAPEUTIC DRUG MONITORING

- There has not been a correlation demonstrated between TGB plasma concentrations and clinical effect. Therefore, routine TDM is not recommended. TGB plasma concentrations in clinical trials range from 1 to 234 ng/ml.

DRUG INTERACTIONS[13]

- TGB does not affect the pharmacokinetics of other drugs.
- TGB metabolism is increased by coadministration of enzyme-inducing drugs resulting in TOP 50 to 65% lower compared to TGB monotherapy.

SPECIAL POPULATIONS

- Children — The weight-corrected Cl of TGB was twofold higher in children than in adults not receiving concurrent enzyme-inducing drugs. In children receiving enzyme-inducing drugs (CBZ, PHT), the weight-corrected Cl of TGB was similar to adults.
- Elderly People — There is no difference in the pharmacokinetics of TGB in elderly patients compared to young adults.[60]
- Renal Disease — There is no effect of renal disease on the Cl of TGB.[61]

- Liver Disease — The Cl of TGB is reduced by 65% in patients with liver disease and dosage reductions may be necessary.
- Pregnancy — There is no information on the effect of pregnancy on the pharmacokinetics of TGB.
- Lactation — Based on the high-protein-binding characteristic of TGB, infants may not have measurable plasma concentrations when the mother is receiving doses of TGB. However, there are no data available.

TOPIRAMATE

INDICATION

- Topiramate[58,62,63] (TOP) is FDA approved for adjunctive treatment of adults and children $\geq$12 years old for partial seizures or primary generalized tonic–clonic seizures.

ABSORPTION AND DOSAGE FORMS

- $F = 0.80$
- Dosage forms available: oral tablets and sprinkles
- Time to peak: 3 to 4 h
- Food delays the rate, but not the extent of absorption

DISTRIBUTION

- $V_d = 0.6$ to 0.8 l/kg
- Protein binding: TOP is 15% bound to proteins ($f_p = 0.85$)

CLEARANCE

- TOP is eliminated by a combination of excretion of unchanged drug and hepatic metabolism.
- $T_{1/2} = 24$ h

DOSING

- Loading doses are not recommended.
- Initial dose: 50 mg/day, increasing from 50 mg/day q 7 days to 200 to 400 mg/day. An even slower increase in the initial dose may improve tolerability to central nervous system side effects (i.e., initial dose 25 mg/day and increased by 25 mg/day q 7 days).

THERAPEUTIC DRUG MONITORING

- There has not been a correlation demonstrated between TOP plasma concentrations and clinical effect. Therefore, routine TDM is not recommended.

Drug Interactions[13]

- TOP metabolism is increased by coadministration of inducing drugs. Plasma concentrations are decreased by 50% and TOP doses need to be adjusted.
- TOP increases the plasma concentrations of PHT in approximately 1/2 of patients receiving the combination, presumably by inhibiting the CYP2C19 minor pathway of PHT.
- TOP decreases the plasma concentrations of the ethinyl estradiol component of oral contraceptives. Women should use an alternative method of birth control or use an oral contraceptive with a higher estrogen content (>35 μg of ethinyl estradiol).

Special Populations

- Children — The weight-adjusted Cl of TOP is 50% higher in children than in adults resulting in TOP plasma concentrations approximately 33% lower in children than in adults.[64]
- Infants — The weight-adjusted Cl of TOP is slightly higher than that observed for children and significantly higher than in adults resulting in a requirement for an increased dose (mg/kg). Titration to effect and not dose is recommended.[65]
- Elderly People — There is not a significant effect of age on the Cl of TOP in the absence of renal impairment.
- Renal Disease — In moderate to severe renal disease, the $T_{1/2}$ of TOP is approximately 50 to 100% longer. Dosage reduction is necessary in patients with moderate to severe renal disease.
- Liver Disease — The Cl of TOP is decreased 25 to 30% in patients with moderate to severe liver impairment. Moderate dosage adjustment with liver disease may be indicated.
- Pregnancy — There is no information on the pharmacokinetics of TOP in pregnancy.
- Lactation — Based on the protein-binding characteristic of TOP, infants would be expected to have clinically significant plasma concentrations if the mother is taking therapeutic doses of TOP. The risk/benefit of breast-feeding needs to be discussed with the mother.

VALPROATE

Indication

- Valproate[66,67] (VPA) is FDA approved for monotherapy and adjunctive therapy in the treatment of simple and complex absence seizures, adjunctively in patients with multiple seizure types including absence seizures,

and as adjunctive therapy in the treatment of patients with complex partial seizures.
- VPA is also FDA approved for the treatment of manic episodes associated with bipolar disease and for use in prophylaxis of migraine headaches.

ABSORPTION AND DOSAGE FORMS

- $F = 1.0$
- Dosage forms available: tablets, capsules, syrup, sprinkles, injectable
- Time to peak for various dosage forms:
  - Syrup: 0.5 to 1.0 h
  - Regular capsules (Depakene, and generics): 1 to 2 h
  - Enteric-coated tablets (Depakote): 3 to 8 h
  - Depakote sprinkles: 3 to 8 h

DISTRIBUTION

- $V_d = 0.09$ to 0.17 l/kg
- Protein binding:
  - Saturable, 85 to 93% bound
  - [VPA] < 50 µg/ml, f_p = 7 to 11%
  - [VPA] 50 to 100 µg/ml, f_p = 11 to 15%
  - [VPA] 100 to 150 µg/ml, f_p = 15 to 25%

CLEARANCE

- VPA is eliminated almost entirely by hepatic metabolism: glucuronidation and β-oxidation are major routes along with CYP450-dependent oxidations and desaturations via CYP2C9 and CYP2C19.
- $T_{1/2}$: Adults: 14 h (5 to 22 h); children: 11 h (5 to 15 h)

DOSING

- Loading doses (intravenous only) for 15 to 20 mg/kg
- Maintenance doses: Initial dose: 125 to 250 mg BID or TID
- Increase doses 125 to 250 mg/day q 7 days until target maintenance doses of 15 to 45 mg/kg/day (adults) or 15 to 60 mg/kg/day (children)

THERAPEUTIC DRUG MONITORING

- Therapeutic range: 40 to 100 µg/ml
- Increases in dose result in nonlinear relationship between total plasma concentrations of VPA and dose due to saturable protein binding (plateau effect); however, there is a linear relationship between unbound VPA plasma concentrations and dose.
- Therapeutic range is the same for all indications.

Drug Interactions[13]

- VPA will increase the plasma concentration of drugs eliminated by glucuronidation (i.e., lamotrigine, lorazepam) and epoxide hydrolase (carbamazepine epoxide) (See Table 11.1).
- VPA will increase the plasma concentration of drugs metabolized by CYP2C9 and CYP2C19 enzymes.
- VPA plasma concentrations are decreased with coadministration of drugs that are inducers of UGT or CYP450.

Special Populations

- Children — The weight-adjusted clearance of VPA is higher in children than in adults resulting in VPA plasma concentrations lower in children than in adults.[22,68]
- Elderly People — A decrease in protein binding results in total plasma concentrations underestimating unbound (or active) VPA plasma concentrations. TDM using total plasma concentrations will result in overdosing.[15,16,69]
- Renal Disease — There is no effect due to renal disease.
- Liver Disease — VPA clearance is decreased 25 to 30% in patients with moderate to severe liver impairment. Moderate dosage adjustment with liver disease may be indicated. Decreased protein binding due to decreased albumin will result in total VPA plasma concentrations underestimating unbound (or active) VPA plasma concentrations. TDM using total plasma concentrations will result in overdosing.[70]
- Pregnancy — There is very little effect of pregnancy on the pharmacokinetics of VPA. Because of decreased albumin, total VPA plasma concentrations will underestimate the unbound VPA plasma concentrations. The dose should be adjusted during pregnancy based on clinical need and not TDM.[17]
- Lactation — Infants of mothers receiving therapeutic doses of VPA do not have measurable VPA plasma concentrations.[18]

ZONISAMIDE

Indication

- Zonisamide[25,58,71] (ZON) is FDA approved for adjunctive treatment of adults and children ≥16 years old with partial seizures.

Absorption and Dosage Forms

- F = not known; but "well-absorbed" as reported in the package insert
- Dosage forms available: capsules

- Time to peak: 2 to 6 h
- No effect of food on ZON absorption

DISTRIBUTION

- V_d = 1.45 l/kg
- Protein binding: 40% bound (f_p = 0.60)

CLEARANCE

- ZON is eliminated by both renal excretion of unchanged drug (35%) and hepatic metabolism. ZON undergoes N-acetylation and CYP3A4 catalyzed reduction.
- $T_{1/2}$ = 63 h
- Time to steady state achieved within 14 days
- Cl: 0.30 to 0.35 ml/min/kg on monotherapy or noninduced patients; 0.5 ml/min/kg on polytherapy with inducing drugs

DOSING

- Loading doses not recommended
- Initial doses: 100 mg q/day, which is increased by 100 mg/day every 2 weeks until a maximum recommended dose of 600 mg/day given BID is reached.

THERAPEUTIC DRUG MONITORING

- There is not good evidence for a concentration-effect relationship with ZON. TDM is not routinely recommended. Plasma concentrations of 7 to 40 µg/ml have been reported in clinical studies. A target plasma concentration of 20 µg/ml has been recommended by one investigator.

DRUG INTERACTIONS

- ZON metabolism is increased by inducers of CYP3A4 (CBZ, PHT, PB).
- ZON does not affect the pharmacokinetics of other AEDs.

SPECIAL POPULATIONS

- Elderly People — There is not an effect of age on the pharmacokinetics of ZON.
- Children — Not known.
- Renal Disease — Severe renal disease (Cl_{cr} < 20 ml/min) results in a 35% decrease in ZON CL and requires decreased ZON doses.

- Hepatic Disease — No information is available on the effects of hepatic disease.
- Pregnancy — No information is available on the effect of pregnancy on the pharmacokinetics of ZON.
- Lactation — Based on the protein-binding profile of ZON ($f_p = 0.60$), infants may have clinically significant plasma concentrations if breast-fed by mothers taking therapeutic doses of ZON. The risk/benefit of breast-feeding needs to be discussed with the mother.

REFERENCES

1. Beghi, E. and Perucca, E., The management of epilepsy in the 1990's, *Drugs*, 49:680–694, 1995.
2. Morselli, P. L., Carbamazepine: absorption, distribution and excretion, in *Antiepileptic Drugs*, 4th ed., Levy, R.H., Mattson, R. H., and Meldrum, B. S., Eds., Raven Press, New York, 1995, 515–528.
3. Thomson, A. H. and Brodie, M. J., Pharmacokinetic optimisation of anticonvulsant therapy, *Clin. Pharmacokinet.*, 23:216–230, 1992.
4. Bertilsson, L. and Tomson, T., Clinical pharmacokinetics and pharmacological effects of carbamazepine and carbamazepine-10,11 epoxide, an update, *Clin. Pharmacokinet.*,11:177–198, 1986.
5. Mattson, R. H. et al., Comparison of carbamazepine, phenobarbital, phenytoin and primidone in partial and secondarily generalized tonic-clonic seizures, *N. Engl. J. Med.*, 313:145–151, 1985.
6. Loiseau, P. and Duche, B., Carbamazepine: clinical use, in *Antiepileptic Drugs*, 4th ed., Levy, R. H., Mattson, R. H., and Meldrum, B. S., Eds., Raven Press, New York, 1994, 555–566.
7. Cheshire, W. P., Trigeminial neuralgia: a guide to drug choice, *CNS Drugs*, 7:98–110, 1997.
8. Albani, F., Riva, R., and Baruzzi, A., Carbamazepine clinical pharmacology: a review, *Pharmacopsychiatry*, 28:235–244, 1995.
9. Bell, W. L., Crawford, I. L., and Shiu, G.K., Reduced bioavailability of moisture-exposed carbamazepine resulting in status epilepticus, *Epilepsia*, 34:1102–1104, 1993.
10. Nuwer, M. et al., Generic substitution for antiepileptic drugs, *Neurology*, 40:1647–1651, 1990.
11. Kerr, B. M. et al., Human liver carbamazepine: role of CYP3A4 and CYP2A8 in 10,11 epoxide formation, *Biochem. Pharmacol.*, 47:1769–1779, 1994.
12. Kerr, B. M. and Levy, R. H., Carbamazepine: carbamazepine epoxide, in *Antiepileptic Drugs*, 4th ed., Levy, R. H., Mattson, R. H., and Meldrum, B. S., Eds., Raven Press, New York, 1995, 529–541.
13. Anderson, G., A mechanistic approach to antiepileptic drug interactions, *Ann. Pharmacother.*, 32:554–563, 1998.
14. Battino, D., Estienne, M., and Avanzini, G., Clinical pharmacokinetics of antiepileptic drugs in paediatric patients. Part II. Phenytoin, carbamazepine, sulthiame, lamotrigine, vigabatrin, oxcarbazepine and felbamate, *Clin. Pharmacokinet.*, 29:341–369, 1995.

15. Willmore, L. J., Antiepileptic drug therapy in the elderly, *Pharmacol. Ther.*, 78:9–16, 1998.

16. Faught, E., Epidemiology and drug treatment of epilepsy in elderly people, *Drugs Aging*, 15:255–269, 1999.

17. Lander, C. M and Eadie, M. J., Plasma antiepileptic drug concentrations during pregnancy, *Epilepsia*, 32:257–266, 1991.

18. Hagg, S. and Spigset, O., Anticonvulsant use during lactation, *Drug Saf.*, 22:425–440, 2000.

19. Pisani, F. and Bialer, M., Ethosuximide: absorption, distribution and excretion, in *Antiepileptic Drugs*, 4th ed., Levy, R. H., Mattson, R. H., and Meldrum, B. S., Eds., Raven Press, New York, 1994.

20. Morselli, P. L. and Fracno-Morselli, R., Clinical pharmacokinetics of antiepileptic drugs in adults. *Pharmacol. Ther.*, 10:65–101, 1980.

21. Sherwin, A. L., Ethosuximide: Clinical use, in *Antiepileptic Drugs*, 4th ed., Levy, R. H., Mattson, R. H., and Meldrum, B. S., Eds., Raven Press, New York, 1994.

22. Battino, D., Estienne, M., and Avanzini, G., Clinical pharmacokinetics of antiepileptic drugs in paediatric patients. Part I: Phenobarbital, primidone, valproic acid, ethosuximide and mesuximide, *Clin. Pharmacokinet.*, 29:257–286, 1995.

23. Palmer, K. and McTavish, D., Felbamate: a review of its pharmacodynamic and pharmacokinetic properties, and therapeutic efficacy in epilepsy, *Drugs*, 45:1041–1065, 1993.

24. Graves, N. M., Felbamate, *Ann. Pharmacother.*, 27:1073–1081, 1993.

25. Walker, M. C. and Patsalos, P. N., Clinical pharmacokinetics of new antiepileptic drugs, *Pharmacol. Ther.*, 3:351–384, 1995.

26. Glue, P. et al., Pharmacokinetic interactions with felbamate: *in vitro–in vivo* correlation, *Clin. Pharmacokinet.*, 33:214–224, 1997.

27. Andrews, C. O. and Fischer, J. H., Gabapentin: a new agent for the management of epilepsy, *Ann. Pharmacother.*, 28:1188–1196, 1994.

28. Gidal, B. E. et al., Inter- and intra-subject variability in gabapentin absorption and absolute bioavailability, *Epilepsy Res.*, 40:123–127, 2000.

29. Blum, R. A. et al., Pharmacokinetics of gabapentin in subjects with various degrees of renal function, *Clin. Pharmacol. Ther.,* 56:154–159, 1994.

30. Goa, K. L., Ross, S. R., and Chrisp, P., Lamotrigine, a review of its pharmacological properties and clinical efficacy in epilepsy, *Drugs*, 46:152–176, 1993.

31. Green, M. D., Bishop, W. P., and Tephley, T. R., Expressed human UGT1.4 protein catalyzes the formation of quaternary ammonium-linked glucuronides, *Drug Metab. Dispos.*, 23:299–302, 1995.

32. Posner, J., Holdich, T., and Crome, P., Comparison of lamotrigine pharmacokinetics in young and elderly healthy volunteers, *J. Pharm. Med.*, 1:121–128, 1991.

33. Tomson, T., Ohman, I., and Vitols, S., Lamotrigine in pregnancy and lactation: a case report, *Epilepsia*, 38:1039–1041, 1997.

34. Patsalos, P. N., Pharmacokinetic profile of levetiracetam: toward ideal characteristics, *Pharmacol. Ther.*, 85:77–85, 2000.

35. Benedetti, M. S., Enzyme induction and inhibition by new antiepileptic drugs: a review of human studies, *Fundam. Clin. Pharmacol.*, 14:301–319, 2000.

36. Nicolas, J.-M. et al., *In vitro* evaluation of potential drug interactions with levetiracetam, a new antiepileptic agent, *Drug Metab. Dispos.*, 27:250–254, 1999.

37. Lott, R. S. and Helmboldt, K., Oxcarbazepine: a carbamazepine analogue for partial seizures in adults and children with epilepsy, *Formulary*, 35:1–9, 2000.

38. van Heiningen, P. N. et al., The influence of age on the pharmacokinetics of the antiepileptic agent oxcarbazepine, *Clin. Pharmacol. Ther.*, 50:410–419, 1991.

39. Rouan, M. C. et al, The effect of renal impairment on the pharmacokinetics of oxcarbazepine and its metabolites, *Eur. J. Clin. Pharmacol.*, 47:161–167, 1994.

40. Dodson, W. E. and Rust, R. S., Phenobarbital: absorption, distribution and excretion, in *Antiepileptic Drugs*, 4th ed., Levy, R. H., Mattson, R. H., and Meldrum, B. S., Eds., Raven Press, New York, 1994.

41. Nelson, E. et al., Phenobarbital pharmacokinetics and bioavailability in adults, *J. Clin. Pharmacol.*, 22:141–148, 1982.

42. Wilensky, A. J. et al. Kinetics of phenobarbital in normal subjects and epileptic patients, *Eur. J. Clin. Pharmacol.*, 23:87–92, 1982.

43. Hargraves, J. A. et al., Identification of enzymes responsible for the metabolism of phenobarbital (abstract), *Int. Soc. Stud. Xenobiotics Proc.*, 10:259, 1996.

44. Nau, H. et al., Placental transfer and pharmacokinetics of primidone and its metabolites, phenobarbital, PEMA and hydroxyphenobarbital in neonates and infants of epileptic mothers, *Eur. J. Clin. Pharmacol.*, 18:18–42, 1980.

45. Treiman, D. M. and Woodbury, D. M., Phenytoin: absorption, distribution and excretion, in *Antiepileptic Drugs*, 4th ed., Levy, R. H., Mattson, R. H., and Meldrum, B. S., Eds., Raven Press, New York, 1995, 301–314.

46. Richens, A., Clinical pharmacokinetics of phenytoin, *Clin. Pharmacokinet.*, 4:153–169, 1979.

47. Jung, D. et al., Effect of dose on phenytoin absorption, *Clin. Pharmacol. Ther.*, 28(4):479–485, 1980.

48. Bajpai, M. et al., Roles of cytochrome P4502C9 and cytochrome P4502C19 in the stereoselective metabolism of phenytoin to its major metabolites, *Drug Metab. Dispos.*, 24:1401–1403, 1996.

49. Graves, N. M. et al., Phenytoin clearances in a compliant population: description and application, *Ther. Drug Monit.*, 8:427–433, 1986.

50. Privitera, M. D., Clinical rules for phenytoin dosing, *Ann. Pharmacother.*, 27:1169–1173, 1993.

51. Ludden, T. M. et al., Individualization of phenytoin dosage regimens, *Clin. Pharmacol. Ther.*, 21:287–293, 1977.

52. Levine, M. and Chang, T., Therapeutic drug monitoring of phenytoin. Rationale and current status, *Clin. Pharmacokinet.*, 19:341–358, 1990.

53. Anderson, G. D. et al., Revised Winter-Tozer equation for normalized phenytoin concentrations in trauma and elderly patients with hypoalbuminemia, *Ann. Pharmacother.*, 31:279–284, 1997.

54. Haidukewych, D., Rodin, E. A., and Zielinski, J. J., Derivation and evaluation of an equation for prediction of free phenytoin concentrations in patients co-medicated with valproic acid, *Ther. Drug Monit.*, 11:134–139, 1989.

55. Bachmann, K. A. and Belloto, R. J. J., Differential kinetics of phenytoin in elderly patients, *Drugs Aging*, 15:235–250, 1999.

56. Cloyd, J. C., and Leppik, I. E., Primidone: absorption, distribution and excretion, in *Antiepileptic Drugs*, 4th ed., Levy, R. H., Mattson, R. H., and Meldrum B. S., Eds., Raven Press, New York, 1994, 459–466.

57. Martines, C. et al., The disposition of primidone in elderly patients, *Br. J. Clin. Pharmacol.*, 30:607–611, 1990.

58. Perucca, E. and Bialer, M., The clinical pharmacokinetics of the newer antiepileptic drugs. Focus on topiramate, zonisamide and tiagabine, *Clin. Pharmacokinet.*, 31:29–41, 1996.

59. Luer, M. S. and Rhoney, D. H., Tiagabine: a novel antiepileptic drug, *Ann. Pharmacother.*, 32:1173–1180, 1998.

60. Jansen, S. S. et al., The pharmacokinetics of tiagabine in healthy elderly volunteers and elderly patients with epilepsy, *J. Clin. Pharmacol.*, 37:1015–1020, 1997.

61. Cato, Ar. et al., Effect of renal impairment on the pharmacokinetics and tolerability of tiagabine, *Epilepsia*, 39:43–47, 1998.

62. Privitera, M. D., Topiramate: a new antiepileptic drug, *Ann. Pharmacother.*, 31:1164–1173, 1997.

63. Sachdeo, R. C., Topiramate: clinical profile in epilepsy, *Clin. Pharmacokinet.*, 34:335–346, 1998.

64. Rosenfeld, W. E. et al., A study of topiramate pharmacokinetics and tolerability in children with epilepsy, *Pediatr. Neurol.*, 20:339–344, 1999.

65. Glauser, T. A. et al., Topiramate pharmacokinetics in infants, *Epilepsia*, 40:788–791, 1999.

66. Levy, R. H. and Shen, D. D., Valproic acid: absorption, distribution, and excretion, in *Antiepileptic Drugs*, 4th ed., Levy, R. H., Mattson, R. H. and Meldrum, B. S., Eds., Raven Press, New York, 1995, 605–619.

67. Levy, R. H., Variability in level-dose ratio of valproate: monotherapy versus polytherapy, *Epilepsia*, 25 (Suppl. 1):S10–S13, 1984.

68. Cloyd, J. C. et al., Pharmacokinetics of valproic acid in children: I. Multiple antiepileptic drug therapy, *Neurology*, 33:185–191, 1983.

69. Bauer, L. A. et al., Valpric acid clearance: unbound fraction and dirunal variations in young and elderly adults, *Clin. Pharmacol. Ther.*, 37:697–700, 1985.

70. Klotz, U., Rapp, T., and Muller, W. A., Disposition of valproic acid in patients with liver disease, *Eur. J. Clin. Pharmacol.*, 13:55–60, 1978.

71. Peters, D. H. and Sorkin, E. M., Zonisamide: a review of its pharmacodynamic and pharmacokinetic properties, and therapeutic potential in epilepsy, *Drugs*, 45:760–787, 1993.

Section IV

Research Applications

12 Classical Modeling

David W. A. Bourne

CONTENTS

INTRODUCTION

The study of pharmacokinetics may be pursued at a number of levels. Considering the detail of the mathematical models involved one may use a noncompartmental, classical, or physiologically based approach. All three approaches require some measure of mathematical description or assumptions with the classical, compartmental approach intermediate in complexity. This chapter describes the development and use of compartmental models in pharmacokinetic research. Noncompartmental and physiologically based pharmacokinetic models are discussed in Chapters 13 and 14, respectively.

COMPARTMENTAL MODELS

Classical pharmacokinetic modeling involves the representation of the subject's body as a collection of compartments.[1-24] Drug concentrations within each compartment are considered to be in rapid equilibrium. That is, they are not necessarily equal concentrations throughout the compartment but are in rapid equilibrium. Rapid is defined empirically. Compartments are defined empirically. Rapid means that the data collected do not support more compartments. The investigator starts with the minimum number of compartments for the entities studied and the route of administration. Additional compartments are included only if the data suggest that they are necessary.

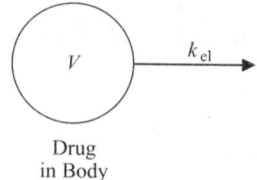

Drug
in Body

FIGURE 12.1 A one-compartment model with first-order elimination after an IV bolus dose.

From a pharmacokinetic point of view, the simplest route of administration is the intravenous (IV) bolus dose. Thus, the simplest compartmental model is the one-compartment model with IV bolus administration and first-order elimination of the drug (Figure 12.1). This model includes an apparent volume of distribution, V. This volume parameter is used to relate the amount of drug in the body with the concentration measured in plasma, serum, or blood (Equation 12.1). As mentioned above, the one-compartment model does not require that the concentration throughout the compartment be the same but that there is an equilibration between the concentrations throughout the body. Therefore, the volume of distribution is not a physical volume and may be many times larger than the size of the subject in cases where the drug is extensively distributed outside the blood.

$$V = \frac{\text{Amount of drug in the body}}{\text{Concentration of drug}} = \frac{X}{C} \qquad (12.1)$$

When elimination follows first-order kinetics, this model can be represented by the differential equation, Equation 12.2.

$$\frac{dC}{dt} = -k_{el} \times C \text{ with the initial condition } C_0 = \frac{D}{V} \qquad (12.2)$$

Using Laplace transforms or other mathematical techniques, Equation 12.2 can be integrated to give Equation 12.3.

$$C = \frac{D}{V} \times e^{-k_{el} \times t} \qquad (12.3)$$

This approach can be expanded to include other routes of administration such as IV infusion (of duration T) and extravascular administrations such as oral, intramuscular, subcutaneous, or topical. Schemes and equations for these models are shown in Table 12.1. The integrated equations in Table 12.1 can be used to calculate drug concentrations after a single IV bolus, IV infusion, or oral dose as shown in Figures 12.2 through 12.4.

TABLE 12.1
One-Compartment Pharmacokinetic Models

ROA	Scheme	Differential Equation	Integrated Equation
IV bolus		$\dfrac{dC}{dt} = -k_{el} \times C$	$C = \dfrac{D}{V} \times e^{-k_{el} \times t}$
IV infusion		$\dfrac{dC}{dt} = \dfrac{k_0}{V} - k_{el} \times C$	$C = \dfrac{k_0}{V} \times [1 - e^{-k_{el} \times t}]$ during $C = \dfrac{k_0}{V} \times [1 - e^{-k_{el} \times T}] \times e^{-k_{el} \times (t-T)}$ after
Extravascular, oral, etc.		$\dfrac{dC}{dt} = \dfrac{k_a \times X_g}{V} - k_{el} \times C$	$C = \dfrac{F \times D \times k_a}{V \times (k_a - k_{el})} \times [e^{-k_{el} \times t} - e^{-k_a \times t}]$

GI = gastrointestinal.

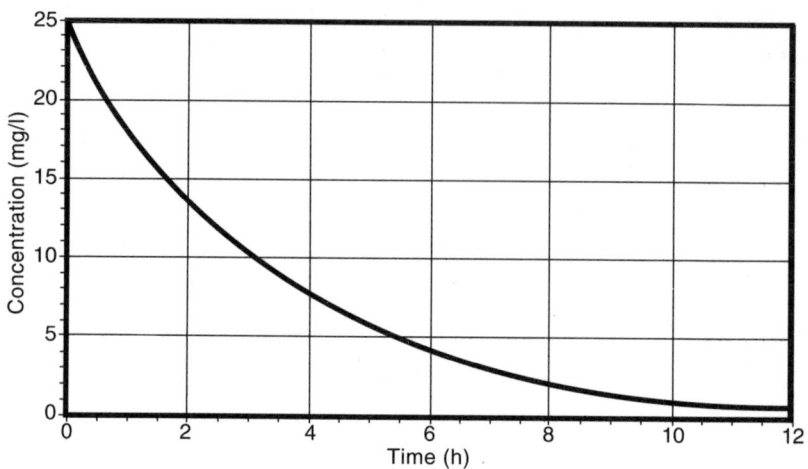

FIGURE 12.2 Linear plot of *C* vs. time after an IV bolus dose.

DATA ANALYSIS AFTER A SINGLE DOSE — ONE COMPARTMENT

Data collected after a single dose can be analyzed using the equations in Table 12.1. Since these equations involve exponential functions, a logarithmic transformation will produce a straight line. The slope will provide information about the value of the rate constants, and the intercept will provide information about the apparent volume of distribution. This can be illustrated readily with data collected after a single IV bolus dose. Taking the natural log of both sides of Equation 12.3 produces Equation 12.4.

$$\ln C = \ln \frac{D}{V} - k_{el} \times t \qquad (12.4)$$

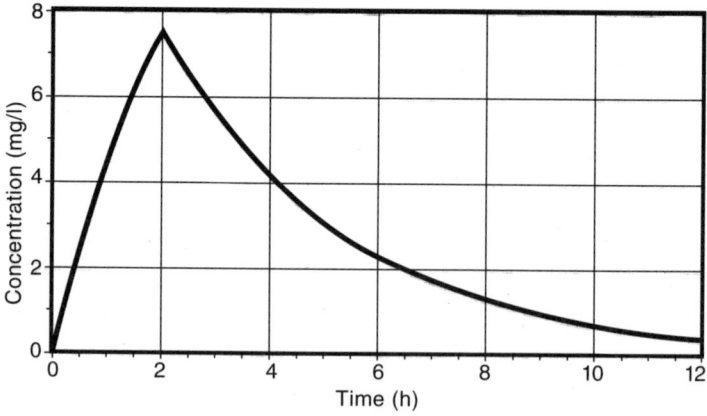

FIGURE 12.3 Linear plot of *C* vs. time after an IV infusion.

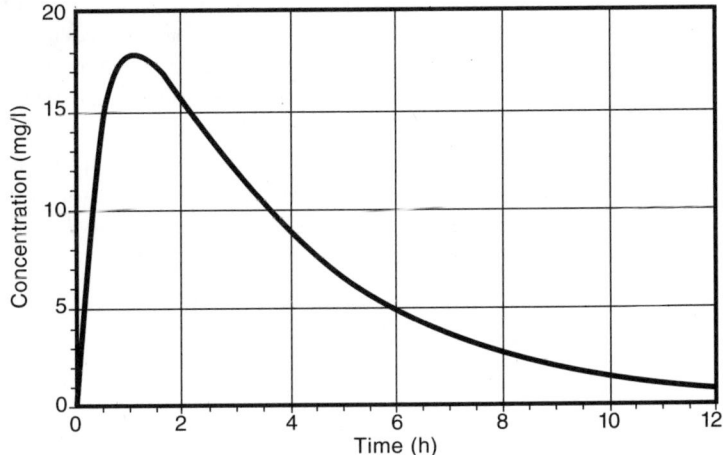

FIGURE 12.4 Linear plot of C vs. time after an oral dose.

Thus, plotting ln C vs. time on linear graph paper or C vs. time on semilog graph paper should produce a straight line, as shown in Figure 12.5. The slope provides a value for k_{el} (Equation 12.5) and the intercept provides a value for D/V. With D known, V can be calculated by Equation 12.6.

$$k_{el} = \frac{\ln C_1 - \ln C_2}{t_2 - t_1} \qquad (12.5)$$

$$V = \frac{D}{C_0} \qquad (12.6)$$

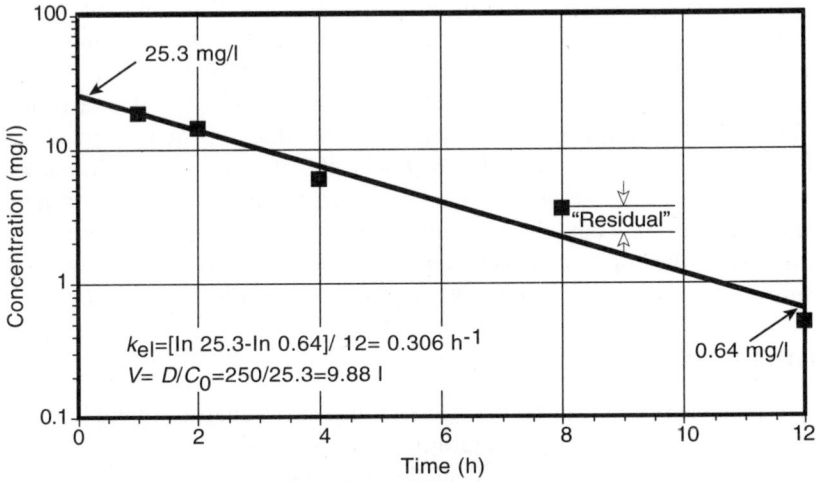

FIGURE 12.5 Semilog plot of C vs. time after a 250-mg IV bolus.

Analysis of data collected following a single IV infusion can be analyzed in a similar fashion. Thus, a semilog plot of the postinfusion data should fall close to a straight line. A value for k_{el} can be determined using Equation 12.5. The calculation of V is similar. However, the data collected after oral (or other extravascular) administration involve more than one exponential term. When these exponential terms are sufficiently different in value, the method of residuals or a curve stripping method can be used to analyze the data. This method is derived from the equation for C vs. time shown in Table 12.1. At later times, assuming k_a is greater than k_{el}, the k_a term becomes close to zero and disappears.

$$C_{later} = A \times e^{-k_{el} \times t}, \qquad \text{where } A = \frac{F \times D \times k_a}{V \times (k_a - k_{el})} \qquad (12.7)$$

Plotting later C values vs. time on semilog graph paper should produce a straight line from which k_{el} can be determined. Combining the original equation with Equation 12.7, the difference between the early C values and C_{later} values can be determined.

$$\text{Residual} = C_{later} - C = A \times e^{-k_{el} \times t} - \left\lfloor A \times e^{-k_{el} \times t} - A \times e^{-k_a \times t} \right\rfloor = A \times e^{-k_a \times t} \quad (12.8)$$

From Equation 12.8 a plot of $C_{later} - C$ vs. time on semilog graph paper provides an estimate of k_a, the absorption rate constant. The technique of curve stripping or method of residuals can be applied to equations involving multiple exponential terms when they differ by at least fivefold.

These data may also be analyzed by nonlinear regression using a computer program such as Boomer (http://www.boomer.org/), SAAM II (http://www.saam.com/), or Win-NONLIN (http://www.pharsight.com/tools/wnl/index.htm). (Information about other pharmacokinetic software can be found at http://www.boomer.org/pkin/soft.html.) These programs are designed to adjust the parameter values until the sum of the weighted square of the residual or difference between the observed and calculated points (WSS) is minimized as shown in the example in Figure 12.5. The advantage of these programs is that the model may be expressed as integrated or differential equations. Multiple lines such as plasma and urine data sets may be fitted simultaneously. The lines are not required to be linear; therefore transformation of the data is not necessary. Also, the data may be weighted to include information about the statistical error in each data point. Nonlinear regression analysis also provides more information in the form of final parameter values and their variability, statistical measures of goodness-of-fit, and plots such as weighted residual plots.

MULTIPLE DOSE ADMINISTRATION

Compartmental models can be used to simulate or to analyze multiple dose data. When uniform doses are given at uniform dosing intervals, T, the concentration after the fourth IV bolus dose can be calculated using Equation 12.9.

$$C = \left(\left(\frac{D}{V} \times e^{-k_{el} \times T} + \frac{D}{V}\right) \times e^{-k_{el} \times T} + \frac{D}{V}\right) \times e^{-k_{el} \times T} + \frac{D}{V}$$

or

$$C = \frac{D}{V} \times e^{-3 \times k_{el} \times T} + \frac{D}{V} \times e^{-2 \times k_{el} \times T} + \frac{D}{V} \times e^{-k_{el} \times t} + \frac{D}{V} \qquad (12.9)$$

Equation 12.9 is the sum of a geometric series and can be simplified to Equation 12.10.

$$C = \frac{D}{V}\left\{\frac{1 - e^{-n \times k_{el} \times T}}{1 - e^{-k_{el} \times T}}\right\}e^{-k_{el} \times t} \qquad (12.10)$$

Equation 12.10 provides the concentration at time t after the nth IV bolus dose. Similar equations can be derived for multiple, uniform IV infusions or extravascular doses given at uniform dosing intervals. When the regimen includes different doses and/or dosing intervals, another approach is necessary. If the disposition processes, distribution, excretion, and metabolism are first order, then, from using the superposition principle, the concentrations from each dose can be added together to give the total concentration from the drug regimen. For example, the concentration at 24 h after a 200-mg dose at 0 hours, 100 mg at 6 h , and 300 mg at 18 h can be given by Equation 12.11.

$$C = \frac{200}{V} \times e^{-k_{el} \times 24} + \frac{100}{V} \times e^{-k_{el} \times 18} + \frac{300}{V} \times e^{-k_{el} \times 6} \qquad (12.11)$$

Data collected after multiple dose administration can also be analyzed by nonlinear regression if the data are collected at appropriate times. One or two sample points during each dosing interval can be combined and fit simultaneously as long as the dosing and sampling regimen are accurately recorded.

MULTICOMPARTMENT MODELS

Earlier in this chapter it was mentioned that the one-compartment model is useful if there is a rapid equilibration between drug concentrations throughout the body. For a number of drugs this is a valid assumption and the one-compartment model can be quite useful. However, for other drugs there are significant observed deviations from this simple model. Drugs given by rapid IV bolus or infusion exhibit concentration–time curves that deviate from a straight line when plotted on semilog graph paper. A "distribution phase" is present in the early data as shown in Figure 12.6. Actually, distribution and elimination are occurring throughout the concentration vs. time profile. It is the slower distribution with these drugs that requires the use of

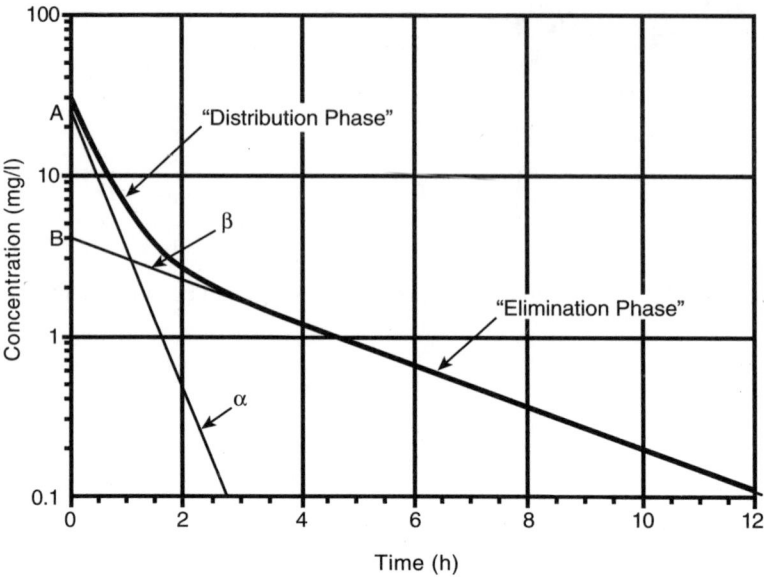

FIGURE 12.6 Semilog plot of C vs. time after an IV bolus dose.

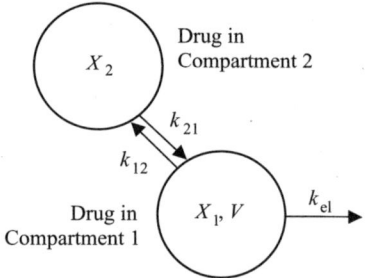

FIGURE 12.7 A two-compartment model in which X_1 and X_2 represent compartments 1 and 2, V represents the volume of compartment 1 with k_{12} and k_{21} representing the first-order rate constants entering and leaving the respective compartments, and k_{el} representing elimination out of the body.

multiple-compartment models. In this case, a second compartment can be included in the scheme as shown in Figure 12.7.

This model can be described mathematically with the differential equation, Equation 12.12.

$$\frac{V \times dC}{dt} = -(k_{el} + k_{12}) \times X_1 + k_{21} \times X_2 \qquad (12.12)$$

Equation 12.12 can be integrated to give Equation 12.13.

$$C = A \times e^{-\alpha \times t} + B \times e^{-\beta \times t} \tag{12.13}$$

where

$$A = \frac{D \times (\alpha - k_{21})}{V \times (\alpha - \beta)} \text{ and } B = \frac{D \times (k_{21} - \beta)}{V \times (\alpha - \beta)}$$

and

$$\alpha + \beta = k_{el} + k_{12} + k_{21}$$

$$\alpha \times \beta = k_{el} \times k_{21}$$

or

$$\alpha, \beta = \frac{k_{el} + k_{12} + k_{21} \pm \sqrt{(k_{el} + k_{12} + k_{21})^2 - 4 \times k_{el} \times k_{21}}}{2} \tag{12.14}$$

Equations for drug concentrations after IV infusion or extravascular single or multiple doses can be derived in a similar fashion. That is, design the model, write the differential equations, and integrate the equations to give useful, working equations. With known values for the model parameters, concentrations after any regimen can be calculated. Analysis of experimental data can be performed to estimate values of these parameters using graphical and nonlinear regression methods. Curve stripping or the method of residuals can be applied to data such as that shown in Figure 12.6. Nonlinear regression analysis can be performed to provide more objective estimates of the parameter values.

SELECTION OF THE BEST MODEL

It is important to recognize that the number of compartments used to represent the drug in the body is determined empirically and in accordance with the principle that simpler is better. Thus, a one-compartment model will be used unless the data can be shown to "require" a more complex model such as a two-compartment model. In practice, there are both subjective and objective criteria for choosing the "best" model. Subjective criteria include looking at the observed and calculated data, weighted residual deviation plots, and variability in the estimated parameter values. The observed and calculated concentration–time curve should be examined to see if there are apparent deviations between the two curves. Systematic deviations between these two curves would suggest another model should be considered. Plots of the weighted residual deviations vs. time or vs. concentration are very useful. A U shape or the appearance of a line through this plot is strong evidence that the model should be expanded. All the nonlinear regression programs will provide estimates of the uncertainty in the values of the parameters. These estimates include standard deviations, coefficients of variation, and confidence intervals.

When the uncertainty in the parameter values becomes too large, the analyst should consider reducing the model. The correlation matrix between parameters can be useful in selecting the parameters that can be removed to make the model smaller. There are statistical criteria that can be used to select the better model. These include the Akaike Information Criterion (AIC) value and the F-test. The AIC value is calculated using the WSS, the number of parameters in the model, and the number of data points. The model with the lower AIC values is usually selected as the better model. The statistical F-test involves the calculation of an F value from the WSS and degrees of freedom from two analyses. The calculated F value is compared with the tabled values and a decision can be made whether the more complex model provides a significant improvement in the fit to the data. The analyst using a combination of subjective and objective criteria can make an educated decision about the best model.

PARALLEL ELIMINATION PATHWAYS

The compartmental model approach can be used to include multiple elimination pathways. For example, unchanged drug excretion and drug metabolism can be accommodated with the addition of suitable components to the model. The diagram shown in Figure 12.8 includes three elimination pathways, excretion of unchanged drug, and two metabolism pathways. If these are the only routes of excretion the overall elimination can be determined from Equation 12.15.

$$k_{el} = k_e + k_{m1} + k_{m2} \tag{12.15}$$

Thus, estimates of k_{el} can be determined from a plot of concentration vs. time on a semilog plot or by nonlinear regression. With information about drug in urine and metabolite 1 in urine, and thus the fraction excreted unchanged or as a metabolite, it is possible to determine k_e, k_{m1}, and k_{m2}. The rate of excretion of unchanged drug can be described using Equation 12.16. This equation can be integrated to give Equation 12.17, which describes the cumulative amount excreted into urine vs. time

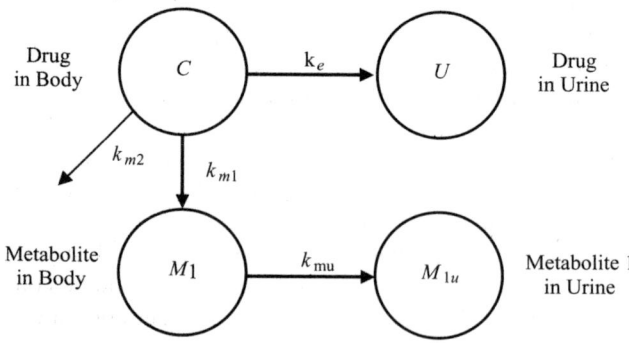

FIGURE 12.8 A one-compartment model illustrating parallel elimination. The subscript e represents elimination of drug, whereas $m1$ and $m2$ represent metabolites 1 and 2, respectively, with mu representing excretion of metabolite from the body into urine.

and can be used with nonlinear regression analysis of urine data. Straight-line equations, more suitable for graphical analysis, can also be derived. These include rate of excretion (Equation 12.18) and amount remaining to be excreted (ARE, Equation 12.19).

$$\frac{dU}{dt} = k_e \times C \times V \tag{12.16}$$

$$U = \frac{k_e \times D}{k_{el}} \times [1 - e^{-k_{el} \times t}] \tag{12.17}$$

$$\frac{dU}{dt} = k_e \times C \times V = k_e \times \frac{D}{V} \times e^{-k_{el} \times t} \times V = k_e \times D \times e^{-k_{el} \times t}$$

$$\ln \frac{dU}{dt} = \ln(k_e \times D) - k_{el} \times t \tag{12.18}$$

$$ARE = U^{\infty} - U = U^{\infty} \times e^{-k_{el} \times t}$$

$$\ln[ARE] = \ln[U^{\infty} - U] = \ln U^{\infty} - k_{el} \times t \tag{12.19}$$

NONLINEAR COMPARTMENTAL MODELS

Compartmental models are not confined to linear systems. It is relatively easy to include nonlinear processes such as saturable metabolism or protein binding. For example, for some drugs one or more metabolism processes may follow Michaelis–Menten kinetics, shown in Equation 12.20. Elimination is described in Equation 12.20 with a nonlinear metabolism process with the parameters V_m (maximum velocity) and K_m (Michaelis constant).

$$\frac{dC}{dt} = -\frac{V_m \times C}{K_m + C} \tag{12.20}$$

At high concentrations the denominator $K_m + C$ approaches C and Equation 12.20 becomes zero order with $dC/dt = -V_m$. This can be seen in Figure 12.9. After the high, 400 mg, dose the concentration follows along a straight line at earlier times. In Figure 12.10 the high dose data fall below the straight line drawn at lower concentrations. At concentrations much lower than K_m the denominator $K_m + C$ approaches K_m and Equation 12.20 becomes a first-order equation with $k' = V_m/K_m$. This can be seen in Figures 12.9 and 12.10. On the linear plot the low-dose data follow an exponential decline, which is a straight line on the semilog plot indicating a first-order process. Nonlinear compartmental models lead to more complex data analysis and dosage regimen calculations. However, the compartmental approach

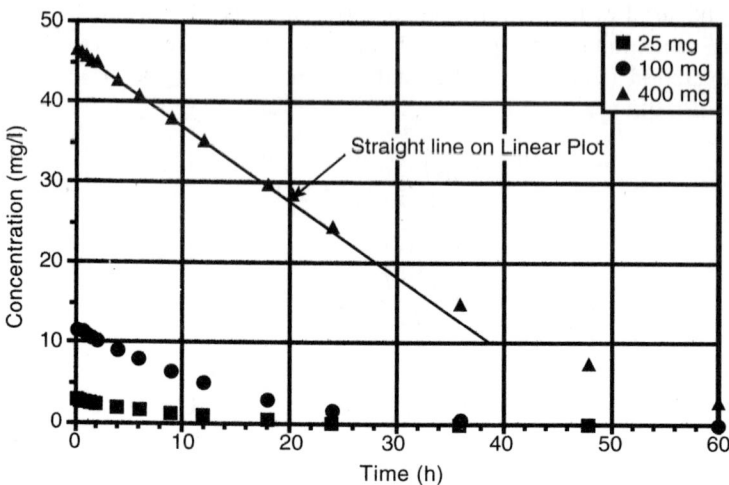

FIGURE 12.9 Linear plot of C vs. time (MM elimination).

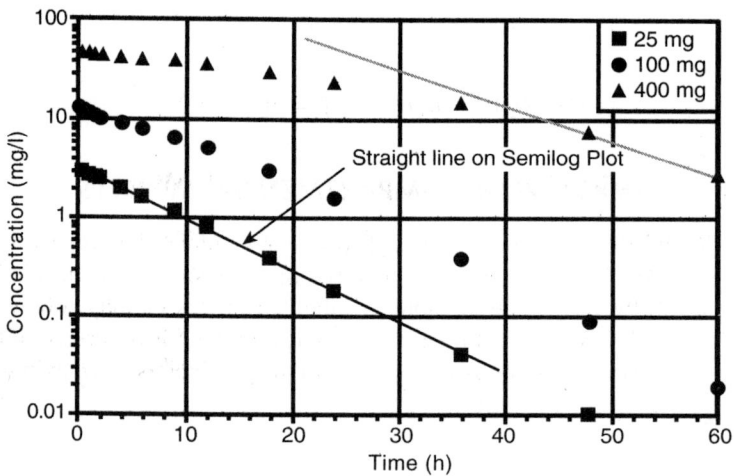

FIGURE 12.10 Semilog plot of C vs. time (MM elimination).

allows these calculations to be made with good data or accurate estimates of the parameter values.

Another example of nonlinear elimination is the case of saturable protein binding. For some drugs, present at higher concentrations, the available binding sites on plasma protein may be saturated. Thus, at high concentrations the fraction free may be higher than at lower concentrations. This can lead to faster elimination at higher concentrations when renal elimination by filtration is a major route of elimination. This is shown in Figure 12.11 with more rapid elimination at higher concentrations and slower, linear elimination at lower concentrations. This phenomenon can be analyzed using compartmental models

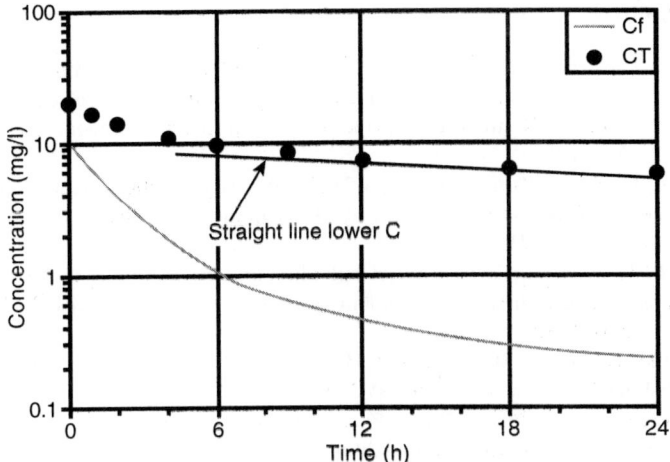

FIGURE 12.11 Semilog plot of C vs. time (saturable binding).

by modifying the elimination process as shown in Equation 12.21, an example with one saturable binding site.

$$\frac{dC_T}{dt} = -k_{el} \times C_f \qquad (12.21)$$

where

$$C_f = \frac{(K_a \times C_T - 1 - K_a \times P) + \sqrt{(K_a \times C_T - 1 - K_a \times P)^2 + 4K_a \times C_T}}{2 \times K_a}$$

In Equation 12.21, C_f is the free drug concentration, K_a is the drug–protein association constant, C_T is the total drug concentration, and P is the total protein concentration.

CONCLUSION

Although compartmental analysis of pharmacokinetic data is empirically based, it can be a powerful method for determining an appropriate dosage regimen, for data analysis and for the investigation of drug disposition processes. Compartmental analysis provides a middle ground between noncompartmental analysis and physiologically based pharmacokinetic analysis. Although the analyses are more involved than noncompartmental analysis, the compartmental models developed can be useful in extrapolating dosage regimen calculations and in the investigation of mechanisms of drug absorption and disposition. In Chapters 13 through 16 these methods will be extended to hybrid models and the modeling of population data.

REFERENCES

1. Benet, L. Z., General treatment of linear mammillary models with elimination from any compartment as used in pharmacokinetics, *J. Pharm. Sci.*, 61:536–541, 1972.
2. Benet, L. Z. and Ronfeld, R.A., Volume terms in pharmacokinetics, *J. Pharm. Sci.*, 58:639–641, 1969.
3. Benet, L. Z. and Turi, J. S., Use of general partial fraction theorem for obtaining inverse Laplace transforms in pharmacokinetic analysis, *J. Pharm. Sci.*, 60:1593–1594, 1971.
4. Bevill, R. F. et al., Disposition of sulfadimethoxine in swine: inclusion of protein binding factors in a pharmacokinetic model, *J. Pharmacokinet. Biopharm.*, 10:539–550, 1982.
5. Bourne, D. W. A., *Mathematical Modeling of Pharmacokinetic Data*, Technomic Publishing Company, Lancaster, PA, 1995.
6. Bourne, D. W. A., Pharmacokinetic Textbooks, available at http://www.boomer.org/pkin/book.html, 2001.
7. Bourne, D. W. A., Triggs, E. J., and Eadie, M. J., *Pharmacokinetics for the Non-Mathematical*, MTP Press, Lancaster, U.K., 1986.
8. Boxenbaum, H., Pharmacokinetics — philosophy of modeling, *Drug. Metab. Rev.*, 24(1): 89–120, 1992.
9. Brown, R. F., Identifiability, its role in design of pharmacokinetic experiments, *IEEE Trans. Biomed. Eng.*, 29(1):49–54, 1982.
10. D'Argenio, D. Z., Optimal sampling times for pharmacokinetic experiments, *J. Pharmacokinet. Biopharm.*, 9(6):739–756, 1981.
11. Dunne, A., Computers: pharmacokinetic parameter estimation — some pitfalls, *Trends Pharmacol. Sci.*, 7(1):11, 1986.
12. Gabrielsson, J. and Weiner, D., *Pharmacokinetic and Pharmacodynamic Data Analysis, Concepts and Applications*, 2nd ed., Swedish Pharmaceutical Press, Stockholm, Sweden, 1998.
13. Gibaldi, M. and Perrier, D., *Pharmacokinetics*, 2nd ed., Marcel Dekker, New York, 1982.
14. Jusko, W. J., Lewis, G. P., and Dittert, L. W., Integral coefficients of multicompartment pharmacokinetic models, *Chemotherapy*, 17:109–120, 1972.
15. Loo, J. C. K. and Riegelman, S., New method for calculating the intrinsic absorption rate of drugs, *J. Pharm. Sci.*, 57:918–927, 1968.
16. Ludden, T. M., Beal, S. L., and Sheiner, L. B., Comparison of the Akaike information criterion, the Schwarz criterion and the F test as guides to model selection, *J. Pharmacokinet. Biopharm.*, 22(5):431–445, 1994.
17. Mayersohn, M. and Gibaldi, M., Mathematical methods in pharmacokinetics — use of the Laplace transform for solving differential rate equations, *Am. J. Pharm. Ed.*, 34(4):608–614, 1970.
18. Mayersohn, M. and Gibaldi, M., Mathematical methods in pharmacokinetics. II — solution of the two compartment open model, *Am. J. Pharm. Ed.*, 35:19–28, 1971.
19. Shargel, L. and Yu, A. B. C., *Applied Biopharmaceutics and Pharmacokinetics*, 4th ed., Appleton & Lange, Stamford, CT, 1999.
20. Sheiner, L. B., Stanski, D. R., Vozeh, S., Miller, R. D., and Ham, J., Simultaneous modeling of pharmacokinetics and pharmacodynamics: application to D-tubocurarine, *Clin. Pharmacol. Ther,*, 25(3): 358–371, 1979.
21. Wagner, J. G., *Pharmacokinetics for the Pharmaceutical Scientist*, Technomic Publishing Company, Lancaster, PA, 1993.

22. Wagner, J. G., Do you need a pharmacokinetic model, and, if so which one? *J. Pharmacokinet. Biopharm.*, 3(6):457–478, 1975.
23. Wagner, J. G. and Nelson, E. Kinetic analysis of blood levels and urinary excretion in the absorptive phase after single doses of drug, *J. Pharm. Sci.*, 53:1392–1403, 1964.
24. Welling, P. G., *Pharmacokinetics, Processes and Mathematics*, American Chemical Society, Washington, D.C., 1986.

13 Noncompartmental Modeling: An Overview

Ronald A. Herman

CONTENTS

INTRODUCTION

Large amounts of research data are often collected, and it is desirable to summarize this information in a meaningful way to explain or understand these data. Mathematical equations are fit to concentration vs. time and/or effect data in an attempt to impart some order to the observations. This modeling is used to describe past behavior. However, an additional very important benefit of modeling data is for prediction of future behavior of the human–drug system interaction. Information from the current data can be extrapolated to other situations by simulating the processes that have been modeled.

1-56676-973-6/02/$0.00+$1.50
© 2002 by CRC Press LLC

APPROACHES TO MODELING
PHARMACOKINETIC DATA

There are three basic approaches to modeling pharmacokinetic data: traditional compartmental models[1,2] (or classical modeling described in Chapters 1 and 12), noncompartmental models[3-7] (examined in this chapter), and physiologically based models.[8] To understand when to use noncompartmental modeling, it is necessary to know the assumptions of each approach and the desired objective of the pharmacokinetic study.

TRADITIONAL COMPARTMENTAL MODELS

Compartments are chosen to represent the body based partially on an empirical or a physiological basis. The number of compartments is determined from the data. Usually a number of models are fit to the data to determine an optimum number of compartments. The route of administration also determines the structure. The model must specify transfer between compartments, including the direction of transfer and the order of transfer (first order, zero order, etc.). If every compartment is connected to a central compartment, then it is referred to as a mammalian model.[9] Structures can be kept very simple (one-compartment instantaneous input), or can be made very complex, such as multicompartment, dual input, or elimination from multiple compartments. A simple model is illustrated in Figure 13.1.

There are a number of assumptions with each approach. The assumptions, and their justifications, for classical pharmacokinetic modeling include:

- Barriers exist between compartments.
- Transfer occurs in a certain direction.
- Transfer occurs at a certain order (e.g., zero order or first order).
- Compartment characteristics
  - It is group of tissues.
  - The drug is homogeneously distributed (ln C vs. t has the same slope for each tissue), which does not mean that each tissue has the same concentration, but that the rate of elimination from each tissue is the same.
  - Drug is instantaneously distributed.
  - Therefore, the compartment is "well-stirred."
- Drug
  - Elimination is only from the central compartment.
  - There is no irreversible tissue binding.

NONCOMPARTMENTAL MODELING

This approach has been incorrectly described as structureless pharmacokinetics, or model-independent pharmacokinetics. These references to noncompartmental modeling really are misnomers. This approach is not "model-less"; instead, it uses a simpler, more general model. Another term used to describe this process is nonparametric pharmacokinetics because a structure with compartments and corresponding

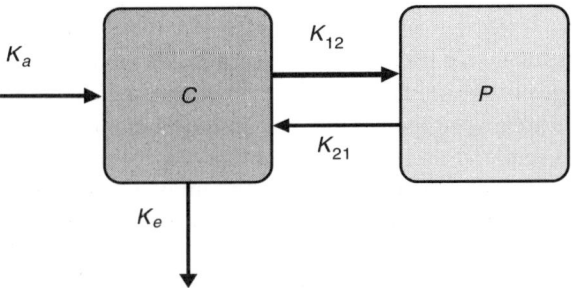

FIGURE 13.1 Classical pharmacokinetic modeling of a two-compartment open model. The symbols C and P represent the central and peripheral compartments, whereas K represents the first-order transfer constants between compartments. The subscripts, a and e, represent absorption and elimination, whereas 12 and 21 represent transfer of drug between compartments 1 and 2 (central to peripheral) or 2 and 1 (peripheral to central).

parameters is not modeled, but instead the response is modeled. The process is based upon a linear systems approach to model the data and this is discussed in more detail in Chapter 16. The simplified model, illustrated in Figure 13.2, assumes the drug is input to a sampling compartment. Some drug irreversibly leaves the sampling compartment and leaves the system (body), whereas additional drug leaves the sampling compartment and goes to the peripheral compartment where it irreversibly leaves the body. For most drugs this is negligible. However, for inhalation anesthetics and drugs with high biliary excretion this is not the case. Some of the drug leaves the sampling compartment and later returns to the sampling compartment after passing through the peripheral system.

The drug is distributed through stochastic (random) processes — convection (drug carried through the blood) and diffusion (through various membranes and tissues). There are two main assumptions inherent to this approach.

Linear Kinetic System

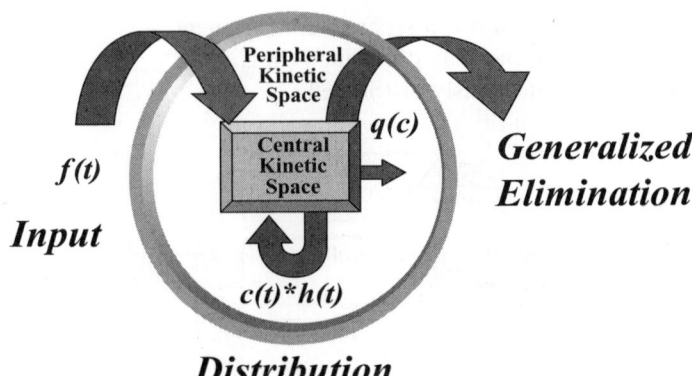

FIGURE 13.2 Noncompartmental modeling.

Superposition

This assumption states that if one input is given followed by a certain response and another input is given and another response is obtained, then under the circumstances when both inputs are given together, the response becomes the sum of the individual responses. For example, if an IV dose and an oral dose were given and the response to each was known, then when both are given simultaneously, the response would be the sum of the two separate responses. This is the principle that is used to determine the response following multiple doses.

Time Invariance

This assumption is that if a certain dose is given, and a certain response obtained, then if that dose is given at any other time, the same response should be observed. For most drugs this is true, but there are some drugs that have time-dependent pharmacokinetics and do not obey this assumption. In these situations, a dose given in the morning may not produce the same result as when it is given at night. Also it is possible that the elimination rate will change if hepatic elimination becomes saturated. Time invariance assumes that a dose given now should have the same response any time later.

PHYSIOLOGICALLY BASED MODELS

For these models, compartments are chosen to represent physiological compartments of the body. Figure 13.3 is an oversimplification of this approach. All physiological compartments should be included in the model. For convenience, just three physiological compartments are illustrated, but there are many more: gastrointestinal tract, lungs, central nervous system, muscles, bone, etc. Many of the same assumptions for the compartments of the traditional models apply here as well. In addition, blood flow must be known or estimated through each compartment. The amount of drug entering and leaving the compartment should be determined.

- Each organ system forms a separate compartment.
  - Drug is homogeneously distributed (ln C vs. t).
  - Drug is instantaneously distributed.

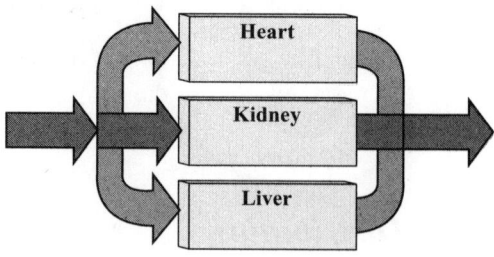

FIGURE 13.3 Simplified physiologically based modeling.

- Therefore, the compartment is "well-stirred" and the concentration of the drug leaving an organ is in equilibrium with the concentration of the drug in the organ. Thus, an equilibrium constant or partition coefficient can be determined from the concentration of the drug in the tissue compared to the concentration in the blood.
- Barriers between compartments (physiological systems):
  - Transfer is dependent on the rate of blood flow through the physiological compartment.
  - Each compartment has a characteristic clearance rate.
  - Therefore, it is necessary to know or determine the rate of blood flow in each organ or tissue system in the model.
- Drug
  - Elimination is only from certain compartments that are specified in the model, for example, the liver and kidneys.
  - There is no irreversible binding of the drug to the tissue.

CHOICE OF MODEL

The choice of which approach to use to model pharmacokinetic data depends primarily on what the study objectives were. However, knowing the assumptions of each modeling approach, the answer to the following questions could also influence the selection of approach.

1. What was done experimentally?
2. What data have been obtained?
3. Can the assumptions of the model be validated from these data?

STUDY OBJECTIVES

There are four main study objectives for analysis of pharmacokinetic data: (1) to summarize the kinetics of the drug (descriptive pharmacokinetics), (2) to quantify one or more kinetic processes of the drug, (3) to explain the pharmacokinetics, or (4) to make pharmacokinetic predictions. The objective of the study will determine which of the three modeling approaches to use. The advantages and disadvantages of each approach are discussed to see how they would fit with these objectives.

TRADITIONAL COMPARTMENTAL MODELS

Generally, these models are simple to deal with conceptually, mechanistically, and mathematically. Equations describing very complex models can be derived in a routine manner using simple "cookbook procedures."[10] One of the difficulties with this approach is to assign structural relevance to the model and the resulting parameters of the model. There is little physiological or anatomical relevance of the compartments. The parameters determined mathematically account for the net effect of the drug distribution, but do little to address the specific structure of the kinetic process. In addition, many of the assumptions are very difficult to verify.[11] This approach can be used to develop descriptive pharmacokinetics for a drug. However, it is not meaningful to summarize in terms of structure-specific parameters that do not have physiological or anatomical significance. Parameters that are not structure specific (T_{max}, C_{max}, AUC, etc.) can be calculated from the fitted model-specific equations. In spite of that, such parameters are more readily obtained by the use of suitable empiric equations in a noncompartmental approach.

A study objective to quantify a pharmacokinetic process or to explain the pharmacokinetics can be accomplished with this approach. However, because the relevant assumptions or the lack of a relevant structure are difficult to verify, these objectives become impractical with this approach. For example, when trying to elucidate hepatic clearance, the classical modeling approach would not be the best choice.

Pharmacokinetic predictions can be easily and more accurately done using the model-independent approach when the structure of the model is not important. However, often a classical modeling approach can be used when model limitations (verifying assumptions, etc.) are met. For example, for gentamicin and vancomycin dosing, a peak and a trough are often measured. For simplicity and clinical practicality, it is assumed that over that sampling interval the drug is eliminated by one-compartment pharmacokinetic behavior. The elimination rate and volume of distribution are calculated for those two levels and are used to make dosing adjustments.[12] However, if adjusting the dose of a cancer chemotherapeutic agent could be accomplished with the use of the area under the curve, then using a more general model-independent approach with fewer verifiable assumptions would be more appropriate.[13]

NONCOMPARTMENTAL MODELING

Utilization of this simplified modeling technique allows problems to be approached in a more general way. This means there are fewer assumptions that usually can be verified easily with a well-designed study. It is also possible to focus on a specific kinetic process (e.g., absorption, distribution, or elimination). When a problem can be approached in a more general way, then stronger predictions about future events can be made. Because this approach attempts to model the response rather than the structure of the process, it does little to explain why the drug exhibits a certain kinetic profile. The mathematical approaches used are slightly different and consequently there are no "cookbook" solutions to problems.[9,14–16] The mathematical and computational approaches are not contained in concise packaged software programs that address all the potential concerns. The total effects of individual kinetic processes are lumped together in functions, such as the unit impulse or characteristic response, the distribution function, and the elimination function. As a result, the individual kinetic processes are not described.

If the study objective is to summarize kinetics, quantify a pharmacokinetic process, or make pharmacokinetic predictions, this approach can be useful because of the more general nature of the approach and the fewer, more readily verifiable assumptions. If the study objective is to explain the pharmacokinetics, then this approach is not very useful. The lumping together of the various kinetic processes does not make it convenient to understand the overall mechanism of the kinetics of the compound studied. Compared to the more structured classical modeling approaches, the degree of complexity and structural differentiation in this approach tends to be a more realistic balance with the quality and quantity of the data typically obtained in pharmacokinetic studies.

PHYSIOLOGICALLY BASED MODELS

Physiological models overcome some of the limitations of the traditional compartmental models.[8] Because blood flow, distribution, and elimination from individual organs and tissue groups are included, the model is more realistic and meaningful. Consequently, the theoretical basis for quantifying and predicting the pharmacokinetic effects is improved. The compartments as well as the model parameters that are determined, such as blood flow, elimination rate, and partitioning coefficients, have a physiological or anatomical meaning. However, it is very difficult to determine many of the physiological or anatomical parameters. The models tend to be overly ambitious or complex and rarely are sufficient data collected. More than urine and blood data is required. It is necessary to know tissue concentrations and it is necessary to know organ blood flow rates.

Because of the complexity of this approach, many parameters and assumptions cannot be verified. As a result, if the study objective is to summarize kinetics, quantify a pharmacokinetic process, or determine pharmacokinetic predictions, this approach is not very useful. However, it is the most useful approach to explain the pharmacokinetics of a compound. Typically, massive sample collection is required and many validation experiments need to be done. As a result, its main application is in the drug discovery process to identify the kinetics and action of a new compound.

The choice of modeling should be dictated primarily by the objective of the study. Sometimes what was actually determined experimentally, as well as what data have been obtained, can influence the choice of modeling approach. Also, if certain assumptions of the model cannot be validated, this may influence the choice of modeling approach. It is important to plan studies appropriately so that the correct data are collected and the appropriate assumptions can be verified.

NONCOMPARTMENTAL MODELING ANALYSIS

This approach is frequently utilized for descriptive pharmacokinetics, to quantify a certain pharmacokinetic process, or to make predictions. The more general nature of this approach and the fewer, more readily verifiable assumptions make it very useful. Drug distribution generally occurs by two stochastic (random) processes:

1. Convection — as the drug is carried through the blood or other fluid with the current
2. Diffusion — as the drug moves through various tissues and membranes[17]

Drug distribution would be nonstochastic if competitive interaction with endogenous substances occurs, or if there is physiological feedback, or if there are biochemical changes. The two assumptions that must be verified for this approach are superposition and time invariance.

Input   **Response**

Superposition

	Input		Response
If:	$f_1(t)$	$\longrightarrow$	$c_1(t)$
And:	$f_2(t)$	$\longrightarrow$	$c_2(t)$
Then:	$f_1(t) + f_2(t)$	$\longrightarrow$	$c_1(t) + c_2(t)$
In general:	$\Sigma_i f_i(t)$	$\longrightarrow$	$\Sigma_i c_i(t)$

Time Invariance

If:	$f(t)$	$\longrightarrow$	$c(t)$
Then:	$f(t - T)$	$\longrightarrow$	$c(t - T)$

An experiment can be designed to verify the assumption of superposition by giving each subject at least two different doses. The time invariance assumption is validated by giving the same dose at two different times, for example, now and again in the evening, assuming preliminary studies indicate there is a sufficient washout of the drug in that time frame. This would help to verify diurnal changes pertaining to the rate of elimination. The same dose could also be given now and again after a significant washout period. Phase I study results should give an approximate idea of elimination half-life so that a washout of at least five half-lives can be designed into the study.

This noncompartmental approach seeks to model the response, not a model structure, so a curve-fitting program is used to fit the data to an appropriate response model. The simplified structure in Figure 13.2 has input occurring into a sampling compartment, with all other activity lumped together in the peripheral system. There is no attempt to describe or ascribe what is happening to the drug outside the sampling compartment. The model for this is shown in the following equation:

$$c(t) = f(t) * c_\delta(t) \tag{13.1}$$

where $c(t)$ is the concentration response function, which is the result of a mathematical convolution of two functions, an input function $f(t)$ and a function that is characteristic of the sampling compartment, $c_\delta(t)$. This latter function is known as the characteristic response or the unit-impulse response. The input function $f(t)$ is not the rate of the release of the drug, but is a composite of at least two processes:

1. Rate of release of the drug
2. The extent of degradation and biotransformation before the drug reaches the sampling compartment

Chapter 16, detailing linear systems analysis, demonstrates the potential for utilizing the characteristic response and the input function to analyze thoroughly the distribution characteristics of the drug.

Pharmacokinetic data that have been collected can be analyzed by fitting the data to a sum of exponentials. A curve-fitting program is used to find the optimal number of exponentials, and the response will be of the form:

$$C(t) = \sum_{i=1}^{n} C_i e^{-\lambda_i t} \tag{13.2}$$

The pre-exponential terms, C_i, and the exponential terms, λ_i, are used in the noncompartmental analysis to calculate a number of descriptive pharmacokinetic terms that describe the disposition of the drug. Each exponential term has a half-life associated with it.

$$t_{1/2} = \frac{\ln 2}{\lambda_i} \tag{13.3}$$

When the drug response is polyexponential, with elimination only from the central compartment, it is possible to calculate area under the curve (AUC) and the area under the moment curve (AUMC) as follows[18]:

$$AUC = \int_{0}^{\infty} c(t)dt = \sum_{i=1}^{n} C_i/\lambda_i \tag{13.4}$$

$$AUMC = \int_{0}^{\infty} tc(t)dt = \sum_{i=1}^{n} C_i/\lambda_i^2 \tag{13.5}$$

And the initial concentration (C_o) is

$$C_o = \sum_{i=1}^{n} C_i \tag{13.6}$$

These terms are used to compute the model-independent pharmacokinetic descriptors that follow. Total body clearance, Cl, which is the rate of drug transfer out of the body, is defined as:

$$Cl = \frac{F \cdot D}{AUC} \tag{13.7}$$

In addition, several peripheral bioavailability terms can be calculated to describe the extent to which the drug is distributed out of the sampling compartment.

PERIPHERAL BIOAVAILABILITIES

V_c	volume of distribution of the sampling compartment
V_{ss}	volume of distribution at steady state
F_{AUC} or $RTPC_{t/c}$	AUC peripheral bioavailability is the ratio of the amount of drug at steady state in the peripheral system to the amount in the sampling compartment; this is also referred to as the residence time partition coefficient

$$V_c = \frac{FD}{C_o} \tag{13.8}$$

$$V_{ss} = \frac{FD \cdot AUMC}{AUC^2} \tag{13.9}$$

$$RTPC_{t/c} = F_{AUC} = \frac{V_{ss}}{V_c} - 1 = \frac{\bar{t}_p}{\bar{t}_s} \tag{13.10}$$

With the noncompartmental model, there are two kinetic spaces of interest, the sampling or central kinetic space and the peripheral kinetic space. The average length of time that all drug molecules spend in a specific kinetic space is the mean residence time of that kinetic space.

MEAN TIME PARAMETERS

$\bar{t}_b (MRT_b)$	mean residence time for the drug in the body — the average total length of time that all molecules spend in the body
$\bar{t}_s (MRT_s)$	mean residence time for the drug in the sampling compartment — the average total length of time that all molecules spend in the sampling compartment before being eliminated from the body
$\bar{t}_p (MRT_p)$	mean residence time for the drug in the peripheral system — the average total length of time that all molecules spend in the peripheral kinetic space

The mean times of these kinetic spaces can be calculated as follows when the drug is input into the central kinetic space and the drug leaving the system (body) from the peripheral kinetic space is negligible.

$$MRT_b = \bar{t}_b = \frac{AUMC}{AUC} = \frac{\displaystyle\sum_{i=1}^{n} C_i/\lambda_i^2}{\displaystyle\sum_{i=1}^{n} C_i/\lambda_i} \tag{13.11}$$

$$\text{MRT}_s = \bar{t}_s = \frac{\text{AUC}}{C_o} \qquad (13.12)$$

$$\text{MRT}_p = \bar{t}_p = \bar{t}_b - \bar{t}_s \qquad (13.13)$$

Please note that the clearance term and the volume of distribution terms include the dose. If the drug is given by an extravascular route and no intravenous dose is given, the bioavailable fraction of the dose, F, will not be known. Consequently, the resulting calculations from Equations 13.7 through 13.9 determine Cl/F, V_c/F, and V_{ss}/F, for which F is not known. Half-lives, mean time parameters, and the residence time partition coefficient are bioavailability independent.

Some pharmacokinetic software packages perform noncompartmental analysis without fitting the entire response curve. These programs compute the elimination rate constant (λ) for the terminal elimination phase of the data, and then use a trapezoidal rule with this elimination rate constant to compute AUC and AUMC. With these terms, the total body clearance, the steady-state volume of distribution, and the mean residence time in the body can be calculated. Without C_o, it is not possible to calculate the volume of distribution of the central compartment or the mean residence time of the sampling compartment. The latter term is therefore critical in accurately determining these parameter values and depends on an unbiased and close fit of the data to Equations 13.2 and 13.6.

The descriptive pharmacokinetic terms illustrated above only begin to touch the surface of the analysis that can be done. When the rate of change of the response function is evaluated, the derivative of $C(t)$, then other clearance terms, peripheral bioavailabilities, and mean time parameters can be computed. These will be examined in Chapter 16 describing linear systems analysis. In addition, it is possible to use intravenous results and extravascular results to deconvolve the input function from the characteristic response function to assess various mean time parameters that examine the arrival of the drug into the sampling compartment.

REFERENCES

1. Godfrey, K., *Compartmental Models and Their Application*, Academic Press, New York, 1983.
2. Jacquez, J., *Compartmental Analysis in Biology and Medicine*, Elsevier, Amsterdam, 1972.
3. Benet, L. Z. and Galeazzi, R. L., Noncompartmental determination of the steady-state volume of distribution, *J. Pharm. Sci.*, 68(8):1071–1074, 1979.
4. Riegelman, S. and Collier, P., The application of statistical moment theory to the evaluation of *in vivo* dissolution time and absorption time, *J. Pharmacokinet. Biopharm.*, 8(5):509–534, 1980.
5. Veng-Pedersen, P., System approaches in pharmacokinetics: II. Applications [Review], *J. Clin. Pharmacol.*, 28(2):97–104, 1988.
6. Veng-Pedersen, P., System approaches in pharmacokinetics: I. Basic concepts [Review], *J. Clin. Pharmacol.*, 28(1):1–5, 1988.

7. Veng-Pedersen, P., Stochastic interpretation of linear pharmacokinetics: a linear system analysis approach, *J. Pharm. Sci.*, 80(7):621–631, 1991.

8. Gerlowski, L. E. and Jain, R. K., Physiologically based pharmacokinetic modeling: principles and applications, *J. Pharm. Sci.*, 72(10):1103–1127, 1983.

9. Veng-Pedersen, P., General treatment of linear pharmacokinetics, *J. Pharm. Sci.*, 67(2):187–191, 1978.

10. Gibaldi, M. and Perrier, D., *Pharmacokinetics*, 2nd ed., Marcel Dekker, New York, 1982.

11. Wagner, J. G., *Pharmacokinetics for the Pharmaceutical Scientist*, Technomic Publishing Company, Lancaster, PA, 1993.

12. Sawchuk, R. J. and Zaske, D. E., Pharmacokinetics of dosing regimens which utilize multiple intravenous infusions: gentamicin in burn patients, *J. Pharmacokinet. Biopharm.*, 4(2):183–195, 1976.

13. Grevel, J., Area-under-the-curve versus trough level monitoring of cyclosporine concentration: critical assessment of dosage adjustment practices and measurement of clinical outcome, *Ther. Drug Monit.*, 15(6):488–491, 1993.

14. Gillespie, W. R., Simple methods for estimation of mean residence time and steady-state volume of distribution from continuous-infusion data, *Pharm. Res.*, 8(2):254–258, 1991.

15. Veng-Pedersen, P., Theorems and implications of a model independent elimination/distribution function decomposition of linear and some nonlinear drug dispositions. I. Derivations and theoretical analysis, *J. Pharmacokinet. Biopharm.*, 12(6):627–648, 1984.

16. Veng-Pedersen, P., Linear and nonlinear system approaches in pharmacokinetics: how much do they have to offer? I. General considerations, *J. Pharmacokinet. Biopharm.*, 16(4):413–472, 1988.

17. Cutler, D. J., Linear systems analysis in pharmacokinetics, *J. Pharmacokinet. Biopharm.*, 6:265–282, 1978.

18. Yamaoka, K., Nakagawa, T., and Uno, T., Statistical moments in pharmacokinetics, *J. Pharmacokinet. Biopharm.*, 6(6):547–558, 1978.

14 Physiologically Based Pharmacokinetic Models

James M. Gallo

CONTENTS

INTRODUCTION

Pharmacokinetics is the study of drug transport, drug distribution, and drug elimination in biological systems. Its roots are in the area of chemical kinetics, which utilize differential equations to describe chemical reactions, typically on a timescale. As depicted in this section, Research Applications, of this text, there are a handful of data analysis and pharmacokinetic modeling approaches that might be applied to any given data set to generate characteristic pharmacokinetic parameters. Physiologically based pharmacokinetic (PB/PK) modeling might be considered a unique approach because it has a singular focus of investigating drug disposition in tissues.

The origins of PB/PK models may be traced to Teorell,[1,2] Bellman,[3] Bischoff and Brown,[4] and Dedrick and Bischoff.[5] The combined investigations of these individuals established the idea of characterizing drug distribution to specific organs in terms of relevant anatomical and physiological variables. In contrast to classical compartmental modeling, tissue compartments represent specific anatomical volumes connected by the blood circulation. Drug uptake into organs is a function of both thermodynamic (i.e., tissue-to-blood partition coefficients, drug protein binding) and membrane transport (i.e., mass transfer coefficient) properties, rather than simple first-order transfer rate constants as used in compartmental models. The PB/PK model parameters are cast into differential mass balance equations that are solved numerically to provide predicted drug concentrations for each organ and compartment.

Individual organs usually have either a blood flow–limited or membrane-limited model structure (Figure 14.1). How any organ is represented depends on the drug and animal that the model depicts. There are three distinct anatomical regions of any organ: vascular, interstitial, and intracellular spaces. The organ structure in a PB/PK model could vary from a single compartment to a three-subcompartment structure depending on the extent of lumping (see Figure 14.1). Any compartment, whether it is a subcompartment or the whole-tissue compartment, is assumed to be a well-mixed homogeneous space yielding a single drug concentration.

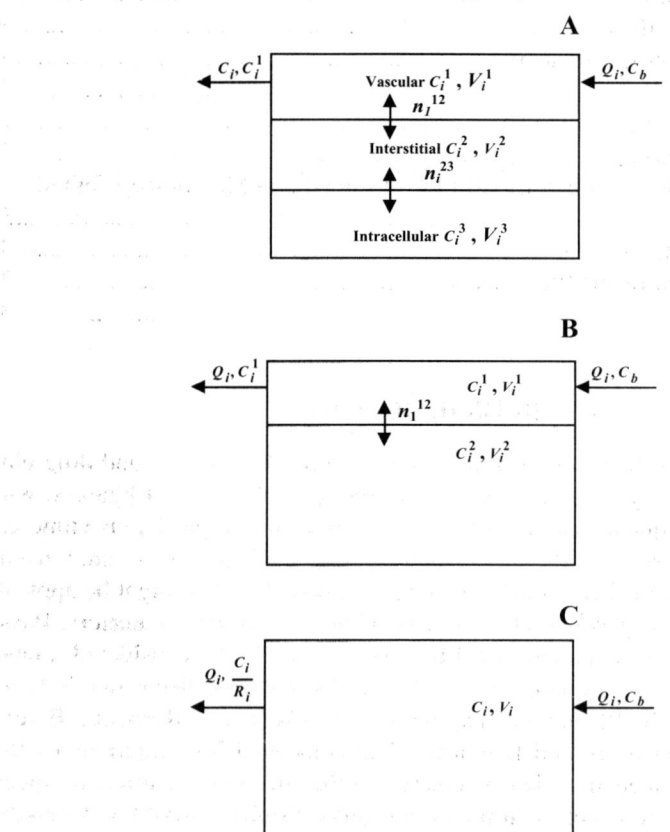

FIGURE 14.1 Individual organ representations for a three-subcompartment (A), two-subcompartment (B), or typical membrane-linked and blood flow–limited (C) PB/PK model. See text for definition of symbols.

A single-compartment tissue structure is referred to as blood flow limited because it assumes blood flow is the rate-limiting step to drug uptake into the tissue. In essence, the vascular barrier presents little resistance to drug uptake and only the rate of blood flow influences tissue drug concentrations. Highly lipophilic drugs and organs of low blood flow would more likely contribute to blood flow–limited organ models.

A membrane-limited organ structure normally is comprised of two homogeneous subcompartments, either an extracellular–intracellular configuration or a vascular–extravascular configuration. In either case, drug transport is rate limited by membrane transport rather than by blood flow to the tissue. A vascular–extravascular arrangement is useful for representing blood–brain barrier transport, whereas extracellular–intracellular arrangements are typical for tissues other than brain. In an extracellular–intracellular organ representation, vascular and interstitial spaces form the extracellular compartment, and thus it is assumed that drug is instantaneously equilibrated between the vascular and interstitial spaces. Vascular–extravascular arrangements lump intracellular and interstitial spaces into a single homogeneous compartment based on the assumption there is no concentration gradient between these milieus. Three-subcompartment organ structures, depicting vascular, interstitial, and intracellular compartments, have also been presented in the PB/PK model format.[6] The level of organ complexity will depend on the number of discrete tissue spaces that can be identified along with the associated drug transport parameters. Division of an intracellular compartment into subcompartments representing different cell types, one possibly serving as a receptor binding site for the drug, is possible in the PB/PK modeling approach. Alternatively, the intracellular compartment could describe free and receptor-bound drug concentrations. Figure 14.2 illustrates some potential organ representations in PB/PK models.

Representation of the whole body by individual organs connected anatomically through the blood circulation is referred to as a "global" PB/PK model. A "hybrid" PB/PK model focuses on a single organ or small group of organs. The organ models are analogous to those found in a global model; yet, the blood concentration of drug entering the compartment is described by an exponential equation or forcing function. The forcing function is normally attained by fitting a polyexponential equation to observed plasma drug concentration–time data, justifying the descriptor *hybrid*. Advantages of hybrid models are that fewer model parameters have to be estimated, and comprehensive tissue drug concentration–time data, somewhat requisite in the global approach, are unnecessary, yet the organ of interest maintains a physiological representation.

Objectives for development of a PB/PK model can be varied. Historically, the end point of model development is subjective or graphical agreement between observed and predicted concentrations, although statistical evaluation of the model is becoming standard. Acceptance of the model implicitly characterizes drug dispositions as linear or nonlinear with respect to clearance, membrane transport, and protein binding. Although this end point is valuable, the models also provide a springboard to enable PB/PK models to reach their full potential as a predictive tool. Model predictions can be conducted under numerous conditions with a driving force to answer "what if" questions (e.g., "What are target tissue drug concentrations if

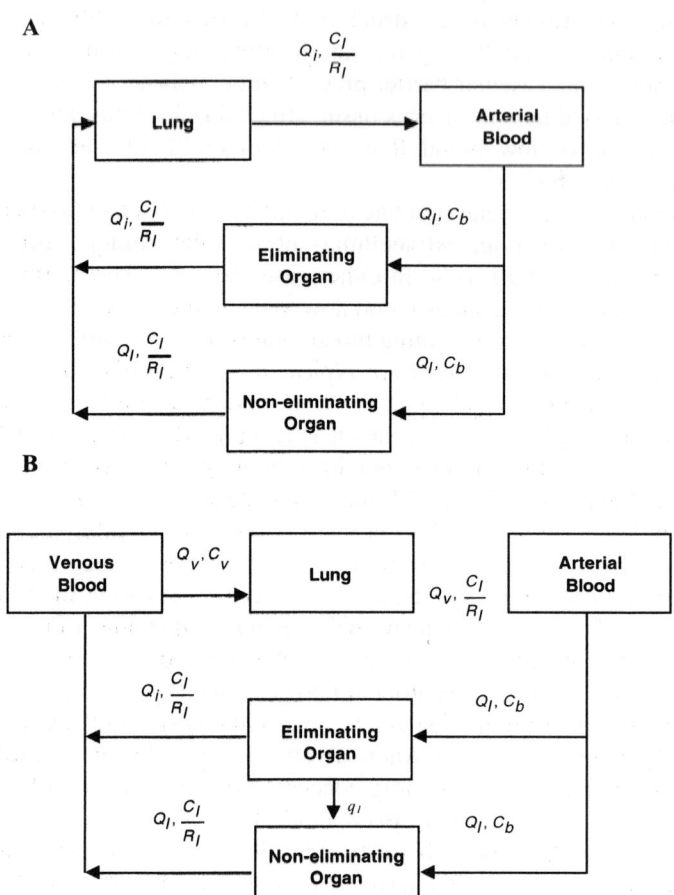

FIGURE 14.2 Possible blood circulation connections in a PB/PK model: (A) venous return incorporated into lung mass balance quation; (B) separate venous blood compartment. See text for definition of symbols.

blood flow or protein binding is altered?"). This type of probing question can be useful in both preclinical and clinical cases that could assist the drug development process. For example, quantitative structure–pharmacokinetics relationships based on PB/PK model parameters have been pursued,[7,8] and could expedite the drug design and discovery processes.

Another application of PB/PK models would be to assess novel drug delivery systems because of their ability to characterize drug disposition at the tissue or cell level. Since tissue uptake and cellular interactions of the drug delivery system will be crucial to successful drug targeting, a PB/PK model capable of characterizing carrier and drug transport can furnish feedback on the design of the delivery system, and can be used as a quantitative tool to evaluate the delivery system.

A hallmark of PB/PK models is the ability to scale up animal-based models to human, thus allowing tissue drug concentrations to be predicted in the absence of data that are difficult or impossible to collect. Initial efforts to apply interspecies extrapolations to anticancer drugs have been extended to chemical risk assessment based on PB/PK models.[9] Empirical allometric equations based on animal body weight have been the mainstay to scale organ weights and blood flow, as well as macropharmacokinetic variables such as clearance and volume of distribution.[10–12] Few have endeavored to study scale-up relationships for drug transport and thermo-dynamic parameters. This can be attributed to the uniqueness of these parameters for each drug and tissue, and also to the perception, possibly unfounded, that large quantities of data are required to develop scale-up relationships. Utilization of *in vitro* systems from different species to measure metabolic rates, partition coefficients, and membrane transport parameters could lead to scale-up relationships, and decrease the number of animal experiments. A shift from global to hybrid PB/PK models will reduce the magnitude of the scale-up effort, and may lead to new paradigms for extrapolating from animal to humans. Establishment of optimized *in vivo* study protocols across different animal species may also limit the number of animal studies necessary to develop allometric or scaling relationships. Extrapolations of PB/PK model to humans is an untapped benefit that should provide a basis for future research.

In summary, PB/PK models should be viewed as a powerful means to represent drug disposition based on mass transport principles, and should be considered as a modeling approach when the emphasis is on understanding the pharmacokinetic properties of the drug in tissues.

FEATURES OF PB/PK MODELS

- Mass balance approach to characterize drug disposition
- Utilize differential equations to describe model systems
- Rationale to consider PB/PK approach if focus is on understanding drug disposition in tissues
- Offer greatest potential to predict drug concentrations under different physiological and pharmacological conditions
- Can be scaled-up from animals to humans

MATHEMATICAL FORMULATION

Figure 14.1A illustrates a non-eliminating tissue compartment, i, divided into three anatomically relevant subcompartments, each homogeneous with respect to drug concentration. The corresponding differential mass balance equations are

$$V_i^1 \frac{d}{dt} C_i^1 = Q_i(C_b - C_i^1) - n_i^{12} \tag{14.1}$$

$$V_i^2 \frac{d}{dt} C_i^2 = n_i^{12} - n_i^{23} \tag{14.2}$$

$$V_i^3 \frac{d}{dt} C_i^3 = n_i^{23} \qquad (14.3)$$

where
C_i^1 = concentration (amount/volume) in compartment 1 of tissue i

C_i^2 = concentration (amount/volume) in compartment 2 of tissue i

C_i^3 = concentration (amount/volume) in compartment 3 of tissue i

C_b = blood concentration entering tissue i

n_i^{12} = net mass flux from compartment 1 to 2, amount/time

n_i^{23} = net mass flux from compartment 2 to 3, amount/time

V_i^1 = volume of compartment 1 in tissue i

V_i^2 = volume of compartment 2 in tissue i

V_i^3 = volume of compartment 3 in tissue i

Compartment 1 represents the vascular space of the tissue and can be identified as a blood, plasma, or serum space. Compartment 2 is the interstitial space and compartment 3 the intracellular tissue volume. Assuming first-order or linear membrane transport for the drug, the flux terms can be expanded as

$$n_i^{12} = h_i^{12}\left(C_i^1 - \frac{C_i^2}{R_i^{12}} \right) \qquad (14.4)$$

$$n_i^{23} = h_i^{23}\left(\frac{C_i^2}{R_i^{23}} - \frac{C_i^3}{R_i^{23}} \right) \qquad (14.5)$$

where h_i^{12} = mass transfer coefficient (volume/time) characterizing drug transport from compartment 1 to compartment 2; h_i^{23} = mass transfer coefficient (volume/time) characterizing drug transport from compartment 2 to compartment 3; $R_i^{12} = f_u^1 / f_u^2$; $R_i^{23} = f_u^2 / f_u^3$, and f_u = the unbound fraction of drug in compartment 1, 2, or 3.

The mass transfer coefficients may also be expressed in units of time^{-1} by multiplying by the appropriate compartmental volume term. Irreversible drug elimination from the tissue requires an expression added to the differential equation representing the subcompartment in which elimination occurs. For example, hepatic drug elimination would be described by a linear or nonlinear expression added to the intracellular liver compartment mass balance equation because this compartment represents the hepatocytes. Formal drug clearance terms are given below for certain models.

The three-compartment tissue model is ordinarily simplified by lumping either all three subcompartments, lumping subcompartment 1 and 2, or lumping subcompartments 2 and 3. These simplifications result in the blood flow–limited (i.e., lumping all three subcompartments) and the membrane-limited (i.e., lumping any two subcompartments) tissue models. Differential mass balance equations for a non-eliminating membrane-limited compartment are

$$V_i^1 \frac{d}{dt} C_i^1 = Q_i(C_b - C_i^1) - n_i^{12} \tag{14.6}$$

$$V_i^2 \frac{d}{dt} C_i^2 = n_i^{12} \tag{14.7}$$

where all terms have the same basic definitions as above. Combining the vascular and interstitial subcompartments defines subcompartment 1 as the extracellular compartment and subcompartment 2 as the intracellular compartment. Lumping interstitial and intracellular subcompartments, as might be done for a brain compartment, defines compartment 1 as the vascular space and compartment 2 as the extravascular compartment. Under conditions of linear membrane transport,

$$n_i^{12} = h_i^{12}\left(C_i^1 - \frac{C_i^2}{R_i}\right) \tag{14.8}$$

Nonlinear or saturable membrane transport is represented as

$$n_i^{12} = \frac{a_i C_i^1}{b_i + C_i^1} - \frac{a_i(C_i^2/R_i)}{b_i + (C_i^2/R_i)} \tag{14.9}$$

where a_i = maximum transport rate, amount/time, and b_i = drug concentration at one-half a_i, amount/volume.

Membrane transport of drugs by both linear and nonlinear mechanisms is represented by addition of Equations 14.8 and 14.9.

Drug elimination from a membrane-limited tissue compartment requires subtraction of the rate of elimination, q_i, from the appropriate mass balance equation, typically from subcompartment 2.

For linear or first-order drug elimination,

$$q_i = k_i V_i^2 C_i^2 \tag{14.10}$$

and for nonlinear elimination,

$$i = \frac{V_m(C_i^2/R_i)}{k_m + (C_i^2/R_i)} \tag{14.11}$$

where k_i = first-order elimination rate constant, time^{-1}
                     V_m = maximum elimination rate, amount/time
                     k_m = Michaelis constant, equal to the concentration at one-half V_m

Lumping compartments 1, 2, and 3 into a single homogeneous tissue compartment implies the blood flow–limited model. The tissue mass balance equation for a non-eliminating organ is

$$V_i \frac{d}{dt} C_i = Q_i \left(C_b - \frac{C_i}{R_i} \right) \tag{14.12}$$

where $R_i = C_i/C_o$, with C_o the effluent venous blood drug concentration.
    Drug elimination terms are now

$$q_i = k_i V_i C_i \tag{14.13}$$

$$q_i = \frac{V_m (C_i / R_i)}{k_m + (C_i / R_i)} \tag{14.14}$$

for linear and nonlinear elimination, respectively. Equation 14.13 is frequently represented as

$$q_i = \frac{\text{Cl}_i^1 C_i}{R_i} \tag{14.15}$$

where Cl_i^1 = total intrinsic organ clearance.
    A complete or global tissue distribution model consists of individual tissue compartments connected by the blood circulation. In any global model, individual tissues may be blood flow limited, membrane limited, or more complicated structures. Connection of the venous and arterial blood circulations can be done in a number of ways depending on whether separate venous and arterial blood compartments are used, or if right and left heart compartments are separated. The two most common methods are illustrated in Figure 14.2 for blood flow–limited tissue compartments. The associated mass balance equations for Figure 14.2A are,

$$V_i \frac{d}{dt} C = \sum_{i=1}^{n} \frac{Q_i C_i}{R_i} - \frac{Q_1 C_1}{R_1} \tag{14.16}$$

$$V_b \frac{d}{dt} C_b = \frac{Q_1 C_1}{R_1} - Q_b C_b \tag{14.17}$$

where 1 and b subscripts refer to lung and arterial blood compartments, and n equals the number of compartments providing venous return. For Figure 14.2B the equations are

$$V_c \frac{d}{dt} C_v = \sum_{i=1}^{n} \frac{Q_i C_i}{R_i} - Q_v C_v \tag{14.18}$$

$$V_1 \frac{d}{dt} C_1 = Q_v \left(C_v - \frac{C_1}{R_1} \right) \tag{14.19}$$

$$V_b \frac{d}{dt} C_b = \frac{Q_v C_1}{R_1} - Q_b C_b \tag{14.20}$$

where the subscript v indicates the venous blood compartment. Whatever representation is used for the venous-arterial system, balance of blood flow has to be maintained, thus above

$$\sum_{i=1}^{n} Q_i = Q_1 \tag{14.21}$$

$$\sum_{i=1}^{n} Q_i = Q_v = Q_b \tag{14.22}$$

for Figures 14.2A and 14.2B, respectively.

Membrane transport of drugs by passive diffusion, although a function of free or unbound drug concentrations, is most often expressed in terms of total drug concentration, as presented in Equations 14.8 and 14.9. Casting these equations in terms of total drug concentrations requires reparameterization of the flux expressions using unbound drug concentrations.[13] For example, Equation 14.8 is arrived at by

$$n_i^{12f} = h_i^{12f} (C_i^{1f} - C_1^{2f}) \tag{14.23}$$

$$C_i^{1f} = C_i^1 / R_i^1 \tag{14.24}$$

$$C_i^{2f} = C_i^2 / R_i^2 \tag{14.25}$$

where superscript f refers to the free drug, $R_i^1 = 1/f_u^1$ and $R_i^2 = 1/f_u^2$. Substituting Equations 14.24 and 14.25 into 14.23 gives

$$n_i^{12f} = h_i^{12f} (C_i^1 / R_i^1 - C_i^2 / R_i^2) \tag{14.26}$$

Setting $h_i = h_i^{12f} / R_i^1$ and $R_i^{12} = R_i^2 / R_i^1$ permits Equation 14.26 to be written as

$$n_i^{12} = h_i (C_i^1 - C_i^2 / R_i^{12}) \tag{14.27}$$

which is equivalent to Equation 14.4.

Although drug fluxes and tissue concentrations are for practical purposes presented in terms of total drug concentrations, division into unbound and protein-bound concentrations is possible. This approach may be beneficial for a drug that undergoes nonlinear protein binding over the concentration range of interest or when pharmacodynamic issues are of concern. Problems encountered in representing free and bound drug mass balance equations have been discussed and center on the use of an effective protein fraction capable of binding drug in tissues.[14]

Input functions, i.e., $I(t)$, describing the rate the administered dose enters a compartment, may have various forms depending on the administration schedule. The input function $I(t)$, is added to the appropriate mass balance equation and can describe any drug administration pattern. First-order absorption may be expressed as

$$I(t) = (F \text{ Dose } k_a)e^{(-k_a t)} \tag{14.28}$$

where F = fraction of the administered dose entering compartment
          k_a = first-order absorption rate constant

Constant rates or zero-order infusions are

$$I(t) = k_0 u(t) \tag{14.29}$$

$$u(t) = \begin{pmatrix} 1, & \text{for } (t \leq T) \\ 0, & \text{for } (t > T) \end{pmatrix} \tag{14.30}$$

where k_0 = zero-order infusion rate constant, amount/time
          T = duration of infusion

Bolus administration is usually characterized by a normalized injection function that effectively inputs the dose in a bell-shaped curve manner. Since bolus administration is not instantaneous, this approach is considered more realistic. $I(t)$ for bolus administration is

$$I(t) = Mg(t) \tag{14.31}$$

where M = dose
          $g(t)$ = normalized injection function and is $30\lambda(\lambda t)^2(1 - \lambda t)^2$
          λ = reciprocal of the injection duration

System analysis techniques have been used to generate input functions for PB/PK models. Oral administration of carbon tetrachloride in different vehicles was successfully described by absorption input functions obtained by deconvolution and disposition decomposition methods.[15,16]

Comparisons of predicted and observed drug concentrations are ordinarily based on total drug concentrations in the tissue because most analytical methods make use of total tissue homogenates. Differentiation of cellular and extracellular drug concentrations is not possible with the destructive tissue sample homogenization procedures routinely employed. Measurements of subcompartment drug concentrations will become more realistic as noninvasive quantitation methods become refined and accessible. A volume-averaged total tissue drug concentration is calculated from predicted subcompartment drug concentrations to allow comparisons to observed total tissue drug concentrations. For a three-subcompartment organ structure as depicted in Figure 14.1, the predicted total tissue concentration C_t is

$$C_t = \frac{V_i^1 C_i^1 + V_i^2 C_i^2 + V_i^3 C_i^3}{V_t} \tag{14.32}$$

where V_t = total tissue volume. Total drug concentrations in a two-subcompartment structure are

$$C_t = \frac{V_i^1 C_i^1 + V_i^2 C_i^2}{V_t} \tag{14.33}$$

Tissue volumes may be expressed in mass or volume units. When volume units are used, a tissue density of 1 is normally assumed. Predicted concentrations from a blood flow–limited compartment may be compared directly to observed values.

Representation of a hybrid model structure employs features of both classical compartmental models and PB/PK models. The target organ or organs in a hybrid model can assume any PB/PK structure, such as depicted in Figure 14.1. The differential mass balance for the target organ or organ systems would be analogous to those presented above. The key difference in the hybrid approach is that the incoming blood drug concentration is characterized by an empirical function. This function is readily characterized by a polyexponential equation as

$$C = \sum_{i=1}^{n} A_i e^{-\lambda_i t} \tag{14.34}$$

where
C = blood or plasma drug concentration
A_i = y-axis intercepts
λ_i = disposition rate constants for ith compartment

The function describing the input drug concentration (i.e., Equation 14.34), C, as a function of time is substituted into the target organ mass balance equations for C_b, as in Equation 14.1, permitting the target organ equations to be solved analytically or numerically. The input function is referred to as a forcing function since predicted target organ drug concentrations will be highly dependent on the nature of the forcing function.

MODEL DEVELOPMENT

No doubt one of the most important and controversial issues connected to PB/PK modeling is parameter estimation. The often large number and diverse types of parameters certainly has contributed to the problem of parameter estimation. Another source of concern, and in part a philosophical one, is the question of whether parameter estimates should be based on *in vitro* or *in vivo* data. An idealistic, and perhaps historical, perspective would utilize *in vitro* methodologies (i.e., partition coefficients, protein-binding parameters, mass transfer coefficients) and literature or tabulated values (i.e., organ blood flows and volumes) to estimate model parameters. Thus, model predictions can be made in lieu of extensive animal experimentation. One would be left with the task of validating or verifying the model predictions. The alternate model development strategy, and the one currently used most often, relies on *in vivo* data. *In vivo* studies in which tissue drug concentrations are measured serve two purposes: parameter estimation and model validation. There has been no consensus on how *in vivo* studies should be conducted for these purposes. Detractors of this model development strategy would indicate that a single *in vivo* study in which tissue drug concentrations are measured serves not only to build the model but also to claim it as a validated model. A pitfall is that collection of drug concentration–time data at one dose level cannot possibly depict nonlinearities that might occur at different doses, doses at which model predictions could be made. It would be sounder to develop the model based on data collected at different doses, and then to use a unique data set to compare to the predicted concentrations. Of course, the limitations or difficulties with this type of approach are related to the rigor of collecting the large amount of tissue concentration–time data. It would seem that strategies that make use of both *in vitro* and *in vivo* investigations may prove advantageous in terms of both model development and model validation.

In summary, although improvements in model development and validation procedures would be beneficial, it is unlikely that a universal protocol will be adopted due to the diversity of PB/PK models. As with any modeling endeavor, the modeler should have a clear idea of which parameter estimation procedures will be employed so that experiments may be designed properly. Methods to estimate specific model parameters are given below.

ORGAN BLOOD FLOWS AND VOLUMES

Numerous compilations of organ blood flows and volumes in different animals are available.[17–20] Organ blood flows may be measured by radioactive test substances, microspheres, or from the organ clearance of highly extracted compounds.[21–23] For example, renal plasma flow can be equated to the clearance of para-aminohippurate. Quantitative autoradiography has been used to measure organ blood flow as well as capillary permeability using standard radiolabeled markers.[24] This method has been particularly useful to obtain regional measurements in brain and brain tumors.

Total tissue volumes are simply determined by weight, assuming a density of 1, and fractionated into subcompartment volumes depending on the model structure.[4] There are tables available to estimate blood and water volumes of organs to permit corrections of blood contamination and to assist in determining effective protein-binding fractions.[17,23,25] There have been sufficient independent investigations of organ blood flow and volumes to develop allometric relationships based on animal body weight.[26] Most PB/PK modeling investigators choose to use literature estimates for organ blood flows, in particular, and at times for tissue volumes. Total tissue volumes can be readily obtained from experiments in which drug concentrations are measured. One should realize that estimates for these parameters, particularly organ blood flows, vary greatly, and should not be overlooked as sources of error in the model.

PARTITION COEFFICIENTS

Estimation methods for tissue-to-blood partition coefficients (i.e., R_i) have been the most prolific, no doubt, due to the need for this parameter in most organ models. Both *in vitro* and *in vivo* parameter estimation techniques are available.

Lin et al.[27] derived an equation for blood flow–limited compartment partition coefficients based on *in vitro* equilibrium dialysis binding studies of a diluted tissue homogenate. Assuming Langmuir-type drug-protein binding, and equality of unbound plasma and tissue drug concentration, the partition coefficient, R_i, is

$$R_i = \frac{1 + \alpha_i}{\gamma(1 + \alpha_p)} \tag{14.35}$$

where $\quad \alpha_{i,p}$ = bound to free drug concentration ratio in tissue i or plasma

$\quad\quad\quad \lambda$ = ratio of whole blood to plasma drug concentration

The authors found that estimates of R obtained by their *in vitro* method agreed with estimates obtained by an *in vivo* technique for the drug ethoxybenzamide.

Chen and Gross[28] derived equations to calculate partition coefficients for blood flow–limited compartments from either constant-rate infusion (i.e., steady-state conditions) or from intravenous bolus regimens. For a non-eliminating organ under steady-state conditions,

$$R_i = \frac{C_i^{ss}}{C_p^{ss}} \tag{14.36}$$

and following an intravenous bolus dose,

$$R_i = \frac{Q_i C_i^0}{Q_i C_p^0 + \alpha V_i C_i^0} \tag{14.37}$$

where $\quad C_i^{ss}$ = steady-state drug concentration in tissue i

$\quad\quad\quad C_p^{ss}$ = steady-state drug concentration in plasma

C_i^0 = terminal phase y-axis intercept from tissue drug concentration–time plot

C_p^0 = initial plasma drug concentration at $t = 0$

α = terminal phase slope from tissue drug concentration–time plot

Gallo et al.[29] have developed the area method for calculation of partition coefficients for both blood flow–limited and membrane-limited compartments. Following intravenous bolus administration, the partition coefficient for a non-eliminating blood flow–limited compartment is

$$R_i = \frac{\text{AUC}_i}{\text{AUC}_p} \tag{14.38}$$

where AUC_i = total area under the tissue drug concentration–time curve

AUC_p = total area under the drug concentration–time curve

Under the same conditions for a membrane-limited compartment, R_i is

$$R_i = \frac{V_i(\text{AUC})_i}{V_i^2 \text{AUC}_p} - \frac{V_i^1}{V_i^2} \tag{14.39}$$

where V_i = total tissue volume

V_i^1, V_i^2 = subcompartment 1 or 2 volumes for tissue i

Both the Chen and Gross[28] and the Gallo et al.[29] methods have been applied to eliminating compartments. Both methods are derived from the specific mass balance equations for the given model structure. Monte Carlo investigations demonstrated that both methods provide reasonably accurate and precise estimates of partition coefficients from concentration–time data sets containing error, data one is likely to encounter from *in vivo* studies.

MASS TRANSFER COEFFICIENTS

Mass transfer coefficients (i.e., h_i) may be obtained from *in vitro* experiments utilizing various membrane permeability methodologies.[30] These include rapid mixing and separation, cell volume, and spectroscopic techniques. A general approach is to add the drug to a cell suspension, rapidly mix and separate the cells, and measure drug concentration in the suspending medium after different mixing times. This experiment is designed to measure the rate of influx into the cell, and if linear bidirectional membrane transport is assumed, the rate of drug efflux would be equal. Specific efflux rates could be obtained by performing the study with drug-loaded cells. Different starting drug concentrations should be used to examine the possibility of nonlinear membrane transport.

Another *in vitro* method was demonstrated by Shah et al.[31] who used a donor–receptor compartment apparatus separated by a cell monolayer to estimate membrane transport parameters. Permeability coefficients, P, were calculated as

$$P = \frac{KV}{AC_o} \tag{14.40}$$

where
K = flux rate of solute, amount/volume
V = volume of donor chamber
A = cross-sectional area of cell surface
C_o = initial solute concentration in the donor chamber

The flux rate is obtained from the slope of receptor chamber solute concentration vs. time plot. The mass transfer coefficient, h_i, is equal to PA, where A is the surface area of the membrane separating the two subcompartments.

Mass transfer coefficients can also be estimated from *in vivo* experiments. Under the condition of steady-state drug concentrations in blood, a method was derived to estimate a linear mass transfer coefficient.[5] Another method, referred to as the moment method, was derived by Gallo et al.[32] to estimate linear mass transfer coefficients for non-eliminating organs. By using Monte Carlo methods it was demonstrated that the method was both accurate and precise.

CLEARANCES

Organ clearances are usually based on *in vivo* rather than *in vitro* studies. Knowledge of total drug clearance and the contribution each elimination pathway has to the total clearance readily allows calculation of organ and intrinsic drug clearances. In general,

$$Cl_i = f_i Cl \tag{14.41}$$

where
Cl_i = organ clearance
f_i = fraction of dose eliminated by organ i
Cl = total systemic clearance

For many drugs, this information is available in the literature, and independent studies may be unnecessary. Intrinsic organ clearance, Cl_i^I, as presented in Equation 14.15, is obtained from

$$C_i^I = \frac{Cl_i Q_i}{Q_i - Cl_i} \tag{14.42}$$

Drug metabolic parameters, V_m and K_m, can be estimated from *in vitro* hepatic enzyme and hepatocyte preparations.[33]

PROTEIN BINDING

Protein-binding parameters are normally obtained from *in vitro* experiments utilizing filtration, centrifugation, and dynamic dialysis or equilibrium dialysis methods. These techniques have been reviewed elsewhere.[34,35]

MISCELLANEOUS

Forcing function methods may be used to estimate any parameters associated with the system mass balance equations.[32,36-38] The procedures may be applied in the context of either a hybrid or global PB/PK model. The method requires nonlinear regression analysis based on tissue and blood drug concentration–time data. The blood or plasma concentration–time data would serve to obtain a forcing function, usually a polyexponential equation, which is then used as an input function in the mass balance equation for the tissue. The model describing the drug disposition of the individual tissue is then fit to the observed tissue concentration–time data with one or more parameters to be estimated. The forcing function procedure would be particularly useful for estimating parameters in which literature estimates are unavailable. Application of the method to a global PB/PK model would require sequentially fitting individual organ models to obtain parameter estimates, and then reconstructing the global model. For a hybrid PB/PK model, the model-fitting procedures result in the final model.

A modification of the forcing function approach makes use of linear systems analysis for individual tissue compartments.[39] Parametric or nonparametric functions are fitted to observed blood drug concentration–time data, and then combined with tissue drug concentration–time measurements deconvolved to obtain a disposition function for drug kinetics in the tissue. From the tissue disposition function, certain parameters for blood flow–limited organ models can be obtained.

Another approach uses Monte Carlo simulations[40,41] to estimate various model parameters. By defining a range of parameter values, the parameter space can be examined in a random fashion to obtain the best model and associated parameter set to characterize the experimental data. This method avoids difficulties in achieving convergence through an optimization algorithm, which could be a formidable problem for a complex model. Each set of simulated concentration–time data can be evaluated by a goodness-of-fit criterion to determine the models that predict most accurately.

DISTINGUISHING BETWEEN BLOOD FLOW–LIMITED AND MEMBRANE-LIMITED ORGAN MODELS

A membrane-limited two- or three-subcompartment organ model is indicated if tissue drug concentrations do not decline in parallel to drug concentrations in plasma or blood. A formal criterion has been given by Dedrick and Bischoff,[5] such that if

$$\frac{Q_i}{h_i} < \left(1 + \frac{V_i^1}{R_i V_i^2}\right) \tag{14.43}$$

then the organ can be represented as blood flow limited. If this condition is not true, then the organ is best represented as being membrane limited. Equation 14.43 can be approximated as $Q_i/h_i \ll 1$, since V_i^1/R_i V_i^2 is much less than 1. Thus, determining whether a blood flow–limited or membrane-limited compartment is compatible with the data requires estimates for R_i and h_i. Often, it is assumed that a blood flow–limited representation is correct in the absence of a formal analysis.

MODEL VALIDATION

Two issues present themselves when the question of PB/PK model validation is raised. The first issue is the accuracy with which the model predicts actual drug concentrations. The actual concentration–time data have most likely been used to estimate certain model parameters. Quantitative assessment, via goodness-of-fit tests, should be done to assess the accuracy of the model predictions. Too often, model acceptance is based on subjective evaluation of graphical comparisons of observed and predicted concentration values.

Goodness-of-fit tests may be a simple calculation of the sum of squared residuals for each organ in the model,[16] or calculation of a log likelihood function.[40] In the former case,

$$\text{SSR} = \sum_{i=1}^{N} (C_i^0 - C_i^p)^2 \tag{14.44}$$

and in the latter case,

$$\text{LL} = \sum_{i=1}^{N} \frac{n_i}{2} \ln\left[1 + \frac{(C_i^0 - C_i^p)}{S_i^2} \right] \tag{14.45}$$

where   SSR = sum of squared residual

N = number of observations

C_i^o = mean of the observed drug concentration

C_i^p = predicted drug concentration

n_i = number of experimental repetitions

S_i^2 = variance of the observed concentrations at each data point

LL = log-likelihood function

Lower values of SSR and LL indicate closer agreement between observed and predicted concentrations for that organ.

The second model validation issue relates to confirming or accepting prospective model predictions when either limited or no data may be available. The ability of PB/PK models to predict drug concentrations under conditions different than those used to generate the experimental data unleashes the predictive power of PB/PK modeling. However, it may be warranted to confirm such predictions with limited data, particularly if the model was originally derived from a less than optimal data set. Evaluation of the PB/PK model should include an examination of the model assumptions and whether the data can differentiate between linear and nonlinear kinetic processes, and between blood flow–limited or membrane-limited organ structures. Favorable responses to these types of questions at least provide a rational basis to accept inferences made from model predictions.

DEVELOPMENT OF PB/PK MODELS

- Both *in vitro* and *in vivo* techniques are available to estimate PB/PK model parameters.
- Well-designed studies and consistency of approach will likely yield a more robust model.
- Model validation techniques have not been applied rigorously.

REFERENCES

1. Teorell, T., Kinetics of distribution of substances administered to the body. I. The extravascular modes of administration, *Arch. Int. Pharmacodyn. Ther.*, 57:205–225, 1937.
2. Teorell, T., Kinetics of distribution of substances administered to the body. II. The intravascular modes of administration, *Arch. Int. Pharmacodyn. Ther.*, 57:226–240, 1937.
3. Bellman, R., Jacquez, J. A., and Kalaba, R., Some mathematical aspects of chemotherapy. I: One organ models, *Bull. Math. Biophys.*, 22:181–198, 1960.
4. Bischoff, K. B. and Brown, R. G., Drug distribution in mammals, *Chem. Eng. Prog. Symp.*, 62:33–45, 1966.
5. Dedrick, R. L. and Bischoff, K. B., Pharmacokinetics in applications of the artificial kidney, *Chem. Eng. Prog. Symp. Ser.*, 64:32–44, 1968.
6. Gallo, J. M. et al., Physiological pharmacokinetic model of adriamycin delivered via magnetic albumin microspheres in the rat, *J. Pharmacokinet. Biopharm.*, 17:305–326, 1989.
7. Dearden, J. C., Molecular structure and drug transport, in *Comprehensive Medicinal Chemistry*, Vol. 4, *Quantitative Drug Design*, Ramsden, C. A., Ed., Pergamon Press, Oxford, 1990, 375–411.
8. Blakey, G. E. et al., Quantitative structure–pharmacokinetics relationships. I. Development of a whole body physiologically based model to characterize changes in pharmacokinetics across a homologous series of barbiturates in the rat, *J. Pharmacokinet. Biopharm.*, 25:277–312, 1997.
9. Watanabe, K. H. and Bois, F. Y., Interspecies extrapolations of physiological pharmacokinetic parameter distributions, *Risk Anal.*, 16:741–754, 1996.
10. Boxenbaum, H. and Ronfeld, R., Interspecies pharmacokinetic scaling and the Dedrick plots, *Am. J. Physiol.*, 245:R768–774, 1983.
11. Igari, Y. et al., Prediction of diazepam disposition in the rat and man by a physiologically based pharmacokinetic model, *J. Pharmacokinet. Biopharm.*, 11:577–593, 1983.

12. Patel, B. A. et al., Comparative pharmacokinetics and interspecies scaling of 3-azido-3-deoxythymidine (AZT) in several mammalian species, *J. Pharmacobiol. Dyn.*, 13:206–211, 1990.

13. Weissbrod, J. M., A Comprehensive Approach to Whole Body Pharmacokinetics in Mammalian Systems: Applications to Methotrexate, Streptozotocin and Zinc, D.Sc. Eng. thesis, Columbia University, New York, 1979, 110–112.

14. Shen, D. and Gibaldi, M., Critical evaluation of use of effective protein fractions in developing pharmacokinetic models for drug distribution, *J. Pharm. Sci.*, 63:1698–1702, 1974.

15. Gillespie, W. R. et al., Application of system analysis to toxicology: characterization of carbon tetrachloride oral absorbtion kinetics, in *Principles of Route-to-Route Extrapolation for Risk Assessment*, Gererity, T. R. and Henry, C. J., Eds., Elsevier Science Publishers, Amsterdam, the Netherlands, 1990, 285–295.

16. Gallo, J. M. et al., A physiological and system analysis hybrid pharmacokinetic model to characterize carbon tetrachloride blood concentrations following administration in different oral vehicles, *J. Pharmacokin. Biopharm.*, 21:551–574, 1993.

17. Altman, P. L., Ed., *Dittmer Biology Data Book*, Vols. 1 and 3, 2nd ed., Federation of American Societies for Experimental Biology, Bethesda, MD, 1972 and 1974.

18. Gerlowski, L. E. and Jain R. K., Physiologically based pharmacokinetic modeling: principles and applications, *J. Pharm. Sci.*, 72:1103–1127, 1983.

19. Spector, W. S., Ed., *Handbook of Biological Data*, W. B. Saunders, Philadelphia, PA, 1956.

20. Brown, R. P. et al., Physiological parameter values for physiologically based pharmacokinetic models, *Tox. Ind. Health*, 13:407–484, 1997.

21. Jansky, L. and Hart, J. S., Cardiac output and organ blood flow in warm and cold-acclimated rats exposed to cold, *Can. J. Physiol. Pharmacol.*, 46:653–659, 1968.

22. Delp, M. D. et al., Distribution of cardiac output during diurnal changes of activity in rats, *Am. J. Physiol.*, 261L:H1–H7, 1991.

23. Altman, P. L. and Dittmer, D. S., Eds., *Respiration and Circulation*, Federation of American Societies for Experimental Biology, Bethesda, MD, 1971.

24. Molnar, P. et al., Absence of host-site influence on angiogenesis, blood flow and permeability in transplanted RG-2 gliomas, *Drug Metab. Dispos.*, 27:1085–1091, 1999.

25. Chen, C. N. and Andrade, J. D., Pharmacokinetic model for simultaneous determination of drug levels in organs and tissues, *J. Pharm. Sci.*, 65:717–724, 1976.

26. Arms, A. D. and Travis, C. C., Reference Physiological Parameters in Pharmacokinetic Modeling, U.S. Environmental Protection Agency, Washington, D.C., February, 1988.

27. Lin, J. H. et al., *In vitro* and *in vivo* evaluation of the tissue-to-blood partition coefficients for physiological pharmacokinetic models, *J. Pharmacokinet. Biopharm.*, 10:637–647, 1982.

28. Chen, H. S. G. and Gross, J. F., Estimation of tissue-to-plasma partition coefficients used in physiological pharmacokinetic models, *J. Pharmacokinet. Biopharm.*, 7:117–125, 1979.

29. Gallo, J. M., Lam, F. C., and Perrier, D. G., Area method for the estimation of partition coefficients used in physiological pharmacokinetic models, *J. Pharmacokinet. Biopharm.*, 15:271–280, 1987.

30. Stein, W. D., *Transport and Diffusion across Cell Membranes*, Academic Press, Orlando, FL, 1986, 52–68.

31. Shah, M. V., Audus, K. L., and Borchardt, R. T., The application of bovine brain microvessel endothelial-cell monlayers grown onto polycarbonate membranes *in vitro* to estimate the potential permeability of solutes through the blood–brain barrier, *Pharm. Res.*, 6:624–627, 1989.

32. Gallo, J. M., Lam, F. C., and Perrier, D. G., Moment method for the estimation of mass transfer coefficients for physiological pharmacokinetic models, *Biopharm. Drug Dispos.*, 12:127–137, 1991.

33. Brown, R. and Seifried, H., Extrapolation of *in Vivo* Metabolic Rate Constants from *in Vitro* Pharmacokinetic Data, U.S. Environmental Protection Agency, Washington, D.C., 1988.

34. Chignell, C. F., Protein binding, in *Drug Fate and Metabolism*, Garrett, E. R. and Hirtz, J. L., Eds., Marcel Dekker, New York, 1977, 187–228.

35. Meyer, M. C. and Guttman, D. E., Dynamic dialysis as a method for studying protein binding, II: Evaluation of the method with a number of binding systems, *J. Pharm. Sci.*, 59:39–48, 1970.

36. Weissbrod, J. M., Jain, R. K., and Sirotnak, F. M., Pharmacokinetics of methotrexate in leukemia cells: effect of dose and mode of injection, *J. Pharmacokinet. Biopharm.*, 6:487–503, 1978.

37. King, F. G. and Dedrick, R. L., Physiological model for the pharmacokinetics of 2β-deoxycoformycin in normal and leukemic mice, *J. Pharmacokinet. Biopharm.*, 9:519–534, 1981.

38. Gallo, J. M., Hassan, E. E., and Groothius, D. G., Targeting anticancer drugs to the brain, II: Physiological pharmacokinetic model of oxantrazole following intraarterial administration 'to rat glioma-2 (RG-2) bearing rats, *J. Pharmacokinet. Biopharm.*, 21:575–592, 1993.

39. Verotta, D. et al., A semiparametric approach to physiological flow models, *J. Pharmacokinet. Biopharm.*, 17:463–491, 1989.

40. Bois, F. Y., Woodruff, T. J., and Spear, R. C., Comparison of three physiologically based pharmacokinetic models of benzene disposition, *Toxicol. Appl. Pharmacol.*, 110:79–88, 1991.

41. Thomas, R. S. et al., Incorporating Monte Carlo simulation into physiologically based pharmacokinetic models using advanced continuous simulation language (ACSL): a computational method, *Fundam. Appl. Toxicol.*, 31:19–28, 1996.

15 Population Pharmacokinetics

Jeffrey S. Barrell

CONTENTS

INTRODUCTION

Relative to the field of pharmacokinetics itself, population pharmacokinetics is in its infancy. Its origin, which can be traced to the mid-1970s, was largely the result of the efforts of Sheiner and Beal of the University of California, San Francisco (UCSF). Many more academic, industrial, and regulatory scientists have since contributed to this field. Some of the more notable events in the evolution of population pharmacokinetics have been captured in a recent review article.[1] The term *population pharmacokinetics* has been defined in several settings and should not be confined

1-56676-973-6/02/$0.00+$1.50
© 2002 by CRC Press LLC

to pharmacokinetic responses alone, as the field, depending on the agent of interest, may encompass pharmacodynamics as well. Population pharmacokinetics can be defined as the study of the sources and correlates of variability in drug concentrations among individuals who represent the target population that ultimately receives relevant doses of a drug of interest.[2] Its advantages over more traditional pharmacokinetic characterization include the actual study of the population of interest (as opposed to healthy volunteers or special populations only), better estimates of population means and variance, opportunities to assess multiple factors that may influence drug disposition, and at least a theoretical cost–benefit advantage. Its value as a distinct field of study can be measured indirectly from the regulatory acknowledgment provided with the recent FDA guidance on population pharmacokinetics, the increased frequency by which drug manufacturers submit population pharmacokinetic studies/analyses with their NDAs, the increased frequency in publications describing population pharmacokinetic theories and applications, and the emergence of contract research organizations (CROs) that specialize in population pharmacokinetics itself or related technologies that depend on population pharmacokinetics, such as clinical trial simulation (CTS).

The settings in which population pharmacokinetics or, more generally, nonlinear mixed effect modeling can be employed vary greatly and span all phases of drug development from discovery through Phase IV. Specifically, these techniques have been employed to screen compounds for selection as drug candidates, analyze toxicokinetic data, assist with interspecies scaling, pool data from preclinical, nonclinical human, or clinical trials as with a meta-analysis, guide special population trials, examine bioequivalence, provide guidance for dosing regimens for pivotal trials or labeling, and assist with pharmacokinetic/pharmacodynamic (PK/PD) analysis of Phase I, II, III, or IV trials. Although some of these areas are highlighted in the section that follows, the conceptual and statistical framework is defined based on the population pharmacokinetic characterization of data obtained from clinical trials.

CONCEPTUAL AND STATISTICAL FRAMEWORK

Population pharmacokinetic analysis proceeds in stages. The most successful efforts have the benefit of prospective study designs, clearly defined objectives, and data analysis plans that are defined prior to study initiation. Although some deviation in the order of the analysis plan is possible, there is a typical sequence that is followed:

- *Exploratory analysis:* Distribution analysis of covariates under investigation, covariate correlation analysis, and investigation into disease process time course if necessary.
- *Structural model definition:* Comparison of all potential model representations based on available data and parameter identifiability, sensitivity analysis to determine data elements that may affect parameter identifiability (missing values, data collection errors, etc.), and simulations to estimate predictive performance prior to covariate analysis.

- *Parameter distribution analysis:* Determination of Bayes individual parameter estimates from structural model to examine parameter covariance, estimate functional expressions of each parameter, and explore variance model expressions.
- *Covariate selection:* Examine covariate addition to model using generalized additive modeling (GAM) or equivalent stepwise insertion techniques.
- *Population PK refinement:* Obtain population pharmacokinetic parameter estimates, determine influential covariates and outliers, and address "reasonable" residual random error.
- *Validation:* Determine predictive performance of the population pharmacokinetic model.
- *Projection and application:* Examine the usefulness of the population pharmacokinetic model to describe pharmacokinetic and relevant covariate relationships for regulatory submission or query, answer "real-world" questions about the data, and propose additional studies based on findings.

The remainder of this chapter discusses the components of each of these stages in an attempt to give the reader more than a cursory glance at the details. Although the full scope of this field deserves a separate text, the material is presented in an abbreviated format with attention to key elements. Those areas that require additional description have been referenced to provide the reader a road map for further study. Although software and estimation methods are discussed in a separate section, the nomenclature used to define structural and error models is consistent with the parameterized, maximum likelihood approach utilized with the NONMEM algorithm developed at UCSF by Beal and Sheiner.[3–13] Given the historical place of NONMEM and its current prominence in population-based PK/PD analyses, some attention has been given to NONMEM solutions and representations. These have been clearly identified.[14–17]

INDIVIDUALS VS. THE POPULATION

The very nature of research involves, at some point, the collection of data. Individual data elements are organized into common groupings to examine the diversity within and across groups. In the pharmaceutical research context, such experiments typically involve the measurement of responses of interest at several selected or randomly observed values of a related variable (or covariate). More specifically, in pharmacokinetic studies we are interested in examining repeated measures (i.e., drug concentrations) over the variable time following administration of a drug to an individual animal, volunteer, or patient. Each subject in this setting belongs to a population of such individuals. The collective responses of this population are believed to be related to the covariate in a similar manner, but with some measure of variation across individuals.

The study of pharmacokinetics has provided many techniques to analyze complex responses within an individual. From a classical statistical reference, we then seek to make inferences about the central tendency of the collective individual data.

Population pharmacokinetics seeks not only to understand the mean population response profile but also to derive information about an individual, which cannot necessarily be obtained from direct measurement of repeated measures within a well-defined study. This necessitates modeling that incorporates the structural detail necessary to define the individual profile and the manner in which variation of these individual profiles exists about the mean population profile.

The history of pharmacokinetics is laden with attempts to explain biologic processes with mathematical expressions. Current practice for the pharmacokinetic characterization of new chemical entities from well-defined, sufficiently sampled studies usually involves the noncompartmental or systems analysis approach presented previously. Modeling with parameters that can be related to physiological, demographic, and pathophysiological variables can be especially important when analyzing patient data as these parameters may be useful in the guidance of initial dosing as well as maintenance regimens. Also, given the increased interest in stochastic simulation to plan future studies and examine the effects of dose and frequency on concentration–time profiles, such models (those that require structural and statistical components) are often required for the identification of patient subgroups.

MODELING, MODEL BUILDING, AND DATA SUMMARIZATION[18,19]

Prior to initiating a population pharmacokinetic analysis, an accurate representation of the underlying pharmacokinetics, the so-called "structural model," must be made. Kinetic phenomena can be described by differential equations or via implicit or explicit integrated forms, typically with time as the independent variable. As with the estimation of any model, we are concerned with identifying the "best" values for parameters $\theta_1, \theta_2, ..., \theta_k$ of a model by minimizing or maximizing some objective function. For example, some function y that is expressed in terms of several θ terms and an independent variable X, such as

$$y = \theta_1 \bullet \exp -[\theta_2 \bullet X] \tag{15.1}$$

which is "fit" to the observed data via an objective function. The objective function quantifies the difference between the observed and predicted responses for a given parameter set. There are several versions of these functions that differ based on our understanding and assumptions about how random error in the response changes with the response and the manner in which we choose to account for this relationship. Common objective functions utilized for structural model determination are shown in Table 15.1. Some software packages accommodate all three objective functions. Techniques such as weighting allow the minimization of relative errors as opposed to absolute errors or adjust for heteroscedasticity. A more detailed explanation of statistical modeling, optimization algorithms, and measures of association for judging lack-of-fit can be found in most statistical texts that discuss nonlinear regression. The general approach involves the estimation of the partial derivatives of the objective function with respect to each of the parameters (θ). Minimization or maximization of the function is determined by setting the partial derivatives equal to zero.

TABLE 15.1
Objective Function Alternatives

Objective Function	Assumptions about Random Error
Ordinary least squares, OLS $$\sum_{i=1}^{n}(Y_{\text{obs}_i} - Y_{\text{cal}_i})^2$$	Assumes a homoscedastic error structure (common or homogeneous variance regardless of response). The random error is the same for all observations.
Weighted least squares, WLS $$\sum_{i=1}^{n}[w_i(Y_{\text{obs}_i} - Y_{\text{cal}_i})^2]$$	Assumes a heteroscedastic error structure (variance changes with the response). The random error is assumed to be some function of the observed data (i.e., if $w_i = 1/Y_{\text{obs}_i}$, the variance is proportional to the response).
Extended least squares, ELS $$\sum_{i=1}^{n}\left[\frac{(Y_{\text{obs}_i} - Y_{\text{cal}_i})^2}{\sigma_i^2} + \ln(\sigma_i^2)\right]$$	Assumes a heteroscedastic error structure. The variance is expressed as a model parameter in conjunction with the structural model parameters. ELS is designated as a maximum likelihood (as opposed to least squares) if the random effects are assumed to be normally distributed.

The θ terms that fulfill this condition are obtained through an iterative process that solves the system of partial derivatives through numerical integration.

All the objective functions shown in Table 15.1 are derived from a least-squares regression approach as previously described, whereas the estimation method more commonly used in population pharmacokinetics and nonlinear mixed effect modeling in general is based on a maximum likelihood (ML) approach. ML is an alternative to the least-squares objective function; it seeks to maximize the *likelihood* or *log-likelihood* function (or to minimize the negative log-likelihood function). In general terms, the likelihood function is defined as

$$L = F(Y, \text{Model}) = \prod_{i=1}^{n}\{p[y_i, \text{Model Parameters}\,(x_i)]\} \qquad (15.2)$$

By this relationship, the probability (now called *L*, the *likelihood*) of the specific dependent variable values to occur is predicted in the sample data, given the respective regression model. Provided that all observations are independent of each other, this likelihood is the geometric sum (Π, across $i = 1$ to n cases) of probabilities for each individual observation (*i*) to occur, given the respective model and parameters (θ) for the *x* values. The geometric sum means that we would multiply out the individual probabilities across cases. As it is customary to express this function as a natural logarithm, the geometric sum becomes a regular arithmetic sum (Σ, across $i = 1$ to n cases). Given the respective model, the larger the likelihood of the model, the larger is the probability of the dependent variable values to occur in the sample. Therefore, the greater the likelihood, the better the fit of the model to the data. If all assumptions for standard multiple regression are met, then the standard least-squares estimation method will yield results identical to the maximum likelihood method. If the assumption of equal error variances across the range of the x variable(s) is violated, then the weighted least-squares method will yield maximum likelihood estimates.

The main focus of the pharmacokineticist or data analyst at this stage is the identification of ill-conditioned states and assessment of the reliability of the parameter estimates. Many software packages offer diagnostic capabilities that assist in the identification of the appropriate structural model. The initial steps in structural model selection typically involve the plotting (linear and log scale) of individual subject concentration–time profiles (or effect–time profiles in the case of pharmacodynamics) with some assessment of individuals who deviate for the main data trend. Plots may suggest different models for different subjects. All individual subject data should be fit to every model under consideration. Generally, in a data-rich environment, the more complex model is chosen to fit all subjects under evaluation in the model selection phase. This may not be appropriate, however, in a sparse sampling paradigm where sufficient sampling may not have occurred to detect a compartment or phase defined by the more complex model. Table 15.2 provides a list of procedures and diagnostics commonly employed during the structural model definition phase when dense sampling is employed in homogeneous volunteers participating in well-defined trials (i.e., a setting where population approach methodologies would not be required).

The typical structural model is chosen from one of several compartmental models, which incorporate the route of administration as a fixed input into the model with certain assumptions (i.e., linear or zero-order input). Of course, greater flexibility in drug input can easily be accommodated with most algorithms. Compartmental models are, for the most part, empirical even though they may incorporate some mechanistic assumptions so they appear more realistic. Numerically, they are generally easier to handle as opposed to mechanistic models. Complex mechanistic or highly parameterized structural models can be accommodated if there is no requirement for a closed-form expression to define the model. Assumptions and assumption testing are discussed from the standpoint of the broader pharmacostatistical model in the next section.

In a setting in which sampling is not an issue, as in the case of most Phase I studies, the estimation of the unique parameters for each subject or patient represents the first stage in the so-called two-stage approach to population analysis. In the second stage, the mean and standard deviation for each parameter in the structural model is estimated. A third stage may be employed to examine the distribution of these parameters within common groupings (fixed effects) and across other observed factors or subject characteristics (covariates) using a statistical model built with an ANOVA/regression approach. The two-stage approach is difficult to criticize when data are relatively dense for each individual and the model does not involve time or dose dependencies. The usual criticism for the two-stage approach is that it inflates the intersubject variability. However, for the traditional, well-defined pharmacokinetic study with many samples and a simple structural model, this problem is unlikely to be important. This is particularly true for clearance, which is almost always well estimated in traditional pharmacokinetic studies. In any event, the mean and standard deviation for each parameter generated

TABLE 15.2
Model Comparison Procedures and Diagnostics Used in the Assessment of Structural Pharmacokinetic Models

Procedure/Diagnostic	Description and Interpretation
Objective function (OF)	OF measures the difference between observed and predicted responses ~ information content (see Table 15.1 for expressions). Smaller values are better (unadjusted for the number of parameters).
Akaike information criteria (AIC)	$$\text{AIC} = n \bullet \ln\left[\sum_{i=1}^{n} w_i (Y_{\text{obs}_i} - Y_{\text{cal}_i})^2\right] + 2p$$ Representation of the information content of a given set of parameter estimates by relating the coefficient of determination to the number of parameters (p). AIC is dependent on the magnitude of the data and the number of observations (n). The most appropriate models will have smaller AIC values. Models being compared must have the same weighting scheme. AIC is relative measure only (i.e., comparison among models; not an absolute measure of goodness-of-fit).
Model selection criteria (MSC)	$$\text{MSC} = \ln\left[\frac{\sum_{i=1}^{n} w_i (Y_{\text{obs}_i} - \bar{Y}_{\text{obs}})^2}{\sum_{i=1}^{n} w_i (Y_{\text{obs}_i} - Y_{\text{cal}_i})^2}\right] - \frac{2p}{n}$$ Scaled version of AIC so that it is unaffected by the magnitude of the data. The most appropriate models will have larger MSC values.
Shwartz criteria (SC)	$$\text{SC} = n \bullet \ln\left[\sum_{i=1}^{n} w_i (Y_{\text{obs}_i} - Y_{\text{cal}_i})^2\right] + \frac{p \bullet \ln(n)}{2}$$ Used in a manner similar to AIC (same restrictions apply).
Residual analysis	$\text{Residual} = Y_{\text{obs}_i} - Y_{\text{cal}_i}$ Residuals vs. concentration may suggest that a weighting scheme is warranted (e.g., model overestimates large concentrations and underestimates smaller concentrations). $$\text{Relative residual} = \frac{\text{Residual}}{\text{Predicted value}}$$ Normalized residuals, interpreted as above. Residuals vs. time may reveal regions of disposition not well explained by the model.
Parameter errors and confidence intervals (CI)	$$\text{CV}(\%) = \frac{\text{Standard error}}{\text{Parameter Estimate}} \bullet 100$$ Large CV values are generally not favorable, but may imply inadequate n or inappropriate sampling times. Univariate CI = estimate ± Standard error $\bullet$ $t(\alpha/2, df)$ Planar CI takes into account the correlation among the parameters and is always larger than the univariate CI.

(continued)

TABLE 15.2 (CONTINUED)
Model Comparison Procedures and Diagnostics Used in the Assessment of Structural Pharmacokinetic Models

Procedure/Diagnostic	Description and Interpretation
Correlation analysis	Observed vs. predicted values A good fit will generally have high r values; the inverse is not necessarily true (i.e., high r values may not signify a good fit). Between parameters Provides an estimate of the extent to which parameters depend on each other (changes in one parameter could be explained or compensated by changes in another). May or may not be indicative of poor estimation precision. May suggest insufficient information to determine both parameters accurately.

---

from the stage two analysis can be utilized as estimates of the population mean and variability, respectively. The two-stage method is generally not applicable in a clinical setting involving large numbers of patients where there are ethical or practical constraints with the frequency and number of samples taken for pharmacokinetic or pharmacodynamic analysis.

The framework for the mixed effect modeling approach to population pharmacokinetic analysis can be defined from the same two/three-stage approach perspective. For the $i = 1, \ldots, n$ individuals in a population of interest, let $x_{ij}, j = 1, \ldots, n_j$ represent the design points on which the y_{ij} responses are observed. In the pharmacokinetic setting, x_{ij} are typically the sampling time points and y_{ij} are the observed concentrations in the biologic matrix of interest (usually plasma or blood). Hence, the pharmacokinetic response can be described by

$$y_{ij} = f(\theta_i, x_{ij}) + \varepsilon_{ij} \qquad (15.3)$$

where the function f denotes the structural model; θ_i is the $p \times 1$ parameter vector for the ith individual and ε_{ij} are the independently and identically distributed (*i.i.d.*) error terms assumed to be normal random variables with a zero mean and a variance that may depend on the mean concentration, $N(0, \sigma_e^2)$. The ε_{ij} accounts for the intraindividual variability and may incorporate model misspecification or other unresolved (or incorrect) error partitioning. In most population pharmacokinetic software packages the structural model is chosen from a library of compartmental models (ADVAN and TRANS subroutine combinations within the NONMEM $PK block), expressed as a closed-form system of equations or defined via differential equations. Using the same nomenclature, we can define the probability density function that accounts for the within-individual variability only as

$$p(y_{ij} | \theta_i, x_{ij}) \qquad (15.4)$$

The intraindividual variation about the ith individual is completely defined when the model for the distribution ε_i is specified. This is expressed in NONMEM nomenclature by the $ERROR block, $Y = f(F, \varepsilon_{ij})$.

The second stage model defines the between-individual variability in the parameters as follows:

$$\theta_i = \theta \mid \eta_i \tag{15.5}$$

where θ is the mean parameter vector for the population and η_i are the individual deviations assumed to be $i.i.d.$ and normal with zero mean vector and covariance matrix Ω. The expression of θ_i shown (additive) is one of numerous ways that individual θ can be defined. In addition, the population θ can be expressed as a function of covariates (β_i). The integration of covariates into the pharmacostatistical model and the covariate selection process are covered in detail in a later section.

The covariate matrix Ω captures both the variances and covariances among the η terms, for example, a model with three η terms.

$$\Omega = \begin{matrix} \omega_{11} & \omega_{21} & \omega_{31} \\ \omega_{12} & \omega_{22} & \omega_{23} \\ \omega_{13} & \omega_{23} & \omega_{33} \end{matrix}$$

The covariance among the η terms is related to their correlation (say, ρ_{12} for the correlation between η_1 and η_2) by $\mathrm{cov}(\eta_1, \eta_2) = \rho_{12} \bullet \omega_{11} \bullet \omega_{22}$. If the parameters are assumed to have zero covariances, Ω reduces to a diagonal matrix whose elements correspond to the variances of the individual deviations from each of the population parameter means. Although this assumption is often employed during early stages of model development, it seldom reflects the true relationship among the parameters and can contribute to the inflation of the η_i terms. Covariance among parameters from well-defined studies can easily be obtained from general linear modeling (GLM) routines within SPLUS or SAS and can be employed to obtain preliminary estimates of the off-diagonal elements of Ω. The estimation of parameter covariance, when appropriate, often improves convergence and the final parameter estimates. The density of the second stage model can then be defined as

$$p(\theta_i \mid \theta, \Omega, \beta_i) \tag{15.6}$$

where β_i represents the individual patient covariate data (i.e., age, sex, race, etc.). The third stage of the mixed effect model approach would represent a Bayesian analysis in which the model would contain the prior distributions of the population parameters.

ASSUMPTIONS AND ASSUMPTION TESTING

Population pharmacokinetics is defined on a backbone of assumptions about the data, the underlying biologic processes that determine the pharmacokinetics, the pharmacokinetic representations of the data, the complete pharmacostatistical model, the statistical methodology chosen to fit and test the model, and the estimation methods chosen to handle the computational aspects of the analysis. The choices made in each of these categories interact to form a unique response surface, which ultimately represents an analyst's version of the truth. It is not possible to avoid assumptions as they allow us to advance the analysis. The complete details of many of the assumptions listed above can be found in classical texts on pharmacokinetics, statistics, and modeling. What follows is a brief description of assumptions common to most regression-based procedures and a manner of assumption testing in a population pharmacokinetic setting largely derived from the work of Karlsson et al.[21]

Davidian and Giltinan[20] have listed the assumptions implicit in a nonlinear regression framework along with the typical application of these assumptions in nonlinear models with repeated measures data. More importantly, they discuss how departures from this classical setting are accommodated in the generalized hierarchical nonlinear model approach as defined in the previous section. Table 15.3 lists the classical assumptions for a nonlinear regression framework along with their applicability in a population pharmacokinetic setting.

In general terms, the violation of these assumptions is accommodated via the modeling of variance in the response and across individuals and through the definition of correlation among variance partitions. For example, the constant variance assumption in Table 15.3 is handled via the specification of the intraindividual error function ($SIGMA block in NONMEM nomenclature) that incorporates potential

---

TABLE 15.3
Classical Nonlinear Regression Assumptions with Application in a Population Pharmacokinetic Setting

Assumptions: Nonlinear Regression	Population Pharmacokinetic Reality
The errors, e_j, have a mean of zero	Ensures that the model for the population response is correctly specified — reasonable for population pharmacokinetics
The errors, e_j, are uncorrelated	Serial concentrations measured from an individual are likely to be correlated
The errors, e_j, have a common variance, σ^2 ($var(e_j) = \sigma^2$), and are identically distributed for all x_j (independent variable; time in the pharmacokinetic situation)	Constant intraindividual variance is frequently violated and typically accounted for with error models that specify the σ^2 vs. concentration relationship; the distribution of σ^2 over x_j (time) is defined by the underlying structural model
The errors, e_j, are normally distributed	Historical requirement for inference; unrealistic for nonlinear models particularly with biologic data

heteroscedasticity both by its functional form and the magnitude of the overall precision in the response. A by-product of this flexibility in the variance expression is the relaxation of the distributional requirements in the classical nonlinear regression model framework. For example, the application of the common constant coefficient of variation model describes a Poisson-like variance structure. In any event, the violation of classical statistical assumptions associated with nonlinear regression has been thoroughly evaluated for both parametric and nonparametric nonlinear mixed effect model approaches.

Karlsson and Sheiner[21] have provided excellent examples of the varied assumptions that come into play during population pharmacokinetic analysis. The decomposition of the assumptions into common categories and application of assumption testing to actual data sets are important concepts to grasp, and the manuscript itself is an excellent reference. Table 15.4 summarizes the assumptions discussed in Karlsson et al.[22] and lists testing procedures to examine the validity of these assumptions.

The descriptions of several proposed techniques for assumption testing are covered in later sections (i.e., graphical and diagnostic techniques). Discussions on software validation have not been included here. Although others contend that there are no procedures available to validate a nonlinear regression algorithm, this clearly does not address the industrial and regulatory requirement for validation of critical computer systems particularly if the outcomes of the population pharmacokinetic analysis are used to support labeling or dose modifications.

ESTIMATION METHODS AND SOFTWARE

The NONMEM package continues to be the most widely used software for population-based PK/PD analysis. Its limitations lie mostly with its user interface, which, despite the numerous modifications to the code including the NM-TRAN preprocessor, remains a warehouse of FORTRAN 77 subroutines. Several alternatives to NONMEM are currently available and others are still under development. It is beyond the scope of this chapter to compare the details of the software itself, nor is it practical given short lifetime of each release. Aarons[23,24] has recently reviewed and critiqued the currently available software for population pharmacokinetic analysis.

All software alternatives and methods are based on a hierarchical nonlinear mixed effects model approach described in the previous section. The programs can be grouped into three main categories: parametric and nonparametric maximum likelihood and Bayesian. They differ in the manner in which they estimate the parameters of the nonlinear mixed effect model. Although general descriptions of each of the estimation methods are provided, a more rigorous description of these methods can be found in work of Davidian and Giltinan.[20] The selection of software will necessitate a corresponding estimation method and the subsequent assumptions and conditions that will be carried forward in the analysis. Some of these are discussed in the next subsection. In addition, the selection of software will likely dictate that the user become familiar with a pseudo-language to express the structural and error components of the pharmacostatistical model. Although libraries of structural models exist in some packages and error models may be specified from a palate of predefined

TABLE 15.4
Assumptions and Assumption Testing Options for Population Pharmacokinetic Analyses

Category	Assumptions	Assumption Testing Procedures
Nature of data	*Error-free dosing:* Dosing histories are accurately recorded	Perform sensitivity analysis on effect of dosing and/or sampling inaccuracy on key parameters.
	Error-free sampling times: Sampling times are accurately recorded	
	Error-free covariate values: Can reflect inaccuracies in recording covariate information or improper/incorrect transformations	Perform sensitivity analysis with variability in suspect covariates added to data. Check parameters, covariate importance to model, and any inferences made.
Model procedures	*Data exclusion:* Effect of data exclusion on outcomes	Examine data inclusion/exclusion and/or imputation rules and procedures on model outcomes.
	Data invention: Effect of data imputation	
	Structural model adequacy: Does the structural model explain the data? *Application of structural model for all subjects on all occasions:* Are there subjects/occasions for which the model is inadequate?	Examine goodness-of-fit for models in question (see Diagnostics). Examine distribution of empirical Bayes estimates and determine extremes/outliers. Determine if IOV or mixture model improves results.
	Covariate modeling strategy: Is the process adequate to identify important relationships or the magnitude of variance partitions?	Compare GAM or equivalent technique to more rigorous single-step insertion techniques within a model-building approach.

Category	Assumption / Question	Check / Action
Model outcomes	*Shapes of covariate relationship:* Are the shapes of individual covariate relationships adequate?	Plot individual Bayes parameters vs. individual covariate data: examine different functional forms of GAM function.
	Absence of interactions: Does a covariate relationship with a parameter depend on another covariate?	Examine coplots (across parameters and covariates) to check for signs of interaction.
	Distribution of modeled parameters: Are η and κ terms normally distributed? (important for efficiency)	Examine distribution of empirical Bayes estimates (should have mean near zero); determine nonparametric distributions.
	Heteroscedasticity in variance model: Do the variance models properly incorporate the dependence of random components on parameters of interest?	Compare proportional vs. additive random error models; examine plots that assess accounting of heteroscedasticity (see Diagnostics).
	Structure and interaction between intersubject (η) and interoccasion (κ) variability: Based on how the model is defined and the estimation method (more important for FO vs. FOCE, NONMEM)	Plot η and κ terms vs. individual parameters to check for correlation; examine impact of additional covariance terms on model outcomes. Examine the distribution of κ terms on suspect occasions; determine the impact of additional κ terms on model.
	Random change in parameters (θ) between study occasions: Assumed to be random and time invariant	Examine distribution of weighted residuals for zero mean/symmetry; determine if autocorrelation affects parameter estimates; check if log-normal distribution of residual error suggests subject-specific effects.
	Shape and independence of residual errors: ε are assumed *i.i.d.* with a zero mean (no serial correlation over time)	
Estimation/software	*Validity of approximation method:* Assumed that method does not introduce bias into estimation	Consider other methods to check (i.e., Laplacian or FOCE vs. FO method in NONMEM)
	Global minimum found: Parameter estimates for final model coincide with lowest value of objective function (OF)	Vary initial estimates of critical parameter to determine if OF stuck in local minimum.
	Software performance bug-free: Assumed that software performs according to documentation	Perform installation qualification (IQ), operation qualification (OQ), and performance qualification (PQ) per internal SOPs prior to usage.
Application	*Portability of model results to "real" world:* Assumed that simulated data and projections compare to observed data	Examine simulation bias (see validation section)

choices, the analyst will ultimately be obliged to examine more complex models requiring the mastery of a fourth generation (i.e., SAS, SPLUS, Nonlin, NM-TRAN, or Pred file syntax) or a formal (Visual Basic, FORTRAN) language.

Tables 15.5 and 15.6 provide an updated summary of the currently available software for population-based PK/PD analysis.[24] Most of the software listed in Tables 15.5 and 15.6 can be installed and run from a variety of platforms. All can be run from a PC/Windows environment, although some require other programs to operate (i.e., FORTRAN or Visual Basic compiler, SAS, SPLUS, etc.). The most current information can be obtained directly from the manufacturer or parties responsible for code distribution. In any event, software development of population-based PK/PD algorithms appears to be very active relative to the previous 20+ years and the available tools will, hopefully, be superior and more user-friendly than the original NONMEM source code.

COVARIATE ANALYSIS

Covariates may be thought of as factors that explain the variability of specific parameters defined within the context of a population-based pharmacokinetics or pharmacodynamics model. They typically fall into demographic, pathophysiological, or study-specific categories. Based on the recent literature, examples of the most commonly identified covariates and the parameters for which they are related are shown in Table 15.7. The utility of covariate identification lies in the observation of potential differences in patient subgroups that may correlate with observed clinical outcomes, the recommendation of dose or regimen adjustment in certain subgroups based on pharmacokinetic differences, or the support for warning or precautions with dose administration in certain subgroups based on PK/PD differences in certain subpopulations. Many such findings are contained within the package inserts of recently approved medicines.

The primary demographic characteristics (age, gender, and race) defining a trial or collection of trials are necessarily considered in all population-based analyses as they either support or define the "Special Population" section of many package inserts. The most difficult covariates to define and examine are those relating to the pathophysiology of the disease state and the patient population. These may indirectly identify subpopulations at greater risk from a drug safety perspective or those who may benefit to greater extent than the patient population as a whole. Of great concern is the assignment of "meaningful" to a covariate as opposed to "statistically significant" or "clinically relevant." Although such distinctions are best defined *a priori*, this does not often occur given the still exploratory manner in which many of these analyses are undertaken.

Prior to the actual covariate identification procedure, the distribution of all covariates under consideration should be known. The designation of variables as categorical, ordinal, or continuous should be assessed to determine if the existing representation is optimal. Also, variable transformations should be considered prior to modeling. If a continuous variable is to be converted to categorical, the assignment of cutpoints should be justified by statistical, clinical, or practical arguments. Data sets should be evaluated for similarities in demographic characteristics between sampled and target populations. The range of a covariate may impact its likelihood of entering a model.

TABLE 15.5
Software Available for Population Analysis: Parametric Maximum Likelihood Method Packages

Method / Program	Description	Operation, Data, and Model Definition	Primary Output
		Maximum Likelihood	
NONMEM	FORTRAN source code in which the maximum likelihood is evaluated with one of two different first-order expansions (FO or FOCE) and a second-order expansion about the conditional estimates of the random effects (Laplacian)	Concentration and dosing events described in ASCII data file; covariates specified as variables NM-TRAN or Pred ASCII file calls data file, defines error (user-defined) and structural (called or user-defined) model — creates FORTRAN executable	I, P, SE, C, OF, CPM
NLME	S-PLUS algorithm utilizing a generalized least-squares (GLS) procedure and Taylor series expansion about the conditional estimates of the interindividual random effects	Data entered as ASCII file, spreadsheet, or database; covariates specified in data set User code defines error and structural model via SPLUS syntax (several variance models provided) Graphics and summary statistics provided through S-PLUS	P, SE, C, GMF, SS
NLINMIX	SAS macro and algorithm that iteratively fits a set of generalized estimating equations (GEE) using a Taylor series expansion around zero or the conditional estimates of the random effects	Data entered as SAS data set (can be converted from ASCII, Excel, or database files); covariates specified as variables User code defines error and structural model via SAS syntax Graphics and summary statistics provided through SAS	P, SE, OF, IC, GMF, CPM, SS
MIXNLIN	SAS program that implements four different algorithms: estimated generalized least squares, interatively reweighted generalized least squares, pseudo-maximum likelihood, pseudo-restricted maximum likelihood	Data entered as SAS data set (can be converted from ASCII, Excel, or database files); covariates specified as variables User code defines error and structural model via SAS syntax Graphics and summary statistics provided through SAS	I, P, SE, HT, GMF, SS

(continued)

TABLE 15.5 (CONTINUED)
Software Available for Population Analysis: Parametric Maximum Likelihood Method Packages

Method / Program	Description	Operation, Data, and Model Definition	Primary Output
WinNonMix	FO and FOCE with maximum likelihood or FOCE with restricted maximum likelihood estimation provided as well as generalized least squares; separate application that can connect with varied data sources (Enterprise edition)	Data entered as ASCII, SAS data set or any ODBC-compliant database file and can be extracted from NONMEM data files Covariates specified as variables Library of structural PK/PD models available; custom models may be defined based on WinNonlin syntax; error models defined via equation editor Dose events entered separate from concentration events	P, SE, IC, OF, GMF, SS
Kinetica 2000 (PPHARM)	Population pharmacokinetics offered as a feature of a broader PK/PD application within an Enterprise (end-to-end, LIMS to report) solution; population approach uses a parametric expectation-maximization (EM) algorithm to compute maximum likelihood estimates	Data entered as ASCII, SAS data set, or any ODBC-compliant database file Covariates specified as variables Library of structural PK/PD models available; custom models may be defined based on Visual Basic syntax; error models defined via Windows interface	I, P, SE, C, IC, OF, GMF, SS

Output Categories: I = iteration history, P = parameter estimates, SE = standard errors, C = correlation matrix, IC = information criteria, OF = objective function, DP = density plots, GMF = graphics/model fits, HT = hypothesis testing, CPM = convergence process monitoring, SS = summary statistics.

TABLE 15.6
Software Available for Population Analysis: Semi- and Nonparametric Maximum Likelihood and Bayesian Packages

Method / Program	Description	Operation, Data, and Model Definition	Primary Output
		Semiparametric – Nonparametric	
NPML	FORTRAN source code in which nonparametric ML method is used. Given that there are no distribution requirements on the random intercept, the likelihood is reduced to a sum and is not approximated.	Data entered as ASCII file; covariates specified in the data set User-defined structural and error models only	I, P, SE, C, OF
NPEM	Nonparametric EM algorithm within USC*PACK suite of programs. Initially, a defined parameter space with an uninformed parameter joint density is used with individual subject data to calculate a new joint density. An iterative two-stage Bayesian algorithm (IT2B) that computes maximum a posteriori (MAP) individual parameter estimates based on population priors is also provided.	Data entered as ASCII file; covariates specified in the data set but linked to structural parameters or expressed in an unlinked manner (qualitative or numerical) User-defined structural and error models only; the intraindividual error model has to be fully specified Covariate interpolation feature provided	P, SE, C, DP, CPM, SS
NLMIX	FORTRAN source code in which semi-nonparametric ML method is used. The marginal log-likelihood of the data is maximized based on user-defined hierarchy. Varied distributional shapes can be represented to express the random effect density.	Data entered as ASCII file; covariates specified in the data set User must write routine to provide data input User-defined structural and error models only	P, SE, C, IC, OF, CPM

(continued)

TABLE 15.6 (CONTINUED)
Software Available for Population Analysis: Semi- and Nonparametric Maximum Likelihood and Bayesian Packages

Method / Program	Description	Operation, Data, and Model Definition	Primary Output
		Bayesian	
BUGS/WinBUGS	The distributions of structural parameters are used as inputs; random interindividual effects are defined by parametric distributions as well. Markov Chain Monte Carlo (MCMC) methods are used to generate posterior probabilities. WinBUGS is the windows GUI-version of the DOS-based BUGS.	Data entered as ASCII or SPLUS file; covariates specified in the data set	I, P, SE, C, GMF, SS
		A library of PK/PD models is available as well as user-defined models; structural models may be defined via GUI (Doodle BUGS) in the WinBUGS program	
		User-defined error models and prior distribution of interindividual variance model parameters required	

Output Categories: I = iteration history, P = parameter estimates, SE = standard errors, C = correlation matrix, IC = information criteria, OF = objective function, DP = density plots, GMF = graphics/model fits, HT = hypothesis testing, CPM = convergence process monitoring, SS = summary statistics.

TABLE 15.7
Common Covariates Identified in Population PK/PD Models

Covariate	Parameter	Example Relationships[a]
Gender	Cl	Nevirapine in patients infected with HIV[25] $Cl = \theta_1 \bullet \exp(\theta_2 \bullet gender)$
Renal function	Cl	Bisprolol in renally impaired patients[26] $Cl = \left(\dfrac{CRE}{90}\right)^{\theta_1} \bullet \theta_2$
Hepatic function	Cl	Bisprolol in hepatically impaired patients[26] $Cl = \left(\dfrac{AST}{20}\right)^{\theta_1} \bullet \theta_2$ AST = serum aspartate transaminase activity (IU/ml)
Age	Cl	Bisprolol in patients with hypertension[26] $Cl = \left(\dfrac{AGE}{50}\right)^{\theta_1} \bullet \theta_2$
Total body weight	Cl	Warfarin in adult patients[27] $Cl = \left(\dfrac{TBW}{70}\right)^{\theta_1} \bullet \theta_2$
	V_{ss}	Zidouvine in patients with HIV[25] $V_{ss} = \theta_1 + \theta_2 \bullet (TBW - 70)$
Race	Cl	Lamotrigine in patients with epilepsy[28] $Cl = \theta_1 + RACE \bullet \theta_2$ RACE = 0 if Caucasian; 1 if non-Caucasian
	V_d	Naratriptan in patients with migraine[29] $V_d = \theta_1 \bullet (1 + RACE \bullet \theta_2)$ RACE = 1 if Black; 0 otherwise
Formulation	K_a	Tiludronate in normal volunteers[30] $K_a = \theta_1 + FORM \bullet \theta_2$ FORM = 1 if test; 0 if reference
Migraine	K_a	Naratriptan in patients with migraine[29] Separate θ for K_a if drug given during nonmigraine period or migraine attack (theorized to correlate with gastric emptying changes)
Smoking	Cl	Warfarin in adult patients[27] $Cl = SMOKE \bullet \theta_1$ SMOKE = 1 if smoker; 0 otherwise

[a] Prototypical examples for each covariate–parameter pairing have been separated and expressed as unique relationships when they may be part of a more complicated expression in the original citation.

One of the techniques commonly used to aid the selection of covariates in the model-building stage and reveal the data structure in general is GAM, which is an extension of multiple linear regression. A multiple linear regression model for a dependent variable Y and a set of "p" predictor variables X_1, ..., X_p is defined as

$$Y = \alpha + \sum_{j=1}^{p} \beta_j X_j + \varepsilon \qquad (15.7)$$

where β_j is a linear regression coefficient, α is a constant, and ε is a variance term that is normally distributed about a zero mean with constant variance. Additive models replace the linear function $\beta_j X_j$ by a smooth nonlinear function, obviating the assumption about the linear dependence of Y on each of the predictors.

$$Y = \alpha + \sum_{j=1}^{p} f_j(X_j) + \varepsilon \qquad (15.8)$$

Thus, linear regression models are a subset of additive models where $f_j = \beta_j$. The procedure begins by the determination of the structural population pharmacokinetic model without covariates. No covariance is initially assumed between the η at this stage. Empirical Bayes estimates for each parameter are obtained and utilized as the "data" or Y in the previous expressions. A stepwise, generalized additive modeling procedure can then be used to identify covariates most useful in explaining the variability in the parameter of interest. This is accomplished by examining a combination of models of the covariates (X_j) in a stepwise fashion. Model discrimination can be made by comparison of the AIC (or similar statistic). For each covariate, a hierarchy of possible models can be examined. The S-PLUS algorithm is the most common venue to run GAM calculation. There is great flexibility in the functional expression of $f_j(X_j)$ and each covariate can enter the model in any of several functional forms. In each step, for each covariate, the models within the hierarchy are applied with the model decreasing the AIC statistic the most retained to the next step. The search is terminated when no model can decrease the AIC any further or when a predefined boundary is attained. In the final stage, mixed effects models describing the relationship between covariates identified using GAM and the parameters are built and covariance among parameters is examined.

NONMEM itself can be utilized to perform covariate selection without GAM. In this setting, the structural model is utilized as the "base" model and single covariates are added in any manner and examined for their effect on the parameter of interest via a drop in the objective function. Each covariate is ranked against the base model by the change in OF, and ranked covariates (highest to lowest) are introduced onto the base model in a stepwise manner with only "significant" covariates added. Upon completion of the fully saturated model, a backward deletion procedure can be utilized to remove redundant covariates.

As there are no formal bounds on the number of models tested in the model-building process, criteria are necessary to differentiate between useful and redundant models. Grevel et al.[26] have offered the following list:

- The difference in the values of the objective function (OF). Based on the NONMEM OF as interpreted by Sheiner, a difference of eight units is assumed to be significant between two models of which the restricted model has one fewer regression parameter than the complete model.
  - A minimum correlation between parameters
  - Small standard errors of the parameter estimates
- Weighted residuals that are randomly scattered around zero when plotted against predicted concentrations.
  - A minimum in the estimates of the interindividual variances

TIME DEPENDENCIES AND INTEROCCASION VARIABILITY

Time-dependent changes in pharmacokinetic parameters can occur for a variety of reasons (transient changes in binding proteins, changes in pathophysiology such as renal function or disease state, or age-related changes in metabolism and elimination in general as might be seen with neonates). Although normally assumed to be constant across all observations within an individual, covariates that change during an observation period may imply that the regression parameters (θ) are not fixed over all observations within an individual. Depending on our knowledge of the time dependency, these changes may be explored in a variety of ways including data splitting, structural models that accommodate changes in the relevant θ, and modeling interoccasion variability.

Data splitting is fairly straightforward and covered in detail in the next section on validation. It simply implies that data to be modeled are partitioned based on differences in sampling (i.e., windows where suspect θ are believed to be constant). The most common data splits to explore pharmacokinetic time dependencies would be single-dose, chronic non-steady-state, and steady-state conditions. Data subsets are modeled individually with all parameters and variability estimates along with any relevant covariate expressions compared in a manner similar to a validation procedure (see next section). Data can be combined in a leave-one-out strategy (see cross-validation description) to examine the uniformity of data "windows."

If the nature of the time dependency is well understood and sampling is sufficient to identify model parameters, which define such expressions, a modification to the typical structural model can be explored. Specifically, as in the case of enzyme induction, changes in clearance may be expected to occur over typical time windows. By allowing the initial value of clearance, Cl(0), to increase in a monoexponential manner until an asymptotic value, Cl(ss), is reached, the clearance at any time, t, can be expressed as a function of these boundary conditions and an induction rate constant, k_I. Such a function (shown below) was proposed by Levy et al.[31] and utilized by Rostami-Hodjegan et al.[32] with certain assumptions to model methadone pharmacokinetics in opiate users.

$$Cl(t) = Cl(ss) - [Cl(ss) - Cl(0)] \bullet exp[k_I \bullet t] \qquad (15.9)$$

Hence, models with and without time dependencies can be compared to evaluate their predictive performance and ability to explain sources of variation. Rostami-Hodjegan et al.[32] compared induction models and data-splitting techniques in their evaluation of adaptive changes in methadone kinetics.

More commonly, time dependencies are explored by modeling interoccasion variability (IOV). In this approach individual parameters (θ_{ij}) incorporate terms for intersubject (η_i) and interoccasion (κ_{ij}) variability. As mentioned previously, there is great flexibility in the ways these terms are expressed. Equations 15.10 and 15.11 illustrate additive and proportional expressions, respectively:

$$\theta_{ij} = \theta + \eta_i + \kappa_{ij} \text{ (additive)} \qquad (15.10)$$

$$\theta_{ij} = \theta \bullet exp[\eta_i + \kappa_{ij}] \text{ (proportional)} \qquad (15.11)$$

where θ is the typical or population parameter and η_i and κ_{ij} are zero-mean symmetrically distributed variables. Correlation between the η and κ can be accommodated as well.

This structure can be defined in NONMEM nomenclature by creating an occasion index (for example, three dummy variables OCI, OCII, and OCIII) over which the parameters of interest can be examined. Hence, in the $PK block the value for clearance (Cl_{ij}) could be defined by an expression:

```
Cl = TVCL*EXP(ETA(1) + ETA(2)*OCI + ETA(3)*OCII +
ETA(4)*OCIII)
```

The distribution of the κ — ETA(2) though ETA(4) in the expression above — is preserved by using the SAME option in the $OMEGA block.

```
$OMEGA BLOCK(1)  0.1;  IIV

$OMEGA BLOCK(1)  0.05;  OCI

$OMEGA BLOCK(1)  SAME;  OCII

$OMEGA BLOCK(1)  SAME;  OCIII
```

Additional details on NONMEM implementation of interoccasion variability are provided in the NONMEM user guides and help files.

VALIDATION AND MODEL CREDIBILITY

One of the most difficult tasks facing a pharmacostatistical analyst is that of demonstrating whether a model is an accurate representation of the actual system being studied — whether the model is valid. If this cannot be shown, conclusions derived from the model will have little value.[33,34] When a model and its results are accepted by the scientific community as valid and are employed in decision-making processes, it can be called "credible." Validation along with verification can occur at several

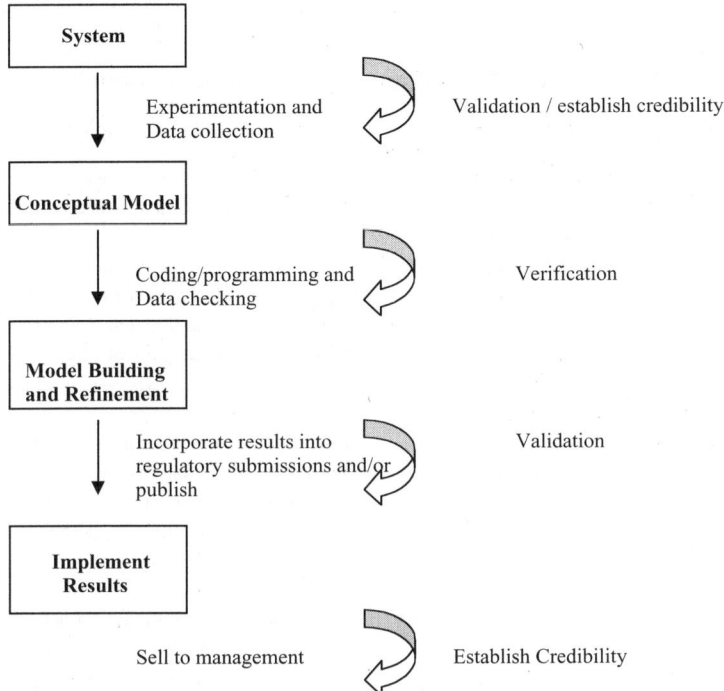

FIGURE 15.1 Schematic of timing and relationship of verification, validation, and model credibility.

places during the establishment of model credibility. In this context, verification is the examination of the translation of the conceptual model into a correctly functioning algorithm (i.e., debugging, data checking, etc.). Figure 15.1 is a modified version of Law and Kelton's proposal for the timing and relationship between validation and verification toward the establishment of model credibility.

Specifically, model validation examines the performance of the population analysis predictions through the projection of the population parameter space onto a new or "different" data set. Model validation is generally divided into internal and external validation categories based on the source of data used to "validate" the model. The data set used to develop the model is called the "index" or "learning" set, whereas that used to examine the performance of the population model is referred to as the "validation" or "test" set. Internal validation relies on subsetting the total data set to be analyzed with the majority of data used in the model-building and refinement stage, whereas external validation implies an external data source (i.e., new experimental study and data). Several techniques useful in the employment of validation strategies are shown in Table 15.8. To some extent, the choice of validation technique is arbitrary but may be constrained by the available data and/or target population.

While the abbreviated details of specific validation techniques are discussed here, these do not represent our total experience with validation and performance

TABLE 15.8
Strategies for Validation of Population Pharmacokinetic Models

Validation Strategy	Typical Setting(s)	Description and Implementation
Data splitting	Large N: single trial; common trials of similar populations; pooled data from Phase I trials (assuming comparable characteristics between index and validation data)	Randomly divide data into two (or more) splits with the significant portion assigned to the index data set. Obtain population parameters from index data set. Examine the "closeness" of index-derived population pharmacokinetic parameters on validation data set (plots of observations vs. predictions, residuals vs. covariates, etc.).
Cross validation resampling	Small data sets (small overall subject N); no test data	Build model with entire data set. Divide data into k subsets; on k occasions, leave one subset from the data and compare the k replicate estimates (may be inefficient because of variation in accuracy estimation).
Bootstrapping resampling	Small data sets (small overall subject N); no test data	Build model with entire data set. Divide data into k subsamples. Each subsample is a simple random sample with replacement. Compare performance of k subsamples. Alternatively, population estimates from the entire data set can be used with Monte Carlo techniques to simulate a large number of subjects from which bootstrapped subsamples can be analyzed and compared as above.
External validation — separate study data	Whenever a separate study is available to be utilized for validation	Build model with data from one or several studies that will serve as the index data. Project population parameters onto data set from a separate trial (validation data set). Compare "closeness" of index-derived population pharmacokinetic parameters on the validation data set as above.
Posterior predictive check (PPC)	The simulated posterior distributions of one or more nonsufficient statistic(s) (NSS_i) are compared with the observed statistics derived from the actual data	Select one or more nonsufficient statistic(s) (NSS_i) from the actual data. Obtain posterior distributions of all fixed and random parameters, which typically requires a full Bayesian method as with POPKAN or BUGS, but can be accomplished with NONMEM under certain conditions and assumptions. Alternatively, posterior distributions can be simulated using a parametric bootstrap. Compute posterior distribution of statistic of interest. Compare NSS from replicates derived from posterior parameter distribution with that from original data set.

measures in general. Although the pharmaceutical literature is generally limited in this area, there are many sources of information in the simulation modeling arena. Balci and Sargent[35] present a comprehensive, albeit somewhat dated, bibliography on validation techniques for simulation models. For example, in a pediatric study setting, the available patient population is likely to be small and the opportunities for additional studies few. Model validation, in this case, is almost exclusively limited to resampling techniques such as bootstrapping or posterior predictive check (see Table 15.8). In both instances, the original data set in conserved and simulation or resampling is used to generate additional data sets for subsequent comparisons.

DATA SPLITTING

As the name implies, data splitting involves the partitioning of the data available for population analysis into data subsets for model building and validation (the index and test data sets, respectively). Typically, this is accomplished by assigning random numbers to each subject in the data set (i.e., a subject's records are kept intact in a particular data set assignment) and then allocating subjects to the appropriate data partition based on predefined percentages (typically 1/3 for validation and 2/3 for the index data in a two-way split). Excel (RAND worksheet function), SAS (normal, rannor, or ranuni functions; others can be expressed), SPLUS (Data menu, then Random Number option in SPLUS 2000 for interactive dialog or use runif, rnorm, or rexp when coding), and many other packages can be used to accomplish this using a variety of distributions. Several of the new software packages (WinNonmix and Kinetica 2000) offer data-splitting routines as well. Once a final model has been built from the index data set, parameter estimates are fixed and then applied to the validation data. Goodness-of-fit is examined and compared to the index data fit. Additionally, the importance of covariates identified as significant in the model-building stage is reexamined with the validation data.

BOOTSTRAPPING AND CROSS-VALIDATION

Bootstrapping and cross-validation are both methods for estimating generalization error based on "resampling." The resulting estimates of generalization error are often used for choosing among various models. In the context of validation, these estimates are used to assess the generalizability of model predictions. The term *bootstrapping* is derived from the idea of pulling oneself up by one's own bootstraps and refers to the process of starting something with no foundation. Bootstrapping is a general technique for estimating sampling distributions. Although we strive to determine the exact sampling distribution of an estimation procedure for computing confidence intervals and for making inferences, we seldom accomplish this and are forced to use asymptotic methods for interval estimation and hypothesis testing. Bootstrapping provides a simple means of obtaining an approximate sampling distribution of a statistic of interest conditional on the observed data. The distribution is approximate and similar to the asymptotic distribution although often preferred (over the asymptotic distribution) as it is based on the observed data. Because of the computational nature of the procedure (computer-generated pseudorandom numbers), the same situation may yield similar, but possibly different results.

Both nonparametric and parametric bootstrap approaches can be pursued depending on whether we are willing to assume we know the "true form" of the distribution of the observed sample (parametric case). The parametric bootstrap is particularly useful when the sample statistic of interest is highly complex (as one might expect when trying to bootstrap a pharmacokinetic parameter derived from a nonlinear mixed effect model) or when we happen to know the distribution, since the additional assumption of a known distribution adds power to the estimate.

The basic procedure for the parametric bootstrap is as follows:

- Determine the "true" distribution for the parameter of interest, i.e., $N(0,1)$, exponential, poisson, etc. This is actually accomplished by assuming the estimates calculated from the data are the "true" values.
- Use one of many computer algorithms to randomly generate a set of values from this distribution (it should be the same number of observations as the original data set). This set of numbers is referred to as the bootstrap sample.
- Evaluate the statistic of interest (typically, the sample mean, variance, or correlation).
- Construct a confidence interval if desired (two methods shown).
  1. Normal approximation method: $\hat{\theta} \pm 1.96\hat{\sigma}$, where $\hat{\theta}$ is the mean of the parameter of interest (actual data), B is the number of bootstrap samples, and $\hat{\sigma}$ is the estimate of the standard error computed by

$$\hat{\sigma} = \frac{1}{B-1} \sum_{i=1}^{B} (\theta_i - \bar{\theta})^2 \qquad (15.12)$$

where

$$\bar{\theta} = \frac{1}{B} \sum_{i=1}^{B} \theta_i \qquad (15.13)$$

  2. Percentile method: Order the θ_i values. The bounds for the $(1 - \alpha) \bullet 100$ confidence interval are defined as follows

$$\text{Lower Bound} = ((\alpha/2) \bullet B)\text{th value}$$

$$\text{Upper Bound} = ((1 - (\alpha/2)) \bullet B)\text{th value}$$

The nonparametric bootstrap is useful when distributions cannot be assumed as "true" or when the sampled statistic is based on few observations. In this setting, an observed data set, for example, $X_1, \ldots, X_n$, where X could be vector-valued (i.e., concentrations at fixed sampling times) can be summarized in the usual way by a mean, median, and variance. An approximate sampling distribution can be obtained drawing a sample of the same size as the original sample from the original data with replacement, for example, $X_1^{*i}, \ldots, X_n^{*i}$, where i is the index of the bootstrap sample that extends through the total number of bootstrap samples (B) to be drawn (typically around 1000 or higher). The statistics of interest can be calculated across the B bootstrap samples and confidence intervals can be constructed as above.

Cross-validation is a "leave-one-out or leave-some-out" validation technique in which part of the data set is reserved for validation. Essentially, it is a data-splitting technique. The distinction lies within the manner of the split and the number of data sets evaluated. In the strict sense a k-fold cross-validation involves the division of available data into k subsets of approximately equal size. Models are built k times, each time leaving out one of the subsets from the build. The k models are evaluated and compared as described previously, and a final model is defined based on the complete data set. Again, this technique as well as all validation strategies offers flexibility in its application. Mandema et al.[36] successfully utilized a cross-validation strategy for a population pharmacokinetic analysis with oxycodone in which a portion of the data was reserved for an evaluation of predictive performance. Although not strictly a cross-validation, it does illustrate the spirit of the approach.

EXTERNAL VALIDATION

In an external validation, data from a study other than that used to build the model are used as the validation or test data. The FDA guidance states that this is the most stringent means of validating a population pharmacokinetic model. Indeed, the performance of the final model developed on a given data set when projected onto data from a separate study goes a long way toward the establishment of a "credible" model and the confidence in the recommendations based on such models.

The procedure may require modifications to the data sets, which, on the surface, may appear heavy-handed, but are necessary to confer a valid projection of a parameter space onto a new data set. Specifically, the design points of both studies need to be compared. The comparability of sampling locations and their densities should be assessed along with dosing frequencies and regimens. The goal is to evaluate whether the detail/complexity of the structural model can be accommodated by the new study data. Of course, the converse is true as well. A second check involves the comparability of the study demographics and the covariate data space in general. Specifically, the ability to project a model containing expressions for parameters, which are dependent on certain covariates, requires those covariates to be present in the new data set. This is not always a guarantee, based on differences in study collection procedures, and may be precluded completely because the new data set is based on a study population in which these covariates are not prevalent. Again, the converse situation in which the new study population may be defined by suspect covariates not previously evaluated in the model-building data set can occur as well. By assuming these issues are solvable, the actual comparison (index to validation) proceeds as with internal validation procedures (see Diagnostics, Graphical Techniques, and Outliers section for details).

POSTERIOR PREDICTIVE CHECK

Few published studies exist in which a posterior predictive check (PPC) has been employed as a validation technique for a population pharmacokinetic analysis.

Despite this, PPC is inherently suited for nonlinear mixed effect modeling of pharmacokinetic data as it utilizes the posterior distribution of parameter estimates to examine whether the salient features of the original data are observed in the derived (replicated) data. Berin and Rubin[37] applied this approach to mixture models of reaction time based on visual tracking experiments conducted in patients with and without schizophrenia. Their application of PPC was directed in the model-building stage and not implemented to "validate" their final model per se. This technique has appeal in both settings as it can serve as a metric in the establishment of a "credible" model.

The general approach requires the identification of one or more features/statistics of the data that are relevant but would not necessarily be fit by the proposed model. These have been defined as nonsufficient statistics (NSS) in the FDA guidance. Examples of NSSs would include C_{max}, T_{max}, trough levels, and baseline or trough effects depending on the pharmacodynamic model — basically, any parameter or pharmacokinetic or pharmacodynamic metric that is observed in the data but not defined in the structural model. Depending on the data density within subjects, it may be possible to interpolate or otherwise estimate these metrics within an individual for the purpose of the PPC. Multiple NSSs are preferable as salient data features are seldom captured by a single metric. The chosen NSSs should also be relevant to the overall modeling objectives. For example, maintaining trough concentrations of antiviral agents above the IC_{90} for resistance development represents a major hurdle for these agents and a measure of effectiveness by which they are compared.

Once the NSSs have been identified, the next step is to generate replicate data sets from the posterior distribution of parameters used to define the model. The number of data replicates is arbitrary but can be guided by the sample size of the original data set (i.e., more replicates for smaller "n" values). Once the replicates have been generated, the NSS can be calculated for each replicate and the distribution of each NSS can be compared to those of the original data. Attention should be paid to the extremes of each NSS (i.e., does the 90th percentile of the original data compare to that of the generated data?). Histograms of the NSS from replicate data are helpful in this regard.

Simulating posterior distributions via bootstrapping can be accomplished in the following manner:

- Simulate a new set of observations (Y^*) using parameter and variability estimates (THETA and ETA in NONMEM) from a multivariate, normal distribution (0, OMEGA) with errors sampled from a multivariate normal distribution (0, SIGMA).
- Fit the final model to the new set of observations and obtain new estimates of THETA, OMEGA, and SIGMA (THETA*, OMEGA*, and SIGMA*).
- Simulate a new set of observations (Y^{**}) using THETA* and ETAs sampled from a multivariate, normal distribution (0, OMEGA*) and errors sampled from a multivariate normal distribution (0, SIGMA*).
- Compute $S(Y^{**})$.
- Repeat first four steps many times (at least 100).

Diagnostics, Graphical Techniques, and Outliers

Diagnostic tests including graphical representations are used during the model-building process and as criteria for validation. The more common tests and procedures are discussed below, along with methods and criteria for outlier identification.

Regression analyses are typically judged by the degree of bias and precision in model outcomes. Bias assesses the degree to which the typical prediction is either too high or too low, and precision is a measurement of the typical magnitude of the error about a true value. Other diagnostics more suited for population pharmacokinetics can be derived from these conceptual metrics. The estimated precision of a population parameter estimate can be expressed in general terms by Equation 15.14.

$$CV = \frac{\sqrt{\text{Estimated variance}}}{\text{True parameter value}} \bullet 100\% \qquad (15.14)$$

where CV represents the coefficient of variation of a parameter estimate, and its estimated variance is calculated based on a large sample or simulation approach. The bias of a parameter estimate can be generally expressed as shown in Equation 15.15.

$$\text{Bias} = \frac{(\text{Mean parameter value} - \text{True parameter value})}{\text{True parameter value}} \bullet 100\% \qquad (15.15)$$

Prediction error can be calculated to examine how well the population model predicts the mean response. The log prediction error for a specific measurement is defined in Equation 15.16.

$$\text{LPE}_{ij} = \log(Y_{ij}) - \log(C_p) \qquad (15.16)$$

where Y_{ij} is the jth measured concentration in the ith individual and C_p is the population mean concentration predicted from the population pharmacokinetic model. C_p at each time point is typically calculated via Monte Carlo simulation as the logarithmic mean of all simulated concentrations using the parameter estimates from the final model. The population model prediction error (pe_j) for each parameter can be defined when estimates of the individual parameters are obtained (as with the POSTHOC option when using NONMEM). The pe_j for a parameter θ would be defined as shown in Equation 15.17.

$$\theta pe_j(\%) = \frac{(\hat{\theta}_j - \theta_j)}{\hat{\theta}_j} \bullet 100 \qquad (15.17)$$

where θ_j is the empirical Bayes estimate of the jth individual true parameter based on population parameters and observed concentrations of individual j.

Graphical analysis techniques are extremely valuable in the examination of the sampling distributions of response and covariate data, the inspection of model fits during the model-building phase, and the examination of covariate correlation with structural model parameters and other covariates.

Table 15.9 lists the common graphical representations with the typical usage and interpretation in a population approach setting. Some specific examples of representative plots are shown in Figures 15.2 through 15.5. Ette and Ludden[38] have provided an informative example of how these and other diagnostic techniques are integrated into the data analysis plan of a population pharmacokinetic analysis.

MISSING VALUES

One of the most common difficulties in the analysis of population pharmacokinetic data is that of missing data. Missing patient identifiers or details of drug administration including compliance occur frequently and are best handled via prospectively defined procedures when possible. Typical solutions to this problem include case/subject deletion and imputation (inserting sample mean or median or estimation via linear regression or an EM procedure). Each of these approaches carries methodological requirements and potential limitations given the observed pattern of the missing data. Analytical limitations in conjunction with sample collection errors can also yield missing observations.

The handling of data that are below the quantification limit (BQL) of the assay is a subject of much discussion. Disparity in the density of observed concentrations across patients can result in a biased estimate of the mean prediction error. Discussions on the treatment of BQL data can be found in the NONMEM network archive at http://www.phor.com/nonmem/nm/index.html. The approach to handling BQL data should be guided by the value of data near the quantification limit (QL) in general. Current analytical methodologies are typically extremely sensitive and often capable of detecting drug levels close to background. Such levels may not be pharmacologically relevant or represent a significant contribution to the underlying pharmacokinetic structural model. In these instances, case deletion is an easy choice for data handling. When this is not the case, more thought must be given to the procedure. After all, the true value of the BQL observation is less than the QL but greater than 0. As such, the distribution of the BQLs is truncated or left censored and hence is not normal. The method proposed by Sheiner (Table 15.10) is simple and preserves the information that may be obtained from such data without modifying the likelihood (based on a NONMEM analysis). Although more heavy-handed approaches have been proposed, these often affect the stability of the objective function. Careful simulation should be employed to test the sensitivity of the objective function to elaborate BQL rules especially when discontinuities are introduced.

Strategies for handling missing values are varied and can depend on the nature and reasons for the missing information. It should be appreciated that missing values do not mean missing information, and the most common and simplest approach, dropping subjects or observations, is the least informative. Table 15.10 lists some typical scenarios for missing data occurrences in a population pharmacokinetic setting as well as possible strategies for data handling. Although compliance can be thought of as missing data,

TABLE 15.9
Common Graphical Representations, Usage, and Interpretation Utilized during Various Phases of a Population Pharmacokinetic Analysis

Plot Type	Usage and Interpretation
Scatter Plots	
Population predicted vs. observed response (PRED vs. DV)[a]	Goodness-of-fit; data points should be clustered uniformly and closely around the line of unity throughout the range of data if fit is good (see Figure 15.2A)
Individual predicted vs. observed response (IPRED vs. DV)	Goodness-of-fit (should appear better than PRED vs. DV as IPRED incorporates η necessary to fit individual data); data points should be clustered uniformly and closely around the line of unity throughout the range of data if fit is good (see Figure 15.2B); can aid in identification of IOV and appropriateness of structural model
Observed, predicted vs. time (PRED, DV vs. TIME)	Goodness-of-fit; predicted data should be consistent with observed data (i.e., shape and amplitude of data signature should be captured by predicted data)
Residual Plots	
Residuals vs. predicted response (RES vs. PRED)	Examination of residual error model; shape of distribution about residual = 0 line can suggest error model representation (i.e., additive for uniform distribution; proportional for cone shape)
Weighted residual vs. predicted response (WRES vs. PRED)	Examination of residual error model; weighted residuals have unit variance and are uncorrelated so that the shape of distribution about residual = 0 line should be uniform if error model is adequate; WRES scale can be viewed as an approximate SD scale (i.e., outside the ±3 unit range should be further explored as a potential outlier, see Figure 15.3)
Residuals vs. time (RES vs. TIME)	Goodness-of-fit and examination of time dependencies; can be used to discern tolerance or induction phenomenon
Weighted residuals vs. time (WRES vs. TIME)	Goodness-of-fit and examination of time dependencies; residuals should be randomly dispersed about the residual = 0 line
Correlation Plots	
Correlation between parameters	Evaluation of structural model; parameters should not be highly correlated
Correlation between η	Diagnostic for need to specify covariance between parameters
Correlation between covariates	Determination of covariate correlation; can determine covariate ranking or support inclusion or omission of certain covariates
Correlation between parameters and covariates	Evaluation of functional relationship between parameters and covariates; can assist in model expression
Distributional Plots	
Histogram of covariates	Evaluation of functional form of covariates; used to define expressions for θ
Histogram of parameters	Check for normality; used to define expressions for θ
Histogram of individual η	Check for normality; used to define expressions for θ
Histogram of partial residuals	Diagnostic of model fit (parameter bias)

[a] NONMEM keywords are shown in parentheses.

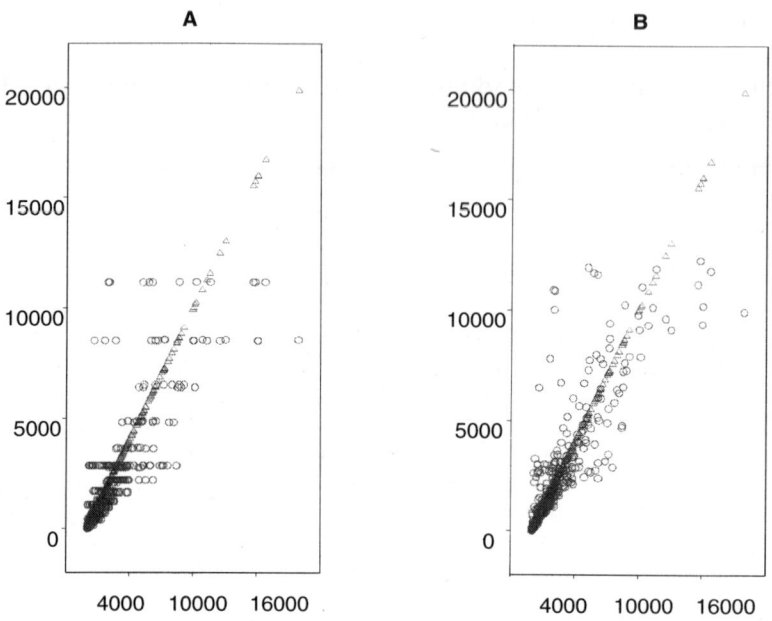

FIGURE 15.2 Population (A) and individual (B) predicted vs. observed concentrations.

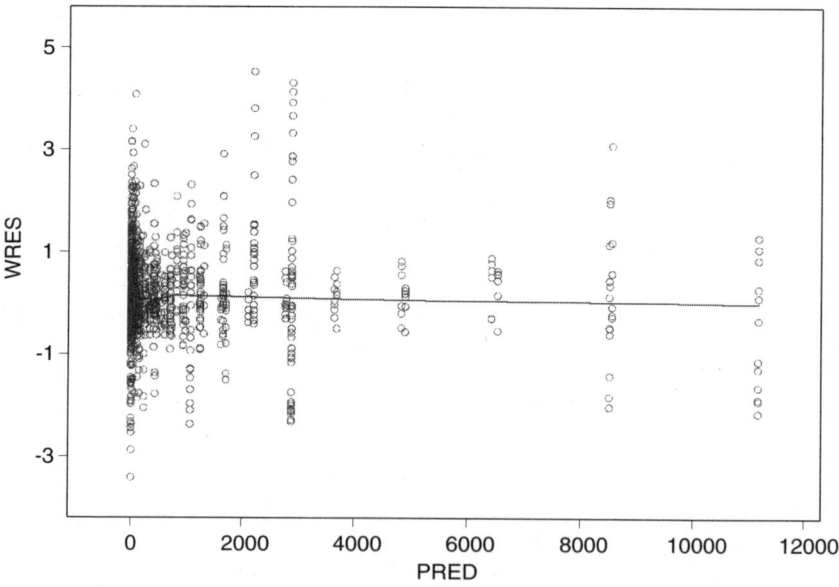

FIGURE 15.3 Weighted residuals vs. predicted concentrations.

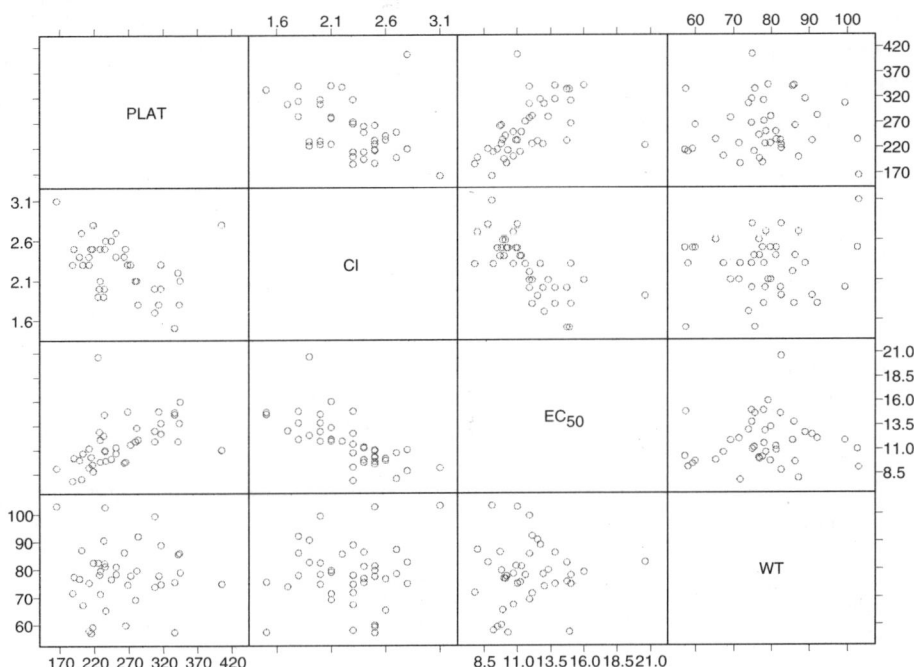

FIGURE 15.4 Correlation plot of parameters (Cl and EC_{50}) and covariates (platelet count and body weight).

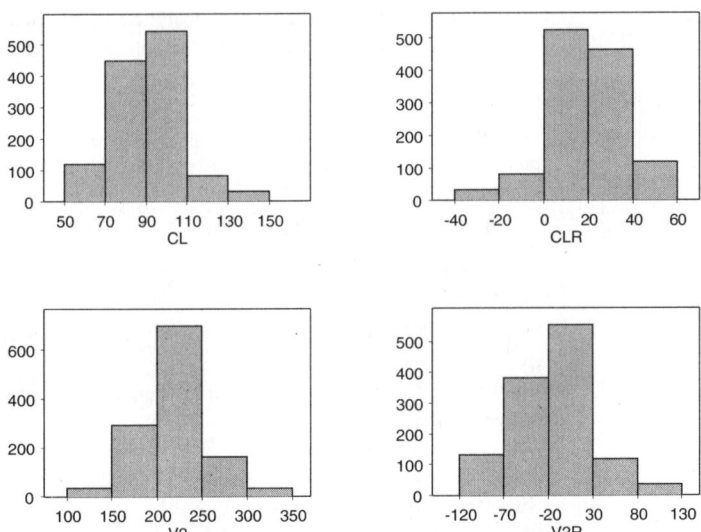

FIGURE 15.5 Histograms of individual predicted parameter values and their partial residuals.

TABLE 15.10

Situations Yielding Missing Values in a Population Pharmacokinetic Setting with Strategies for Data Handling

Situation	Data Treatment Strategies
	Missing Covariate
Categorical variable (i.e., missing subject's race)	Case deletion: Eliminate subject from data set Model/predict covariate: Given other data (covariate, outcomes, etc.), model covariate response and predict missing values Indicator variable for missingness

Continuous variable (i.e., missing subject's age)	Case deletion
	Imputation:
	  Mean/median: Straightforward; can be fine-tuned by averaging across subgroups if highly correlated
	  Linear regression: Based on known or established relationship with other available predictors
	  EM estimation
	Indicator variable for missingness

Ordinal variable (i.e., missing risk assessment)	Case deletion Model/predict covariate Indicator variable for missingness
	Compliance/Dosing Information
Compliance records (i.e., pill count data, etc.)	Case deletion: Can be guided by compliance data (i.e., deletion rule defined based on severity of noncompliance) Compliance model: Compliance as an outcome modeled as a response or utilized as a covariate expressed in the nonlinear mixed effect model Indicator variable: Compliance as a quantal (i.e., 0 = compliant; 1 = noncompliant) or ordinal (0 = fully compliant; 1 = >75%; 2 = 50 – 75%; 3 = <50%) response
Missing dosing records (i.e., infusion rate not recorded)	Case deletion Per protocol value used (most subject/patient protocols will have predefined subject allocation schedules providing dosing details for every subject enrolled) Model/predict dose
	Missing Response Data
No information	Case deletion Interpolation: Depending on data density within subject (linear, spline, etc.)

BQL data	Case deletion: Can be refined based on location and order of BQL data
	Sheiner method:
	  Delete all but the first BQL observation in a continuous series of BQLs
	  Assign the first BQL to 50% of the assay quantification limit (i.e., in NONMEM, DV = QL/2)
	  Use an additive plus proportional error model with the standard deviation of the additive component greater than or equal to 50% of the assay quantification limit

it often can be incorporated directly into the pharmacostatistical model as either a covariate or an outcome depending on how it is estimated. Recent works have utilized varied compliance markers in a population-based analysis with great success. Hence, the reader seeking to incorporate compliance data/expressions into a mixed effect model is best directed to the studies by Girard et al.[39,40] and Vrijens and Goetghebeur.[41]

Study Design

More so than traditional clinical pharmacology studies, proof-of-concept studies, or dose-finding studies, population PK/PD studies have been scrutinized because of the logistical, data accuracy, covariate assessment, and protocol design problems they present. Several meetings have been held to discuss these issues as well as specific issues relating to the design of population-based PK/PD studies. From these meetings, general guidelines have emerged:

- A population PK/PD study should not compromise the main objectives of the clinical trial.
- It is important to communicate the purpose of the population PK/PD analysis to the investigators and to convince them of the importance of accurate timing.
- Some prior knowledge of the PK/PD models and covariate relationships is necessary for the analysis of sparse Phase III data.
- Computer simulation and optimal design measures may be useful in defining sparse sampling time.
- Population methods must be specified as completely as possible in the protocol.

SIMULATION

The utility of simulations to forecast formulation, regimen, and subpopulation "what-if" scenarios has been widely appreciated, and such simulations have found their way into population PK/PD reports and manuscripts, NDA application summaries, and the background sections of protocols. Simulation can occur at multiple stages of a population pharmacokinetic project. For a prospectively designed study, simulation can examine design features such as sampling schemes and dosing regimens. The goal of such simulations is to assess the impact of input and sampling on the information content (i.e., is there sufficient sampling to estimate k_a?) and the assumptions made in a given study (i.e., the trough concentrations of greater than 80% of the patients studied should be greater than 40 nM at the lowest dose).

Simulation techniques vary based on the degree of randomness desired. If one wants the typical subject only, a deterministic solution is achieved by solving the structural model for a given input at defined sampling times. Error (intra- and inter-individual) partitions are not necessary. Similarly, the extremes of a certain parameter can be simulated by fixing the estimate of that parameter to its upper or lower bound and proceeding with a deterministic solution. More informed simulations will, of course, specify the degree of stochasticity required to answer the questions posed. For example, simulation can be used during the model development and refinement process through a combination of simulation and fitting. This technique is valuable for deciding whether a model is too complex for the data collected or if a specific covariate is important for the accurate prediction of structural model parameters. In the more common setting, the fully specified model is utilized allowing all random variables to roam within their likely limits. This will permit the most informative examination of regimen or subpopulation influence on the response.

NONMEM provides a great deal of flexibility in simulation options. The user manuals contain most of the details required. In short, the $SIMULATION record and the seed number initiates the process. The order of the simulation statement before or after the $ESTIMATION record will dictate whether data supplied are to be fit or simulated. The subproblems option defines the number of unique profiles to be simulated for a given set of subject identifiers and design points. The data file essentially contains the skeleton of the profile to be simulated (i.e., sampling points and covariate details are specified) with the DV field essentially a placeholder for the simulated

NMTRAN Control File

```
$PROBLEM Simulation Example;
$INPUT ID TIME AGE SEX WT DV MDV
EVID
$DATA Simex.csv
$SUBROUTINE ADVAN 1
$PK
TVCL=THETA(1)
TVV=THETA(2)
CL=TVCL+ETA(1)
V=TVV+ETA(2)
$THETA (.005) (10)
$OMEGA (0.1) (0.1)
$ERROR
Y=F + EPS(1)
$SIGMA (0.02)
$SIMULATION (1985) SUBPROBLEMS=50
ONLYSIMULATION
$TABLE ID TIME CL V NOPRINT
FILE=101.TAB NOHEADER
```

Data File, Simex.csv

| ID | TIME | AMT | AGE | SEX | WT | DV |
MDV	EVID					
1	0	250	55	1	50	.	1
1						
1	1	0	55	1	50	.	0
0						
1	2	0	55	1	50	.	0
0						
1	3	0	55	1	50	.	0
0						
1	4	0	55	1	50	.	0
0						
1	6	0	55	1	50	.	0
0						
1	8	0	55	1	50	.	0
0						
1	12	0	55	1	50	.	0
0						
1	16	0	55	1	50	.	0
0						
1	24	0	55	1	50	.	0
0						
1	48	0	55	1	50	.	0
0						
1	72	0	55	1	50	.	0
0						

FIGURE 15.6 NONMEM NMTRAN control and data files necessary to simulate 50 profiles from a specified subpopulation.

response. The $TABLE record defines the simulated output to be captured. Figure 15.6 shows a typical NMTRAN control and data file pairing to obtain 50 simulated profiles over a 72-h time course from a single-dose administration.

The extension of this work has resulted in attempts to simulate clinical trial outcomes based on "what-if" scenarios governed by treatment and study design possibilities. In short, the marriage of pharmacostatistical modeling (as with population-based analyses) and Monte Carlo techniques has created a simulation environment that permits the evaluation of drug and design features on safety and efficacy consequences, which have not been previously obtained in the pharmaceutical industry. Clinical trial simulation (CTS) is an emerging technology with few examples existing in the literature. Many industrial, academic, and regulatory groups are investing resources to pursue the areas that may best benefit from this technique. The Pharsight Corporation Trial Simulator is the current industry standard for CTS. The user interface is reasonably user-friendly and has flexibility to accommodate any design nuance. Other competitive programs are in development. NONMEM itself is very capable of simulating clinical trials. This is done using the $SIMULA-TION statement and the $TABLE statement for output as previously discussed. The difficulty with using NONMEM in this regard lies in the fact that the user must set up a data set (possibly a large data set) that includes all the events (including "other events" EVID = 2 for samples, randomizations, etc). Also, the event list is not adaptive, such that the collection of an extra sample based on some event in the trial (toxicity, lack of efficacy, etc.) cannot be accommodated. Dose adjustments need to be handled via a change in bioavailability. Hence, while possible, the number of "workarounds" to use NONMEM in this regard often precludes its use for CTS.

CONCLUSION

SUPPORT AND EDUCATION

Despite three decades of research on population-based PK/PD approaches and numerous examples in which these techniques have proved valuable, there remains some skepticism about the value of these efforts particularly in support of the registration of new medicines. Unfortunately, the paradigm shift advocated by Sheiner[13] moving clinical research toward learning-based vs. the traditional confirmatory (hypotheses-testing) type trial design is not widely embraced at present. This may change with time, but more common today is an environment in which those supporting these efforts (population pharamcokinetics) must continually educate the clinical community and occasionally their own management. In some instances the bias against population pharmacokinetics is specific to a therapeutic area and may be related to the perceived "failures" of past efforts. Regardless, organizations that have critiqued their internal efforts such as La Roche,[42] Parke-Davis,[43] and the U.S. FDA[44] have generally found their experiences with population pharmacokinetics valuable.

Academic, industrial, and regulatory groups with strong expertise in population-based techniques and analyses have habits and traits in common that make them effective. These include cross-disciplinary support of population pharmacokinetic programs including membership from pharmacokinetics, clinical, and statistical groups, in-house programming, data management, and application development support, multiple project leaders (i.e., no single guru), ongoing education in population pharmacokinetic techniques and methodology, and recognized clinical expertise and influence. These features are not easy to come by and those seeking to build such groups may find it difficult to secure the required head count in an environment in which such skills are not initially appreciated.

DOCUMENTATION AND "TELLING THE STORY"

The requirements for the U.S. regulatory submission of population-based analyses have been codified by the FDA guidance on population pharmacokinetics.[2] In general, the submitted report must include the statement of the objectives, design issues, a discussion of relevant assumptions, analyses hypotheses, methods, results and discussion, and commentary on the application of the results. The objectives section typically defines the desire to characterize the pharmacokinetics or pharmacodynamics based on sparse data, the major covariates to be investigated, the quantification of unexplained sources of variability, the discrimination of models, and the provision of meta-analyses of pooled study data if appropriate. The design section should discuss whether the analysis is based on a prospectively designed study or constitutes a retrospective analysis. It should also cover inclusion/exclusion criteria, describe the study population, explain or defend the choice of sampling times, discuss the measured responses, and provide commentary on possible sources of bias.

Documentation on assumptions should address those assumptions implicit in the pharmacokinetic or pharmacodynamic model and the statistical methodology chosen to evaluate the data. It should also state the assumed sensitivity of the parameters required to define the model relative to the data space being evaluated as well as any preconceived notions regarding biomarkers or surrogate markers evaluated as responses or covariates in the analysis. Hypotheses should be defined based on what was held *a priori* as true before the study or analysis, what was developed from preliminary or exploratory data analysis, and what would constitute a difference or equivalence in an effect or outcome. In some instances the criteria for difference as opposed to equivalence can be defined from a statistical viewpoint independent of the actual study design. This approach does not always confer regulatory acceptance, however.

The methods section must address the logistical, quality assurance, and data analysis specifications utilized during the conduct of the study or studies, the handling of the samples and data, and the subsequent data analysis and interpretation. It should contain the details of the analytical rigor including pertinent information on assay validation and measurement error and the sample-handling procedures, which may contribute to this error. An assessment of compliance should be provided along with any preconceived assumptions about the effect of compliance on outcomes. Finally, the details of the data analysis including model-building and reduction procedures along with criteria for outlier identification and data subsetting should be described in the methods section. The results and discussion section should be straightforward, but attention to the description of the data should be given as well as model-building and validation results. Specifically, dependent and independent variables should be characterized by distribution analysis and correlated with responses and covariates. The adequacy of the sampled distributions relative to the objectives should be discussed (i.e., Is the sampling adequate for the structural model proposed? Are the demographic characteristics of the study population consistent with the target population for which the drug is intended?). The guidance lists the desire of the FDA to be able to reconstruct analyses through the provision of electronic files (data, control files, complete output for key intermediate and final steps).[2] This is more easily managed and documented if the results section contains a flow diagram that identifies the analysis stages and indexes the various stages for referencing with electronic files and appendices. Finally, the results section must contain commentary on the reliability of the findings through some of the procedures discussed previously (accuracy and precision of estimates, validation, diagnostics, etc.). The results application statement should discuss any recommendations made based on the results of the population analysis, particularly if there are dose modification implications that are to be reflected in the labeling.

Although the documentation procedures discussed above are decidedly from an industrial perspective, they have universal appeal as indicators of scientific rigor and thoroughness. Documentation aside, an often unaddressed requirement for any scientist supporting these efforts is the attention paid to "telling the story." Indeed, some of the skepticism among our clinical colleagues is based on the ineffective or complete lack

TABLE 15.11
Common Practices for Effective Communication of Population Pharmacokinetic Ideas, Outcomes, and Best Practices

Tips for "Telling the Story"

Internal efforts	Form cross-disciplinary working group to discuss regularly current internal efforts, manpower, best practices, and review past performance.
	Solicit the assistance of discovery stage scientists to understand possible mechanism-based reasons for observed but unanticipated effects or correlations — explain what you're doing!
	Take advantage of internal symposia to educate scientists in related fields.
	Pursue integration of population pharmacokinetic efforts into the clinical development plans of early stage compounds.
	Create internal repository of complete population pharmacokinetic efforts (code, model development chronology, documentation, etc.)
	Identify external consultants (academic and CRO-based) to review internal efforts and deliver an external opinion of past efforts and the "added-value" of the population approach in general.
External efforts	Participate in local users groups (MUFPADA, East Coast Population Approach Group, PAGE, etc.).
	Publish; present at national or international meetings (helpful to present/publish in therapeutic area–specific venues in addition to settings where population pharmacokinetic methods and data analyses are well understood).
	Maintain a current appreciation of the population pharmacokinetic literature and be aware of regulatory (U.S., EU, etc.) changes in opinion and requirements.

of explanation regarding the approach in general, potential benefit of the effort, demonstrated return on investment, and historical results with related compounds or with compounds in relevant therapeutic areas. The utility of pharmacokinetic and more recently pharmacodynamic data in the drug development process is broadly held today. This came only after years of demonstrated benefit and a great deal of education. The broader acceptance of population pharmacokinetics will only come after sustained education and demonstration of benefit to therapy or to the drug development process itself in terms of time and cost savings. "Buy-in" among pharmacokinetic, statistical, and clinical groups is typically not achieved unless there is representation from each of these groups on the design and analysis team. Similarly, the communication of outcomes, benefits, pitfalls, and best practices must become part of the corporate memory of all stakeholder groups. Some useful practices for effective communication of population pharmacokinetics are shown in Table 15.11.

Population approach techniques are powerful tools useful in the characterization of sources of variation in the patient population particularly when applied to studies used for the registration of new chemical entities. The correct usage of these techniques requires significant investment on behalf of the analyst or scientist. The only way to master these techniques is to apply oneself in the field of study.

REFERENCES

1. Hochhaus, G., Barrett, J. S., and Derendorf, H., An evolution of pharmacokinetics and pharmacokinetic/dynamic correlations during the 20th century, *J. Clin. Pharmacol.*, 40:908–917, 2000.
2. Population Pharmacokinetics Guidance, U.S. Food and Drug Administration, Rockville, MD, 1999.
3. Sheiner, L. B., Rosenberg, B., and Melmon, K. C., Modelling of individual pharmacokinetics for computer-aided drug dosage, *Comp. Biomed. Res.*, 5:441–449, 1972.
4. Sheiner, L. B., Rosenberg, B., and Marathe, V. V., Estimation of population characteristics of pharmacokinetic parameters from routine clinical data, *J. Pharmacokinet. Biopharm.*, 5:445–479, 1977.
5. Beal, S. L. and Sheiner, L. B., NONMEM Users Guide, Parts I to IV, Division of Clinical Pharmacology, University of California, San Francisco, 1985.
6. Sheiner, L. B. and Beal, S. L., Evaluation of methods for estimating population pharmacokinetic parameters, I. Michelis-Menten model: routine clinical data, *J. Pharmacokinet. Biopharm.*, 8:553–571, 1980.
7. Sheiner, L. B. and Beal, S. L., Evaluation of methods for estimating population pharmacokinetic parameters, II. Biexponential model and experimental pharmacokinetic data, *J. Pharmacokinet. Biopharm.*, 9:635–651, 1981.
8. Sheiner, L. B. and Beal, S. L., Evaluation of methods for estimating population pharmacokinetic parameters, III. Monoexponential model and routine clinical data, *J. Pharmacokinet. Biopharm.*, 11:303–319, 1983.
9. Beal, S. L. and Sheiner, L. B., Estimating population pharmacokinetics, *CRC Crit. Rev. Biomed. Eng.*, 8:195–222, 1982.
10. Sheiner, L. B. and Beal, S. L., Bayesian individualization of pharmacokinetics: simple implementation and comparison with non-Bayesian methods, *J. Pharm. Sci.*, 71:1344–1348, 1982.

11. Beal, S. L., Population pharmacokinetic data and parameter estimation based on their first two statistical moments, *Drug. Metab. Rev.*, 15:173–193, 1984.

12. Sheiner, L. B. and Benet, L. Z., Pre-marketing observational studies of population pharmacokinetics of new drugs, *Clin. Pharmacol. Ther.*, 38:481–487, 1985.

13. Sheiner, L. B., Learning vs. confirming in clinical drug development, *Clin. Pharmacol. Ther.*, 61:275–291,1997.

14. Peck, C. et al., Opportunities for integration of pharmacokinetics, pharmacodynamics and toxikinetics in rational drug development, *Clin. Pharmacol. Ther.*, 51:465–473, 1992.

15. Peck, C., Population approach in pharmacokinetics and pharmacodynamics: FDA view, in *Proceedings of the COST B1 Conference*, 1992, 157–168.

16. Temple, R., The clinical investigation of drug use by the elderly: Food and Drug guidelines, *Clin. Pharmacol. Ther.*, 42:681–685, 1987.

17. Holford, N. H. G. and Peace, K., The effect of tacrine and lecithen in Alzheimer's disease. A population pharmacodynamic analysis of five clinical trials, *Eur. J. Clin. Pharmacol.*, 47:17–24, 1994.

18. Steimer, J. L. et al., The population approach: rationale, methods, and applications in clinical pharmacology and drug development, in *Pharmacokinetics of Drugs, Handbook of Experimental Pharmacology*, Vol. 110, Welling, P. G. and Balant, L. P., Eds., Springer-Verlag, Berlin, 1994, 404–451.

19. Sanathanan, L. P., Random effects modeling in population: kinetic/dynamic analysis, *Drug Inf. J.*, 25:307–318, 1991.

20. Davidian, M. and Giltinan, D. M., Hierarchical nonlinear models, in *Nonlinear Models for Repeated Measurement Data*, Chapman & Hall, London, 1995, 1–11, 98–115.

21. Karlsson, M. O. and Sheiner, L. B., The importance of modeling interoccasion variability in population pharmacokinetic analyses, *J. Pharmacokinet. Biopharm.*, 21(6):735–750, 1993.

22. Karlsson, M. O. et al., Assumption testing in population pharmacokinetic models: illustrated with an analysis of moxonidine data from congestive heart failure patients, *J. Pharmacokinet. Biopharm.*, 26(2):207–246, 1998.

23. Aarons, L., Population pharmacokinetics: theory and practice, *Br. J. Clin. Pharmacol.*, 32:669–670, 1991.

24. Aarons, L., Software for population pharmacokinetics and pharmacodynamics, *Clin. Pharmacokinet.*, 36(4):255–264, 1999.

25. Zhou, X.-J. et al., Population pharmacokinetics of nevirapine, zidovudine, and didanosine in human immunodeficiency virus-infected patients, *Antimicrob. Agents Chemother.*, 43(1):121–128, 1999.

26. Grevel, J., Thomas, P., and Whiting, B., Population pharmacokinetic analysis of bisprolol, *Clin. Pharmacokinet.*,17(1):53–63, 1998.

27. Mungall, D. R. et al., Population pharmacokinetics of racemic warfarin in patients, *J. Pharmacokinet. Biopharm.*, 13(3):213–227, 1985.

28. Grasela, T. H. and Donn, S. M., Neonatal population pharmacokinetics of phenobarbital derived from routine clinical data, *Dev. Pharmacol. Ther.*, 8:374–383, 1985.

29. Fuseau, E. et al., The integration of the population approach into drug development:a case study, naratriptan, in *European Cooperation in the Field of Scientific and Technical Research: The Population Approach, Measuring and Managing Variability in Response, Concentration and Dose*, Aarons, L. et al., Eds., Brussels, 1997, 203–214.

30. Maier, G. A. et al., Characterization of the highly variable bioavailability of tiludronate in normal volunteers using population approach methodologies, *Eur. J. Drug Metab. Pharmacokinet.*, 24(2):249–254, 1999.

31. Levy, R. H. et al., Pharmacokinetic interactions of chronic drug treatment in epilepsy: carbamazepine, in *The Effects of Disease State on Drug Pharmacokinetics*, Benet, L., Ed., Academy of Pharmaceutical Sciences, Washington, D.C., 1976, 87–95.

32. Rostami-Hodjegan, A. et al., Population pharmacokinetics of methadone in opiate users: characterization of time-dependent changes, *Br. J. Clin. Pharmacol.*, 48:43–52, 1999.

33. Ette, E. I., Population model stability and performance, *J. Clin. Pharmacol.*, 37:486–495, 1997.

34. Bruno, R. N., et al., A population pharmacokinetic model for Docetaxel (Taxotere): model building and validation, *J. Pharmacokinet. Biopharm.*, 24:153–172, 1996.

35. Balci, O. and Sargent, R. G., A bibliography on the credibility assessment of and validation of simulation and mathematical models, *Simuletter*, 15:15–27, 1984.

36. Mandema, J. et al., Characterization and validation of a pharmacokinetic model for controlled-release oxycodone, *Br. J. Clin. Pharmacol.*, 42:747–756, 1996.

37. Berin, T. R. and Rubin, D. B., The analysis of repeated-measures data on schizophrenic reaction times using mixture models, *Stat. Med.*, 14:747–768, 1995.

38. Ette, E. I. and Ludden, T. M., Population pharmacokinetic modeling: the importance of informative graphics, *Pharm. Res.*, 12(12):1845–1855, 1995.

39. Girard, P. et al., A Markov mixed effect regression model for drug compliance, *Stat. Med.*, 17:2313–2334, 1998.

40. Girard, P. et al., Do we need full compliance data for population pharmacokinetic analysis? *J. Pharmacokinet. Biopharm.*, 24(3):265–282, 1996.

41. Vrijens, B. and Goetghebeur, E., The impact of compliance in pharmacokinetic studies, *Stat. Methods Med. Res.*, 8(3):247–262, 1999.

42. Reigner, B. G. et al., An evaluation of the integration of pharmacokinetics and pharmacodynamic principles in clinical drug development: experience within Hoffman La Roche, *Clin. Pharmacokinet.*, 33:142–152, 1997.

43. Olson, S. C. et al., Impact of population pharmacokinetic-pharmacodynamic analyses on the drug development process: experience at Parke-Davis, *Clin. Pharmacokinet.*, 38(5):449–459, 2000.

44. Sun, H. et al., Population pharmacokinetics: a regulatory perspective, *Clin. Pharmacokinet.*, 37:41–58, 1999.

16 The Linear Systems Approach

Peter Veng-Pedersen

CONTENTS

1-56676-973-6/02/$0.00+$1.50
© 2002 by CRC Press LLC

INTRODUCTION

The relative advantages and disadvantages of linear system analysis (LSA) and noncompartmentally based pharmacokinetic (PK) modeling to other modeling approaches have previously been discussed by several authors.[1-10] A modern interpretation of LSA refers to modeling that makes use of LSA principles or, additionally, modeling that focuses on response relationships more than structure relationships.[9] *Systems analysis* is the preferred terminology in this work rather than noncompartmental modeling because some of the so-called noncompartmental methods do in fact make use of a "compartment" (e.g., sampling space). This work is mainly aimed at covering the basic principles of LSA-based PK modeling. Nonlinear systems analysis modeling is dealt with to a much lesser extent.

LINEAR SYSTEMS ANALYSIS

LSA is a modeling approach in pharmacokinetics/pharmacodynamics (PK/PD) that applies general linear principles such as convolution and deconvolution to simplify and generalize linear PK/PD relationships.

It would be incorrect to characterize LSA approaches generally as empirical because LSA is often based on a mechanism, e.g., molecular stochastic independence (MSI) that is a physiologically and mechanistically realistic, basic mechanism (see the appendix to this chapter).

Also, the LSA approach cannot generally be characterized as a "black box" approach because LSA offers various degrees of structure differentiation. In fact, it offers an excellent opportunity to analyze specific kinetic mechanisms (e.g., drug absorption and drug elimination) unconfounded by kinetic elements that theoretically or practically may be difficult or impossible to resolve by traditional modeling means.

THE LINEARITY CONCEPT

The concept of linearity is differently defined in different contexts and has often been the cause of confusion in PK/PD. For example, in regression analysis, linearity often refers to linear regression equations, which have the following general form, linear in the parameters, p, to be determined:

$$y = p_1 g_1(x) + p_2 g_2(x) + \dots + p_n g_n(x) \tag{16.1}$$

where $g_j(x)$ are arbitrary functions of the independent variables. In pharmacokinetics, "linear pharmacokinetics" or "linear drug disposition" have often been interpreted traditionally in a narrow sense as kinetics described by linear compartmental models or first-order processes. Linearity in this case does not refer to linear regression equations, but refers to the fact that linear compartmental models involve first-order compartmental transfer and accordingly are described by linear differential equations. More generally, it is recognized that the solution of systems of linear differential equations guarantees certain general linear properties such as superposition and dose linearity. Thus, in a more modern interpretation linearity of compartmental models refers to such general properties. There are many other types of models such as physiological models that may also be described by systems of linear differential equations and, as such, result in linear system properties. However, linear differential equations are *not* a required foundation for the general linear properties of a linear kinetic system. As discussed in this work, a kinetic system exhibiting general linear properties may be described in terms of purely statistical kinetic principles (molecular stochastic independence, MSI) without any connections to linear differential model equations or other structure-specific mathematical formulations. LSA considers linearity at a higher, more general level than other modeling approaches. Linearity in LSA is based on operational principles not limited to the specificity of linear compartmental systems or other structured modeling systems such as physiological models or recirculation models.

Linear Systems: A linear system is a system having two or more kinetic variables that are linearly related through a linear relationship.

Linear Relationships: A linear relationship (L) between two kinetic variables x and y,

$$y = L(x) \tag{16.2}$$

is a relationship that follows the general property of the linear operator (L).

In the following presentation, "L" denotes a general linear operational relationship (defined in section on Mathematical Element of LSA).

Nonlinear Relationships: A relationship between two kinetic variables x and y is nonlinear if it does not follow the property of a linear relationship.

Rationale for Using LSA: LSA may be a useful alternative to other PK/PD analysis approaches for several reasons:[9]

- The analysis may be simplified by LSA through a simpler model structure.
- LSA is largely based on generalization. Linearity is considered in a general way not specific to a particular structure such as compartmental, physiological recirculation, etc.
- The analysis may be done by LSA with fewer assumptions.
- The assumptions, e.g., general linear relationships between kinetic variables, are more readily verified and tested than structure assumptions of more structured modeling approaches.
- Structure identifiability problems may be avoided or significantly reduced by the use of the LSA approach.[6] Specific kinetic processes such as drug absorption and elimination may be analyzed in a more direct way by removing some structure complexity that otherwise may confound the analysis.
- Numerically, the LSA approach may be implemented by calculations that involve estimation by (general) linear regression, e.g., use of cubic spline functions.[11-14] The intrinsic problems of nonlinear estimation common in more structured methods can thereby be avoided or significantly reduced.
- Through its greater generality, the LSA approach allows a wider interpretation of the kinetics being evaluated, and makes more apparent what is required experimentally to elucidate the kinetics more conclusively.

MATHEMATICAL ELEMENTS OF LSA

THE LINEAR OPERATOR L

A linear operator is a mathematical operator with the following property:

$$L(k_1 x_1 + k_2 x_2) = k_1 L(x_1) + k_2 L(x_2) \tag{16.3}$$

where x_1 and x_2 are two arbitrary cases of the variable x, and k_1 ure k_2 arc two arbitrary constants.

It is readily shown that several simple mathematical operations such as differentiation, integration, and linear transformations (scaling and translation), as well as more complex operations such as convolution, deconvolution, and Laplace transformations (and inverse Laplace transformation) have the above linear operator property.

Variable x in the definition of a linear operator is typically a time-dependent variable, $x(t)$. The specific case $k_1 = k_2 = 1$ results in the well-known simple superposition property of the linear operator.

$$L(x_1 + x_2) = L(x_1) + L(x_2) \tag{16.4}$$

The specific case $k_2 = 0$ results in the "proportionality property" of the linear operator.

$$L(kx) = kL(x) \tag{16.5}$$

This property is exemplified in the well-known dose linearity in linear pharmacokinetics. Compounded linear operations are simplified as a linear operation, i.e.,

$$L_n(\ldots L_2(L_1(*)) \ldots) = L(*) \tag{16.6}$$

Kinetically, this corresponds to a preservation of linearity in a chain of linear kinetic steps. A linear combination of linear operators is simplified as a linear operation, i.e.,

$$k_1 L_1(*) + k_2 L_2(*) + \ldots + k_n L_n(*) = L(*) \tag{16.7}$$

for all values of the constants $k_1, k_2, \ldots, k_n$. The inverse linear operation L^{-1} defined as

$$L^{-1}(L(*)) = * \tag{16.8}$$

is a linear operator. For example, convolution is a linear operation. Since deconvolution is the inverse operation of convolution, it is a linear operation. If two kinetic variables y_1 and y_2 are both linearly related to a third kinetic variable x, then they are linearly related; from

$$y_1 = L_1(x) \tag{16.9}$$

$$y_2 = L_2(x) \qquad (16.10)$$

it follows that y_1 is linearly related to y_2 (or y_2 is linearly related to y_1),

$$y_1 = L(y_2) \qquad (16.11)$$

Most linear operators operating on a time-dependent variable, $x(t)$, produce a time-dependent result. However, some linear operators (e.g., integration type of operators) may result in a constant. For example, the total integration of the drug concentration profile, $c(t)$, produces a constant, the area under the curve (AUC). The AUC is accordingly linearly related to the drug concentration. More surprising is the fact that a linear operation performed on a constant may result in a time-dependent result. For example, the linear operation convolution will always have a time-dependent result.

LINEAR RELATIONSHIPS CONSIDERED AS RESPONSE SYSTEMS

Theorem 16.1 If x and y are linearly related and y_1 is the response to x_1 and y_2 is the response to x_2, then for arbitrary constants k_1 and k_2 the response to $x = k_1 x_1 + k_2 x_2$ will be

$$y = k_1 y_1 + k_2 y_2 \qquad (16.12)$$

Theorem 16.2 Let y_1 and y_2 denote the responses to arbitrary cases of x: $x = x_1$ and $x = x_2$, respectively. The variable y is linearly related to x if, for arbitrary constants k_1 and k_2, the response to $x = k_1 x_1 + k_2 x_2$ is

$$y = k_1 y_1 + k_2 y_2 \qquad (16.13)$$

IDENTIFICATION OF LINEARITY

IDENTIFICATION OF LINEARITY FROM MODEL EQUATIONS

Linearity between variables in model equations that describe a kinetic system may be readily identified even for complex equations presented in implicit or explicit form, including differential or nondifferential forms, according to the following theorem.

Theorem 16.3 Consider a mathematical relationship between variables x and y defined by some equations, E. Let x_1 and x_2 denote arbitrary cases of x, and let y_1 and y_2 denote the corresponding solutions; i.e., (x_1, y_1) and (x_2, y_2) each satisfies E. The relationship between x and y is linear, $y = L(x)$, if (x_3, y_3) given by

$$x_3 \equiv k_1 x_1 + k_2 x_2 \qquad (16.14)$$

$$y_3 \equiv k_1 y_1 + k_2 y_2 \qquad (16.15)$$

is a solution to E and k_1 and k_2 are arbitrary constants.

As an example of the application of Theorem 16.3, consider the differential equations for a two-compartment linear model

$$x_1' = -E_1x_1 + k_{21}x_2 + I_1 \tag{16.16}$$

$$x_2' = k_{12}x_1 - E_2x_2 \tag{16.17}$$

where E_1 and E_2 are the sum of the first-order exit rate constants from compartments 1 and 2, respectively, and I_1 is an arbitrary input into compartment 1 (the analysis also applies in the cases of a separate or simultaneous input, I_2, into compartment 2, but this is not considered for the sake of simplicity). Consider an arbitrary case, denoted by (1) as subscript, i.e., by definition let $x_{1(1)}$, $x_{2(1)}$, and $I_{1(1)}$ satisfy the differential equations:

$$x'_{1(1)} = -E_1x_{1(1)} + k_{21}x_{2(1)} + I_{1(1)} \tag{16.18}$$

$$x'_{2(1)} = k_{12}x_{1(1)} - E_2x_{2(1)} \tag{16.19}$$

Similarly, consider another arbitrary case denoted by subscript (2).

$$x'_{2(2)} = -E_1x_{1(2)} + k_{21}x_{2(2)} + I_{1(2)} \tag{16.20}$$

$$x'_{2(2)} = k_{12}x_{1(2)} - E_2x_{2(2)} \tag{16.21}$$

Multiplying the two equations of case (1) by the arbitrary constant k_1 and the two equations of case (2) by the arbitrary constant k_2 followed by pairwise adding produces the following equations after simple rearrangement:

$$x'_{1(3)} = -E_1x_{1(3)} + k_{21}x_{2(3)} + I_{1(3)} \tag{16.22}$$

$$x'_{2(3)} = k_{12}x_{1(3)} - E_2x_{2(3)} \tag{16.23}$$

where

$$x_{1(3)} \equiv k_1x_{1(1)} + k_2x_{1(2)} \tag{16.24}$$

$$x_{2(3)} \equiv k_1x_{2(1)} + k_2x_{2(2)} \tag{16.25}$$

$$I_{1(3)} \equiv k_1I_{1(1)} + k_2I_{1(2)} \tag{16.26}$$

From this it is concluded that:

- Variables x_1 and x_2 are linearly related according to Theorem 16.3.
- Variables x_1 and x_2 are also each linearly related to the arbitrary input I_1. The linearity between the input (I_1) and the variables x_1 and x_2 is verified

similarly to above. Consider, for example, the relationship between x_1 and I_1 (x_2 is then considered an auxiliary variable). Accordingly, $(x_{1(1)}, I_{1(1)})$ and $(x_{1(2)}, I_{1(2)})$ are solutions leading to $(x_{1(3)}, I_{1(3)})$, which is also a solution, where $x_{1(3)} = k_1 x_{1(1)} + k_2 x_{2(1)}$ and $I_{1(3)} = k_1 I_{1(1)} + k_2 I_{1(2)}$. Thus, x_1 and I_1 are linearly related according to Theorem 16.3. Linearity between x_2 and I_1 is shown in an identical manner.

IDENTIFICATION OF LINEARITY FROM MEASURED RESPONSES

Linearity between measured pharmacokinetic response variables in a kinetic system may be readily identified in a manner similar to above, according to the following theorem.

Theorem 16.4 Consider a kinetic system with responses R_1 and R_2 measured at the same or distinct sampling sites S_1 and S_2 resulting from a drug input, I, at an input site P. Let $(R_{1(1)} R_{2(1)})$ and $(R_{1(2)} R_{2(2)})$ denote the responses resulting from two different arbitrary inputs $I_{(1)}$ and $I_{(2)}$ at P, respectively. Let $(R_{1(3)} R_{2(3)})$ denote the responses resulting from a third input $I_{(3)}$ at P that is an arbitrary linear combination of the first two inputs

$$I_{(3)} = k_1 I_{(1)} + k_2 I_{(2)} \tag{16.27}$$

with k_1 and k_2 as arbitrary constants. R_1, R_2, and I are linearly related (L) if

$$R_{1(3)} = k_1 R_{1(1)} + k_2 R_{1(2)} \tag{16.28}$$

$$R_{2(3)} = k_1 R_{2(1)} + k_2 R_{2(2)} \tag{16.29}$$

A PK response may be measured as a concentration typically denoted by $c(t)$, but can as well be an amount, often denoted $A(t)$. For the sake of generality PK responses/variables, which typically are time-dependent functions, will therefore be denoted by $R(t)$ or R.

LINEAR DRUG DISPOSITION

Linear drug disposition is defined as a linear relationship (L) between the input, I, of drug into the systemic circulation and the resulting measured concentration, C, of drug in the systemic circulation:

$$C = L(I)$$

In a structured modeling context, linear drug disposition is commonly thought of as a disposition governed by first-order processes. Contrary to this, the LSA approach defines linear disposition independently of the structure of the disposition processes (distribution and elimination). Linear disposition is simply a linear relationship (L) between input (I) and the measured systemic drug concentration, C. A so-called nonlinear absorption is sometimes observed in oral administrations when an increase or decrease in the dose results in a nonproportional change in the drug level, C. Such nonlinearity does not imply that the drug has a nonlinear disposition because the nonlinearity may arise from a nonlinear presystemic (first-pass) biotransformation or degradation.

Disposition linearity implies proportionality between intensity of input and the systemic drug concentration. The input intensity proportionality is defined in the following general manner: If an arbitrary input (I) results in the concentration $C = L(I)$, then an increase or decrease in the input (new input = kI) results in the same proportional change in the drug concentration. This is due to the general proportionality property of the linear operator $L(kI) = kL(I) = kC$. This proportionality is generalized to any type of input, continuous as well as discontinuous and bolus input.

The LSA linear drug disposition definition does not impose any restrictions on the mathematical form of the drug input response. This is in contrast to linear compartmental models. For example, a two-compartment model implies a biexponential response to an IV bolus injection:

$$c = A_1 e^{-\alpha_1 t} + A_2 e^{-\alpha_2 t} \tag{16.30}$$

where A_1, A_2, α_1, and α_2 are parameters that are always positive for this disposition model. In contrast, LSA does not impose this structure constraint on the parameters. For example, in LSA, when representing the IV bolus response by a biexponential equation, the A_2 parameter may be allowed to become (slightly) negative, which for $A_1 + A_2 \geq 0$, and $\alpha_2 > \alpha_1$ still ensures non-negativity for $t \geq 0$ as physiologically required, but allows for a more versatile set of curve shapes. Also, strictly speaking in reality $c(0) = 0$, since the concentration is not measured at the injection site. This physiological constrain is readily considered in a structure-independent way by the LSA approach.

MOLECULAR/PROBABILITY BASIS FOR LINEARITY IN PHARMACOKINETICS

LINEAR INPUT–RESPONSE LINEARITY

Kinetic linearity has traditionally been associated with specific kinetic mechanisms, e.g., first-order processes for drug distribution and elimination. In contrast, in LSA it is not necessary to assume such physiologically problematic kinetic mechanisms. This is well exemplified in dealing with input–response systems in PK. Such systems can be considered in LSA in a very fundamental way on the basis of sound physiological principles involving largely statistical kinetic concepts, namely, molecular stochastic independence (MSI).

MOLECULAR STOCHASTIC INDEPENDENCE

Consider a drug molecule introduced into a body at point P at the arbitrary time T. Let $\Pr(t)$ denote the probability that the drug molecule, in its original or biotransformed form, is present at a given sampling space, S, at an arbitrary later time $T + t$. The drug exhibits *molecular stochastic independence* (MSI) if $Pr(t)$ is independent of the presence in the body of other such drug molecules in original or biotransformed form.

Theorem 16.5 MSI results in a linear input (I)–response (R) relationship.

$$R = L(I) \tag{16.31}$$

The linearity (L) is a convolution-type linearity involving a unit impulse response (UIR) function ("characteristic response"):

$$R = \int_o^t UIR(t - u)I(u)du \equiv UIR \ (*) \ I \tag{16.32}$$

The UIR is given by

$$UIR \equiv L(I_{bolus})/D_{bolus} \tag{16.33}$$

$$I_{bolus} \equiv \text{bolus input of a dose } D_{bolus} \tag{16.34}$$

(see the appendix). The UIR of the linear input–response relationship is directly related to the MSI probability function, $Pr(t)$ (see the appendix) and is proportional to this function when the response, R, is the amount or concentration of drug, in original or biotransformed form, present in the sampling space. It is not required that the concentration in the sampling space be uniform. In that case, the concentration response is defined as an "average" concentration in the sampling space, i.e., the amount in the sampling space divided by the constant volume of the sampling space. This will be the proper concentration when the amount in the sampling space is linearly related to the input. Note that the derivation (see the appendix) is based on the amount in the sampling space. The derivation, therefore, makes no assumption of concentration uniformity in the sampling space.

LINEAR RESPONSE–RESPONSE LINEARITY

Theorem 16.6 MSI leads to linearity between kinetic response variables, R_1 and R_2, resulting from the same input at the same input site.

$$R_1 = L_1(I) \tag{16.35}$$

$$R_2 = L_2(I) \tag{16.36}$$

The MSI probability function $Pr(t)$ and thus the UIR is characteristic for the transport of the drug molecules from the given input point, P, to the sampling space, S (see the appendix). Depending on the definition of the sampled response as a biotransformed form of the drug or unchanged drug, the transport from P to S may involve biotransformation, in addition to physical transport. Strictly speaking, the drug input site, P, must be an entry via a point. An input site spread out over a region will not result in a simple convolution-type input–response relationship if the probability function $Pr(t)$ is different from different points of the input site region. The sampling space, S, can be any physiologically fixed space. In blood drawings, this space will simply be the space in the body occupied by the blood drawn in the sampling. Accordingly, in simple blood drawings the sampling space will, strictly speaking, differ for different volumes of blood drawn. This becomes important to consider if the drug is not uniformly distributed in the sampling space, which may happen since the drug does not need to be evenly distributed in the sampling space for the convolution relationship to true (see the appendix). Accordingly, the sampling space can include heterogeneous biopsy tissue samples or whole organs or regions of the body analyzed noninvasively by electronic scanning. In fact, the sampling space can consist of disjoint, widely different spaces in the body. This is in strong contrast to the usual compartmental modeling assumption that considers compartments to be homogeneous.

The convolution derivation assumes a time-invariant system. This follows from the fact that the probability function $Pr(t)$ in the derivation (see the appendix) does not depend on T, the time the drug molecules enter at the input point, P. The convolution derivation also assumes an instantaneous sampling procedure.

CONVOLUTION AS A LINEAR OPERATOR

The convolution linear operator met in linear input–response systems (Theorem 16.5) is of great importance in LSA and is encountered in many different contexts. This linear operator consists of a "kernel function," $g(t)$, and a specific integration operation,

$$L_g(*) \equiv \int_0^t g(t-u)(*)du \tag{16.37}$$

The operand (*) can be another function, $h(t)$, in which case the operation simply becomes

$$L_g(h(t)) = \int_0^t g(t-u)h(u)du \tag{16.38}$$

The operand can also be a constant, k, in which case the operation only involves a simple integration:

$$L_g(k) = k\int_0^t g(u)du \tag{16.39}$$

When the operand is an impulse function, $D\delta(t)$, $\delta(t) \equiv$ Dirac delta function, $D =$ amount of impulse, then the operation becomes a simple scaling of the kernel function:

$$L_g(D\delta(t)) = Dg(t) \tag{16.40}$$

The integration operation of the convolution operator is symmetric with respect to the kernel and the operand. Thus, $L_g(h(t)) = L_h(g(t))$. For this reason a convolution involving two functions, g and h, is often written $g * h$ (or $h * g$).

$$g * h = h * g = \int_0^t h(t-u)g(u)du = \int_0^t h(u)g(t-u)du \qquad (16.41)$$

Table 16.1 presents some useful general properties of convolution, and Table 16.2 presents some useful specific convolution expressions.

GEOMETRIC INTERPRETATION OF CONVOLUTION

The convolution operation between two functions is explained in Figure 16.1. The convolution of function g (Figure 16.1, graph A) and function h (Figure 16.1, graph B) results in the convolution function $g * h$ (Figure 16.1, graph F). Five steps are involved in the calculation of the value of the convolution function at the arbitrary time t (vertical line in Figure 16.1, graph F):

- The two functions are separated by the time, t. One function (h) is flipped around the y axis.
- The function flipped around and the other function (g) are overlaid.

TABLE 16.1
Some General Properties of Convolution[a]

Item

1. $g * h \equiv \int_0^t g(t-u)h(u)du$

2. $g * h = h * g$

3. $g * h * q = g * (h * q) = (g * h) * q$

4. $g * k = k\int_0^t g(u)du$

5. $g * (k_1 h_1 + k_2 h_2) = k_1 g * h_1 + k_2 g * h_2$

6. $\int_0^\infty g * h \, dt = (\int_0^\infty g \, dt)(\int_0^\infty h \, dt)$

7. $\int_0^t g * h \, dt = (\int_0^t g \, dt) * h = g * (\int_0^t h \, dt)$

8. $\int_0^\infty t(g * h)dt = (\int_0^\infty t g \, dt)(\int_0^\infty h \, dt) + (\int_0^\infty g \, dt)(\int_0^\infty t h \, dt)$

9. $\lim_{t \to \infty}(\int_0^t g \, dt) * h = (\int_0^\infty g \, dt)(\int_0^\infty h \, dt)$

10. $g * h = 0$ at $t = 0$ if g and h are not impulse functions

11. $g * h \to 0$ for $t \to \infty$ if g and $h \to 0$ for $t \to \infty$

12. $[f * g]' = f(t)g(0) + f(t) * g'(t) = f(0)g(t) + f'(t) * g(t)$

13. $[g * k]' = kg$

14. $g * h > 0$ if $g > 0$ and $h > 0$

15. $g * \delta(t) = g$, where $\delta(t)$ if the Dirac delta function

[a] k, k_1, and k_2 are constants.

TABLE 16.2
Some Specific Convolutions[a]

Item	g	h	$g * h \equiv \int_0^t g(t-u)h(u)\,du$
1	$e^{-\alpha t}$	$e^{-\beta t}$	$(e^{-\alpha t} - e^{-\beta t})/(\beta - \alpha)$ for $\alpha \neq \beta$
2	$e^{-\alpha t}$	$e^{-\alpha t}$	$te^{-\alpha t}$
3	k	$e^{-\alpha t}$	$k(1 - e^{-\alpha t})/\alpha$
4	k_1	k_2	$k_1 k_2 t$
5	k	t^n	$kt^{n+1}/(n+1)$ $\qquad n = 0, 1, 2, \ldots$
6	t^n	t^m	$n!m!\,t^{n+m+1}/(n+m+1)!$
			$n = 0, 1, 2, \ldots\ m = 0, 1, 2, \ldots$
7	t	$e^{-\alpha t}$	$[t - (1 - e^{-\alpha t})/\alpha]/\alpha$
8	t^2	$e^{-\alpha t}$	$[t^2 - 2t/\alpha + 2(1 - e^{-\alpha t})/\alpha^2]/\alpha$
9	t^3	$e^{-\alpha t}$	$[t^3 - 3t^2/\alpha + 6t/\alpha^2 - 6(1 - e^{-\alpha t})/\alpha^3]/\alpha$

[a] Constants are denoted by k.

- The overlaid functions are multiplied; i.e., corresponding values of the overlaid curves are multiplied from time = 0 to time = t to form the product curve seen in Figure 16.1, graph E.
- The product curve is integrated from time = 0 to time = t.

The resulting integral (AREA) is equal to the value of the convolution of the two functions at time t, i.e., AREA = $g(t) * h(t)$. This geometric interpretation is

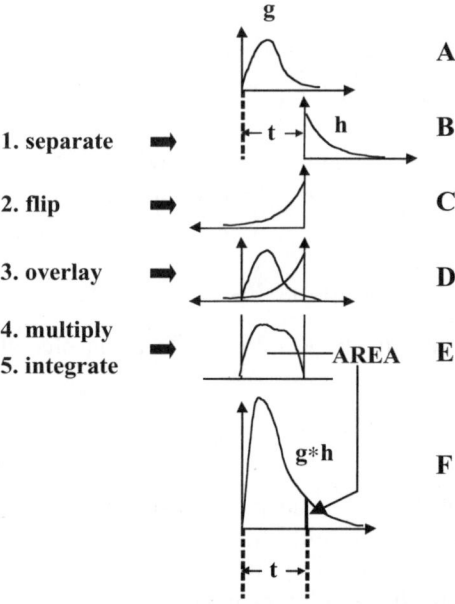

FIGURE 16.1 Geometric interpretation of convolution.

not only useful for understanding the convolution operation, but is also useful in deriving important relationships for more complex convolution operations, e.g., convolution of discontinuous functions, and in determining bounds for the convolution, etc.

CONVOLUTION AND LAPLACE TRANSFORMS

The Laplace transform, which is a linear operator, is frequently used as a mathematical tool when dealing with linear systems. The Laplace transform is often useful in dealing with more complex convolution relationships. For example, consider the following property of the Laplace transform operation L_L:

$$L_L(g * h) = L_L(g)L_L(h) \tag{16.42}$$

where

$$L_L(*) \equiv \int_0^\infty e^{-st}(*)dt \tag{16.43}$$

Thus, to find the convolution expression for two function g and h, one can form the product of the Laplace transform of g and the Laplace transform of h, and then invert. This is a useful approach if the product of the Laplace transforms is a Laplace transform that readily can be backtransformed to give the convolution expression. Consider, for example,

$$L_L(t^n) = n!/s^{n+1} \tag{16.44}$$

Accordingly,

$$L_L(t^n * t^m) = L_L(t^n)L_L(t^m) = n!m!/s^{n+m+2} \tag{16.45}$$

Inversion of the Laplace transform $n!m!/s^{n+m+2}$ is then readily done using the specific Laplace transform formula for t^n which then gives the expression shown in Table 16.2, item 6.

LSA AND DRUG INPUT–RESPONSE IN PK

As presented above (Theorem 16.5) MSI results in a convolution-type input–response relationship.

$$R = UIR * I \tag{16.46}$$

From the derivation of this relationship, it is evident that the input–response relationship is characterized by the following five items:

1. Position of input (input point, or input site)
2. Position of response (sampling space)

3. Type of response measured (e.g., concentration or amount of the drug or biotransformed drug)
4. Type of input (e.g., rate or amount of input)
5. A convolution-type linear operator (UIR *) that is specific to the subject-drug and items 1 to 4 above

The input, I, is commonly defined as the rate of drug input $I = f(t)$. However, the convolution relationship also allows for bolus-type input: $I = D\delta(t)$ where $\delta(t)$ is the Dirac delta function (see Table 16.1, item 14). The cumulative amount of input obtained by integrating the rate of input is related to the area under the response curve by a convolution relationship identical to the input–response relationship, i.e., see Equations 16.47 through 16.49,

$$A(t) = \int_0^t I \, dt = \int_0^t f(t) \, dt \tag{16.47}$$

$$AUR(t) = \int_0^t R \, dt \tag{16.48}$$

$$AUR(t) = UIR(t) * A(t) \tag{16.49}$$

The response, R, may be expressed as a concentration or an amount and is not limited to the parent drug but can be any biotransformed form of the drug. If the drug and metabolites are linearly related to the input (Theorem 16.5), then the sum of the drug and the metabolite input–response (concentration or amounts) is also linearly related to the input through a simple convolution relationship. This follows from the following derivation:

For the drug (D):

$$R_D = UIR_D * I \tag{16.50}$$

For the metabolites (Mj):

$$R_{Mj} = UIR_j * I \tag{16.51}$$

For response = drug + metabolites:

$$R \equiv R_D + \sum_j R_{Mj} = \left(UIR_D + \sum_j UIR_j \right) * I = UIR * I \tag{16.52}$$

Thus, if the response measured is radioactivity from a labeling of the drug, then the total radioactivity resulting from the drug and metabolites is linearly related to input through a convolution relationship if both the metabolites and the drug are linearly related to the input.

The response, R, of an input–response relationship depends on the whole history of the input. The importance of this fact should not be ignored in dealing with prediction of responses: *If a response variable is related to the input through a convolution relationship and its current value is perturbed from its initial, baseline value by prior input(s), then accurate predictions of the future response require information about the prior input.* For example, the value of the response variable at time $= T$, $R(T)$, could be achieved by a simple bolus injection at time $= T$, or by a constant-rate input (infusion) at a rate that ensures that the response reaches $R(T)$ at time T. The future response resulting from discontinuing the input at time T will be different in these two cases that both result in the same response value at time T.

SIMPLE DERIVATIONS OF COMMON PK EXPRESSIONS USING LSA

Often PK equations for common inputs such as first-order absorption; zero-order input, or constant-rate infusions are derived from the differential equations describing the kinetics. LSA offers a very attractive alternative to such derivations that is more direct and does not require the use of differential equations or Laplace transforms. The LSA derivations can be done simply by elementary convolution operations (see Tables 16.1 and 16.2) in conjunction with the input–response convolution relationship between concentration, $c(t)$, and the rate of input, $f(t)$:

$$c(t) = \text{UIR}(t) * f(t) \tag{16.53}$$

Most drugs with a linear disposition have a polyexponential response to a systemic bolus input. Accordingly, by definition the unit impulse response (IV bolus response normalized with respect to the bolus dose) is then well described by

$$\text{UIR}(t) = \sum_{j=1}^{n} A_j e^{-\alpha_j t} \tag{16.54}$$

where typically $n = 2$ or 3 for most drugs. Thus, the concentration response to a constant input, $I = k = \text{constant}$, simply becomes (see Table 16.2, item 3)

$$c(t) = k * \left(\sum_{j=1}^{n} A_j e^{-\alpha_j t} \right) = K \sum_{j=1}^{n} \frac{A_j}{\alpha_j} (1 - e^{-\alpha_j t}) \tag{16.55}$$

Similarly, the absorption rate of a first-order absorption is given by

$$f(t) = FDk_a e^{-k_a t} \tag{16.56}$$

where

$$FD = \int_0^t f(t)dt = \text{Dose absorbed} \tag{16.57}$$

D is the drug dose administered and k_a = first-order absorption rate constant. The concentration response to the first-order absorption is obtained by a simple convolution (see Table 16.2, item 1).

$$c(t) = (FDk_a e^{-k_a t}) * \sum_{j=1}^{n} A_j e^{-\alpha_j t} \qquad (16.58)$$

$$= FDk_a \sum_{j=1}^{n} \frac{A_j}{k_a - \alpha_j} (e^{-\alpha_j t} - e^{-k_a t}) \qquad (16.59)$$

LSA APPROACH TO DISCONTINUED INPUT AND MULTIPLE INPUTS

Laplace transforms consider discontinued input, but the procedure is unnecessarily complex and tedious. LSA offers a more direct solution to this problem by the use of simple convolution or the use of the superposition principle in a "generalized" fashion, as seen as follows.

CONVOLUTION AND DISCONTINUOUS INPUT

The concentration response to an arbitrary input, $f(t)$, that is discontinued at time = T is simply given by

$$c(t) = \int_0^{\min(t,\,T)} \mathrm{UIR}(t-u)f(u)\,du \qquad (16.60)$$

Consider, for example, a "linear drug" with a polyexponential UIR as given above. Let the input be first order and discontinued at time = T, so that

$$f(t) = FDk_a e^{-k_a t} \qquad o \leq T \qquad (16.61)$$

$$= 0 \qquad t > T \qquad (16.62)$$

The convolution integral gives by simple integration the following concentration response to the discontinued input:

$$c(t) = FDk_a \sum_{j=1}^{n} \frac{A_j}{k_a - \alpha_j} e^{-\alpha_j t} (1 - e^{-(k_a - \alpha_j)\min(t,\,T)}) \qquad (16.63)$$

THE SUPERPOSITION PRINCIPLE APPLIED TO DISCONTINUOUS INPUT

The superposition principle allows any input to be "partitioned" in arbitrary components as long as the sum (superposition) of these components adds to become equal to the original input, i.e.,

$$R = \text{UIR} * I = \text{UIR} * (I_1 + I_2 + \ldots + I_n) = \text{UIR} * I_1 + \text{UIR} * I_2 + \ldots + \text{UIR} * I_n \quad (16.64)$$

Correspondingly, the response is similarly arbitrarily partitioned.

$$R = R_1 + R_2 + \ldots + R_n \qquad (R_j \equiv \text{UIR} * I_j) \quad (16.65)$$

The partitioning may be a purely mathematical operation and does not need to have physiological relevance. This opens up some elegant ways of dealing with discontinuous inputs. For example, consider a constant input that is started at time $t = 0$ and stopped at $t = T$.

$$I = k \text{ for } 0 \leq t \leq T \quad (16.66)$$

$$I = 0 \text{ for } t > T \quad (16.67)$$

This input can be partitioned into two inputs:

$$I = I_1 + I_2 \quad (16.68)$$

where

$$I_1 = k \quad \text{for } t \geq o \quad \text{and} \quad I_2 = -k \quad \text{for } t > T \quad (16.69)$$

Note that both I_1 and I_2 are continuous, nonstop constant input of the same magnitude ($|k|$) with I_1 started at $t = 0$ and I_2 started at time $t = T$. Note also that the I_2 input is negative and not a "meaningful" input (and does not have to be because this is purely a mathematical operation). The responses R_1 and R_2 to the two inputs I_1 and I_2 are readily derived because they are both continuous input. The response to I_1 is very simple to obtain (see example above):

$$R_1(t) = c(t) = k * \left(\sum_{j=1}^{n} A_j e^{-\alpha_j t} \right) = k \sum_{j=1}^{n} \frac{A_j}{\alpha_j} (1 - e^{-\alpha_j t}) \qquad \text{for } t > 0 \quad (16.70)$$

The response to I_2 is simply related to the first response because of the "symmetrical" partitioning.

$$R_2(t) = -R_1(t - T) = -k \sum_{j=1}^{n} \frac{A_j}{\alpha_j} (1 - e^{-\alpha_j(t-T)}) \qquad \text{for } t > T \quad (16.71)$$

Accordingly, by summation and simple rearrangement, the response to the discontinuous input I simply becomes

$$R = R_1 + R_2 = c(t) = k \sum_{j=1}^{n} \frac{A_j}{\alpha_j} (e^{-\alpha_j(t-T)_+} - e^{-\alpha_j t}) \quad (16.72)$$

where subscript "+" is used to denote the so-called truncation function, i.e.,

$$(t - T)_+ \equiv t - T \qquad \text{For } t > T$$
$$\equiv 0 \qquad \text{otherwise} \qquad (16.73)$$

The above examples of discontinuous input are readily extended to consider infusion input that are started and stopped multiple times.

LSA AS AN ALTERNATIVE TO MULTIVARIATE LINEAR COMPARTMENTAL MODELING

A drug and its metabolites are sometimes measured simultaneously, perhaps in one or more tissues. The traditional approach to PK analysis in such cases has been to apply an explicitly structured modeling approach, which is inherently problematic for several reasons. LSA offers an attractive alternative to multivariate PK analysis for drugs with a linear disposition that eliminates many of the problems in traditional modeling. The basic principle of the LSA approach to this problem is very simple. Consider, for example, two PK responses resulting from the same input (I):

$$R_1 = L_1(I) \qquad (16.74)$$

$$R_2 = L_2(I) \qquad (16.75)$$

Both responses are related to the input through a convolution-type linear relationship. Accordingly, a unique inverse relationship to the input exists for the two responses:

$$I = L_1^{-1}(R_1) \qquad (16.76)$$

where, L_1^{-1} denotes the inverse of the L_1 operator. Substitution of I in $R_2 = L_2(I)$ gives the fundamental relationship between the two responses:

$$R_2 = L_2(L_1^{-1}(R_1)) = L(R_1) \qquad (16.77)$$

The relationship between R_2 and R_1 is unique and *independent* of the input. The relationship is independent of the input in the sense that it does not depend on the rate or extent of the input. However, as for any input–response system, the relationship depends on the site of the input. Thus, as long as the drug disposition does not change (time-invariant disposition) and the drug enters the system through the same input site, the relationship between the responses remains unchanged. The R_1, R_2 relationship is a linear operational relationship. In its simplest form it can be a simple convolution-type relationship; otherwise it involves additional linear operations (see examples below). The relationship applies to any two PK responses in a multivariate PK system with a linear disposition and, as such, is an alternative to the traditional linear compartmental multivariate analysis. Perhaps the biggest power of the approach stems from the most remarkable fact that *the response relationships can be determined when the input is not known.*[10] Although the *ad hoc* conceptual derivation above makes use of deconvolution (L_1^{-1}), which requires information about the response to a known input, neither deconvolution nor a known reference input is required to determine the linear relationship between the responses.[10]

THE RESPONSE MAPPING OPERATOR

A theoretical analysis and a computer program have been presented for determining PK relationships by the LSA approach for linear drug dispositions.[10] The general analysis, response mapping operation (RMO) considers as specific examples the common cases where individual PK responses are well approximated by simple polyexponential functions, which is often the case. The procedure is briefly: Let $R_1(t)$ and $R_2(t)$ be two PK responses resulting from an arbitrary input at input site P. The two responses are uniquely related by the following relationship[10]:

$$R_2(t) = \sum_{j=m_2}^{m_1} w^{(j-1)}(0)R_1^{(m_1-1)}(t) + w^{(m_1)}(t) * R_1(t) \qquad (16.78)$$

where

$$m_2 = k_2 + 3 \qquad (16.79)$$

$$m_1 = k_1 + 3 \qquad (16.80)$$

$$R_1^{(j)}(0) = 0 \text{ for } j = -1, 0, 1, ..., k_1 \qquad (16.81)$$

$$R_2^{(j)}(0) = 0 \text{ for } j = -1, 0, 1, ..., k_2 \qquad (16.82)$$

A "derivative number" $k_1 = -1$ or $k_2 = -1$ is here used to denote a case where neither the response function (R) nor its derivatives is initially zero. The "mapping function" $w(t)$ is given by

$$w(t) = L^{-1}\left(L(R_2)\frac{1}{s^{m_1}L(R_1)}\right) \qquad (16.83)$$

where L and L^{-1} denote Laplace transform and inverse Laplace transforms, respectively. More specifically, the derivative elements of the mapping function, $w(t)$, are given by the following relationships:

$$w^{(m_1)}(t) = \sum_{j=1}^{m_1} \Phi^{(j-1)}(0)R_2^{(m_1-j)}(t) + \Phi^{(m_1)}(t) * R_2(t) \qquad (16.84)$$

$$w^{(i)}(0) = \sum_{j=1}^{i} \Phi^{(j-1)}(0) R_2^{(i-j)}(0) \qquad i = 1, 2, \ldots \qquad (16.85)$$

where the auxiliary function $\Phi(t)$ is given by

$$\Phi(t) = L^{-1}\left(\frac{1}{s^{m_1} L(R_1)}\right) \qquad (16.86)$$

Example 16.1 Consider the very simple system, for which

$$R_1(t) = A_1 e^{-\alpha_1 t} \qquad (16.87)$$

$$R_2(t) = B_1 e^{-\beta_1 t} + B_2 e^{-\beta_2 t}, B_1 + B_2 = 0 \qquad (16.88)$$

The two responses could, for example, represent the drug concentration in the blood and some peripheral tissue following a bolus injection. For this case, $k_1 = -1$, $k_2 = 0$. Applying the above response mapping algorithm produces the following relationship between the responses:

$$R_2(t) = w^{(2)}(t) * R_1(t) \qquad (16.89)$$

where

$$\begin{aligned}
w^{(2)}(t) &= u_1 e^{-v_1 t} + u_2 e^{-v_2 t} \\
u_1 &= B_1(\alpha_1 - \beta_1)/A_1 \\
u_2 &= B_2(\alpha_1 - \beta_2)/A_1 \\
v_1 &= \beta_1 \\
v_2 &= \beta_2
\end{aligned} \qquad (16.90)$$

Example 16.2

$$R_1(t) = A_1 e^{-\alpha_1 t} + A_2 e^{-\alpha_2 t}, \qquad A_1 + A_2 = 0 \qquad (16.91)$$

$$R_2(t) = B_1 e^{-\beta_1 t} + B_2 e^{-\beta_2 t}, \qquad B_1 + B_2 = 0 \qquad (16.92)$$

In this case $k_1 = 0$, $k_2 = 0$, and the mapping algorithm produces the following linear operational relationship between the two response variables:

$$R_2(t) = w^{(2)}(0) R_1^{(2)}(t) + w^{(3)}(t) * R_1(t) \qquad (16.93)$$

where

$$w^{(2)}(0) = -(B_1\beta_1 + B_2\beta_2)/(A_1\alpha_2 + A_2\alpha_1) \tag{16.94}$$

$$w^{(3)}(t) = u_1 e^{-v_1 t} + u_2 e^{-v_2 t} \tag{16.95}$$

$$u_1 = B_1(\beta_1 - \alpha_1)(\beta_1 - \alpha_2)/(A_1\alpha_2 + A_2\alpha_1) \tag{16.96}$$

$$u_2 = B_2(\beta_2 - \alpha_1)(\beta_2 - \alpha_2)/(A_1\alpha_2 + A_2\alpha_1) \tag{16.97}$$

$$v_1 = \beta_1 \tag{16.98}$$

$$v_2 = \beta_2 \tag{16.99}$$

DECONVOLUTION

Deconvolution is the inverse operation of convolution and is mainly applied to determine the input, I, in an input–response convolution relationship.

$$R = \text{UIR} * I \tag{16.100}$$

In that case, deconvolution deals with either of two problems:

Deconvolution Problem 1: Given $R(t)$ and $I(t)$, determine $\text{UIR}(t)$
Deconvolution Problem 2: Given $R(t)$ and $\text{UIR}(t)$, determine $I(t)$

The main deconvolution problem, problem 2, requires problem 1 to be solved first.

Problem 1 To determine UIR, it is required that a known input be applied and the response measured. Several possibilities exist.

$$\text{Bolus input: UIR} = R_{\text{BOLUS}}/\text{Bolus Dose} \tag{16.101}$$

$$\text{Constant input: UIR} = (dR/dt)/k \ (I = k = \text{constant}) \tag{16.102}$$

$$\text{Post-steady state: UIR} = -(dR/dt)/k \ (I = k = \text{constant}) \tag{16.103}$$

The response differentiated in the last case (post-steady state) is the response, R, observed after a constant-rate input has led to a steady state. R is then the post-infusion response after establishment of steady state. The general procedure in all the above cases is to fit a suitable function to the (R, t) data to estimate $R(t)$ and then obtain $\text{UIR}(t)$, as given above. Polyexponential functions have worked very well in approximating $R(t)$, resulting in a polyexponential representation of $\text{UIR}(t)$. However, the determination of UIR is not limited to the above three cases but can be determined from the response to any known input function.

Problem 2 Several deconvolution methods exist for determining the input to solve deconvolution problem 2. Basically these can all be classified in two categories, namely the direct methods and the prescribed input function methods.

THE DIRECT DECONVOLUTION METHOD

In this method, which has been extensively applied,[15-19] the UIR(t) is determined as given above. Additionally, a suitable function is fitted to the input–response data from the unknown input to determine $R(t)$. The input function is then determined directly from the R and UIR functions using the following deconvolution formula that is applicable when

UIR$(0) \neq 0$:

$$I(t) = \left[R'(t) - \frac{\text{UIR}'(0)}{\text{UIR}(0)} R(t) - h(t) * R(t) \right] / \text{UIR}(0) \tag{16.104}$$

The auxiliary function, $h(t)$, in this formula, called the distribution function, is obtained from the following Laplace transform inversion:

$$h(t) = L^{-1} \left[\frac{\left(s - \dfrac{\text{UIR}'(0)}{\text{UIR}(0)} \right) L(\text{UIR}) - \text{UIR}(0)}{L(\text{UIR})} \right] \tag{16.105}$$

Consider, for example, a simple case where UIR(t) $= Ae^{-\alpha t}$. In this example $h(t)$ becomes zero, leading to the following one-exponential deconvolution formula.

$$I(t) = [R'(t) + \alpha R(t)]/A \tag{16.106}$$

Consider, as another example, the more common case where UIR $= A_1 e^{-\alpha_1 t} + A_2 e^{-\alpha_2 t}$. The distribution function is then given by

$$h(t) = L^{-1} \left[\frac{\left(s + \dfrac{A_1 \alpha_1 + A_2 \alpha_2}{A_1 + A_2} \right) \left(\dfrac{A_1}{s + \alpha_1} + \dfrac{A_2}{s + \alpha_1} \right) - (A_1 + A_2)}{\dfrac{A_1}{s + \alpha_1} + \dfrac{A_2}{s + \alpha_2}} \right] \tag{16.107}$$

resulting in

$$h(t) = Ge^{\gamma t} \tag{16.108}$$

where

$$G = A_1 A_2 \left(\frac{a_1 - a_2}{A_1 + A_2} \right)^2 \tag{16.109}$$

$$\gamma = -(A_1\alpha_2 + A_2\alpha_1)/(A_1 + A_2) \tag{16.110}$$

Thus, the input rate is

$$I(t) = \left[R'(t) + \frac{A_1\alpha_1 + A_2\alpha_2}{A_1 + A_2} R(t) + Ge^{\gamma t} * R(t) \right] / (A_1 + A_2) \tag{16.111}$$

The solution to the more general case with

$$\text{UIR}(t) = \sum_{j=1}^{n} A_j e^{-\alpha_j t}$$

is as follows:

$$I(t) = \left[R'(t) + \frac{\displaystyle\sum_{i=1}^{n} A_j \alpha_j}{\displaystyle\sum_{j=1}^{n} A_j} R(t) + R(t) * \sum_{j=1}^{n-1} G_j e^{\gamma_j t} \right] / \sum_{j=1}^{n} A_j \tag{16.112}$$

The γ_j are obtained as the $n-1$ roots of the following polynomial:

$$P(x) = \sum_{i=1}^{n} A_i \prod_{j=1 \neq i}^{n} (\alpha_j + x) \tag{16.113}$$

Each corresponding G_j parameter is subsequently obtained from the γ_j parameter, according to

$$G_j = -\left(\sum_{m=1}^{n} A_m \right) \left[\sum_{j=1}^{n} \frac{A_j}{\gamma_i + \alpha_j} \sum_{k=1 \neq j}^{n} \frac{1}{\gamma_i + \alpha_k} \right]^{-1} \tag{16.114}$$

SPECIAL CONSIDERATIONS FOR UIR(0) = 0

The above analytically exact direct deconvolution algorithm does not apply when UIR(0) = 0. However, the method is readily extended to include UIR(0) = 0 as follows. Differentiation of R with respect to time in this case gives

$$R'(t) = \text{UIR}(0)I(t) + \text{UIR}'(t) * I(t) = \text{UIR}'(t) * I(t) \tag{16.115}$$

Thus, the solution becomes

$$I(t) = \left[R''(t) - \frac{\text{UIR}''(0)}{\text{UIR}'(0)} R'(t) - h(t) * R'(t) \right] / \text{UIR}'(0) \qquad (16.116)$$

where

$$h(t) = L^{-1} \left[\frac{\left(s - \frac{\text{UIR}''(0)}{\text{UIR}'(0)} \right) L(\text{UIR}') - \text{UIR}'(0)}{L(\text{UIR}')} \right]$$

$$= L^{-1} \left[\frac{\left(s - \frac{\text{UIR}''(0)}{\text{UIR}'(0)} \right) s L(\text{UIR}) - \text{UIR}'(0)}{s L(\text{UIR})} \right] \qquad (16.117)$$

Consider, for example, $\text{UIR}(t) = A(e^{-\alpha_1 t} - e^{-\alpha_2 t})$, $\alpha_2 > \alpha_1$. This could be the UIR for an oral solution where the input–response is the concentration of the drug in the general systemic circulation. Inserting $\text{UIR}'(0) = A(\alpha_2 - \alpha_1)$ and $\text{UIR}''(0) = A(\alpha_1^2 - \alpha_2^2)$ in the expression for h above gives $h = -\alpha_1 \alpha_2$ leading to

$$I(t) = [R''(t) + (\alpha_1 + \alpha_2) R'(t) + \alpha_1 \alpha_2 * R(t)] / [A(\alpha_2 - \alpha_1)] \qquad (16.118)$$

Thus, if the same subject was given an oral test formulation resulting in a drug concentration response:

$$R(t) = c(t) = B_1 e^{-\beta_1 t} + B_2 e^{-\beta_2 t} + B_3 e^{-\beta_3 t} \qquad (16.119)$$

with the constrain $B_1 + B_2 + B_3 = 0$, then the input function $I(t)$, which in this case is the rate of release of dissolved drug into the gastrointestinal tract, is obtained from the above expression for $I(t)$ with

$$R''(t) = \beta_1^2 B_1 e^{-\beta_1 t} + \beta_2^2 B_2 e^{-\beta_2 t} + \beta_3^2 B_3 e^{-\beta_3 t} \qquad (16.120)$$

$$R'(t) = -\beta_1 B_1 e^{-\beta_1 t} - \beta_2 B_2 e^{-\beta_2 t} - \beta_3 B_3 e^{-\beta_3 t} \qquad (16.121)$$

THE PRESCRIBED INPUT FUNCTION DECONVOLUTION METHOD

The prescribed input function method[11,20,21] is also called "deconvolution through convolution." In this method, the UIR(t) is first obtained in the same way as in the direct method just described. A suitable functional is then chosen to represent the input rate; e.g., this "prescribed function" may be a polyexponential expression, a polynomial function, a cubic spline function, or other empirical function. Let this function be denoted $f(t, p)$, where p is an unknown parameter vector of the input function; then,

$$I(t) \cong f(t, p) \tag{16.122}$$

$$R(t) \cong f(t, p) * \text{UIR}(t) \tag{16.123}$$

The convolution expression $f(t, p) * \text{UIR}(t)$ is then fitted by appropriate means to (R, t) data treating the p parameters as fitting parameters and $\text{UIR}(t)$ as a fixed function. The fitting will, depending on the functional form of f, constitute a linear or nonlinear regression problem that can be solved numerically specifically, or be solved by any curve-fitting program that allows fitting of user-defined functions. Accordingly, let

$$\hat{R}(t) \equiv f(t, \hat{p}) * \text{UIR}(t) \tag{16.124}$$

denote the "best fit" to the (R, t) data; then the input is given by

$$\hat{I}(t) = f(t, \hat{p}) \tag{16.125}$$

The "deconvolution through convolution" method is particularly suitable to use when the functional forms of the prescribed input function, $f(t)$, and $\text{UIR}(t)$ are such that an analytical expression can be found for their convolution. In such cases numerical convolution, which is less exact than analytical convolution, is avoided. For example, let

$$\text{UIR}(t) = A_1 e^{-\alpha_1 t} + A_2 e^{-\alpha_2 t} \tag{16.126}$$

and consider a biexponential "prescribed input function,"

$$I(t) \cong B_1 e^{-\beta_1 t} + B_2 e^{-\beta_2 t} \tag{16.127}$$

then an analytical convolution expression is obtained (see convolution Tables 16.1 and 16.2).

$$R = I * \text{UIR} = \sum_{j=1}^{2} \sum_{k=1}^{2} \frac{A_j B_k}{\alpha_j - \beta_k} (e^{-\beta_k t} - e^{-\alpha_j t}) \tag{16.128}$$

In the fitting of R to input–response data the UIR function is kept constant; i.e., the above A and α parameters are fixed, whereas the B and β parameters of the prescribed biexponential input function are treated as fitting parameters. Most frequently, the prescribed input function method is a two-step procedure where in the first step the UIR is determined separately. The UIR is then subsequently used in the second step in a fixed form in the deconvolution through convolution procedure. Alternatively, the deconvolution can be done in a single-step procedure by *simultaneously* fitting the UIR and the R functions to the reference input and test input data. However, the single-step procedure may only have an advantage over the two-step procedure if there is a significant change in the disposition (UIR) between the reference and test administration. Numerous deconvolution methods have been proposed with diverse names such as "point-area methods," "point-point methods," "explicit methods," and "implicit methods." In spite of this naming diversity, the methods proposed basically fall into the two classes, namely, "direct methods" and "prescribed input function methods," described above. A comprehensive individual discussion of the many methods is beyond the scope of this presentation. However, the relative merit of deconvolution methods is perhaps best judged on the basis of an analysis of the source of errors in deconvolution often disregarded.

SOURCES OF ERRORS IN DECONVOLUTION

The sources of errors (SE) in deconvolution may be classified as intrinsic errors and methodological errors, with the latter type of errors divided into experimental errors and data analysis errors.

INTRINSIC ERRORS

SE1: Nonlinear Disposition

Any significant interaction between the drug molecules or between drug molecules and their biotransformed forms in the transport from the input site to the sampling space creates a deviation from MSI and thus a deviation from linearity (see the appendix). The source of this interaction is most typically competitive, saturable processes such as binding, enzymatic biotransformation, and active transport. Alternatively, the interaction may be caused by pharmacological or toxicological effects of the drug that change its disposition, e.g., change in blood flow, biochemical changes, etc. These errors very often depend on the dose or drug concentration and may be reduced considerably by operating at low concentrations. Use of drug tracers in "tracer amounts" may be useful in this respect.

SE2: Time Variance

Biological systems are constantly changing due to environmental impact, intake of food, normal hormonal fluctuations, biorhythms, etc. Thus, drugs dispositions are expected to change over time, creating a deviation from the time-invariance assumption of deconvolution. This error may be reduced by proper attention to these factors. The use of a simultaneous reference administration of a tracer form of the drug can eliminate this error if the time variance is predominantly a "between experiment" disposition change and little changes occur within the time span of a single administration.[22] Giving the reference close to the test formulation may also reduce this error.[23] The possibility also exists to model the disposition change specifically to reduce the error.[24] For example, the most common disposition change is a change in the drug clearance resulting from a change in the elimination kinetics. The disposition decomposition–recomposition method enables an exact correction in the deconvolution for such changes.[24] The UIR disposition function can be decomposed as follows:

$$UIR' = -k \cdot UIR + h * UIR \qquad (16.129)$$

where the generalized central elimination rate constant k is given by

$$k = Cl/V + \int_0^\infty h(t)dt \qquad (16.130)$$

where Cl and V are the clearance and volume of distribution, respectively, and the distribution function $h(t)$ is as defined previously. A perturbation in the central elimination kinetics will result in a perturbed elimination rate constant.

$$\hat{k} = \hat{Cl}/V + \int_0^\infty h(t)dt \qquad (16.131)$$

leading to a perturbed disposition function $U\hat{I}R$.

$$U\hat{I}R' = -\hat{k} \cdot U\hat{I}R + h * U\hat{I}R \qquad (16.132)$$

The exact clearance correction deconvolution method involves four steps:[24]

1. The distribution function $h(t)$ is obtained from the UIR of the reference administration.
2. The changed clearance, $\hat{Cl}$ is obtained from the terminal log-linear elimination phase of the test administration and from the distribution function of the UIR in step 1.
3. The perturbed unit impulse response $U\hat{I}R$ is determined by recomposition of the disposition kinetics from the distribution function and the perturbed clearance via Laplace transformation.

$$U\hat{I}R = L^{-1}\{[s + \hat{Cl}/V + \int_0^\infty h(t)dt - L(h)]^{-1}\}/V \qquad (16.133)$$

4. The response from the test administration is deconvolved using the perturbed unit impulse response determined in step 3 to give the input functions:

Theoretically the best method to eliminate a time-variance error in deconvolution is to use a simultaneous reference administration of a tracer form of the drug. This will eliminate the error if the time variance is a "between experiments" disposition change. A "within experiments" time variance error will be reduced but not eliminated by this advanced approach.

Methodological Errors

Experimental Errors

SE3: Difference in input site for test and reference (Figure 16.2). The UIR derived from the reference administration is specific for a specific input site (*point of entry*), P, as evident from the deconvolution derivation (see the appendix). The drug in the test preparation will most frequently not enter the system through this reference input point, P, but through some other input point with intrinsically different UIRs (T, Figure 16.2). For example, in regular oral absorption evaluations by deconvolution, the reference input site (point) is in a peripheral large vein. However, the input site from the oral absorption is the entry into the general systemic circulation after the first pass of the drug through the liver and initial entry into the circulation. The ideal, but not practical, experimental setup in this case is to inject the reference at the exit "point" of the liver to obtain a better UIR for the deconvolution. The degree of difference between the ideal and not ideal UIR is directly related to the difference in the corresponding input site–sampling site probability functions (see the appendix). The difference may, in some cases, be pronounced such as, for example, when evaluating gastrointestinal release of drug using an oral solution reference.

SE4: Dispersed reference input site. The deconvolution derivation is based on an input point (see the appendix). Accordingly, if the reference drug is entering the system in a dispersed way over some input region or area, then the different input points over this region may have different UIRs. Effectively, the single UIR obtained from such a reference administration is a weighted average of the UIRs from the different entry points with weights equal to the fraction, f, of the reference dose entering via the points, i.e.,

$$\text{UIR} = \Sigma f_j \bullet \text{UIR}_j \qquad (16.134)$$

For example, in evaluation of "gastrointestinal bioavailability"[17] an oral solution of the drug is administered approximately "momentarily" in a "bolus" fashion and a UIR is obtained from the resulting drug concentration in the blood. This reference input sites area in this case spreads out over the region of the gastrointestinal tract. It is expected that the UIR from a "point" in the small intestine is significantly different from the UIR from a "point" in the colon area. Consequently, when using an oral solution reference to deconvolve the drug concentration response from an

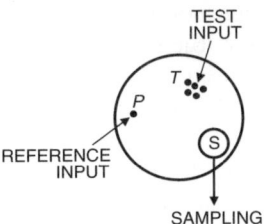

FIGURE 16.2 Difference in reference and test input sites creates potential problems in deconvolution.

oral test preparation the evaluation of the gastrointestinal drug release rate may be quite erroneous. This source of error will be smallest when the test and reference become dispersed in a similar way over the gastrointestinal tract such that similar fractions (f_j) of dissolved drugs are dispersed over the length of the tract. This does not appear very likely in dealing with many oral controlled-release formulations and a drug solution reference, thus introducing significant errors.

SE5: Non-instantaneous response sampling. The derivation of the deconvolution relationship (see the appendix) assumes an instantaneous sampling, which is never done. However, the error from this will be small if the UIR changes very little over the sampling period. This is usually not the case initially in the rapid distribution phase of the drug, making rapid and time-accurate samplings much more important in the initial disposition phase.

SE6: Inconsistent sampling space. The deconvolution derivation assumes identical sampling space for test and reference, which often may be impractical and thus introduce some errors. In blood sampling the sampling space is the somewhat disperse region occupied by the blood volume withdrawn in the sampling. Normally, the concentration of the drug will be homogeneous over this region. If this is not the case, then withdrawing different volumes of blood will essentially mean the sampling is not done from exactly the same sampling space. Also, the repeated blood sampling may not be done from the same site of the body. A potential other problem often not realized in blood sampling may occur when the sampling procedure significantly influences the blood flow in the area of the sampling space, e.g., by some blockage of the local blood flow.

Data Analysis Errors

The merit of deconvolution methods proposed in PK analysis only partially discussed by others[21,25] is appropriately judged by the extent to which data analysis errors discussed in this section are considered.

SE7: Mathematically inexact deconvolution. Numerical procedures such as numerical integration, numerical solution of differential equations, and some matrix–vector formulations of linear systems are numerical approximations and as such contain errors. This type of error is largely eliminated in the direct deconvolution method where the deconvolution is based on a mathematical exact deconvolution formula (see above). Similarly, the prescribed input function method ("deconvolution through convolution") will largely eliminate this numerical type of error if the convolution can be done analytically so that numerical convolution is avoided.

SE8: Response estimation errors. Undoubtedly, the biggest problem in deconvolution is to estimate closely the true input responses resulting from the reference and in particular the test administrations. There are two primary reasons that make this especially difficult.

1. In estimating the response from the reference administration it is not enough to get a good estimation of the response. It is also very important to get a good estimation of the derivative of the response. This is because the input function appears as an integrand in the convolution integral requiring a differentiation of the response to extract the input function.

This is clearly evident from the analytically exact deconvolution formula, which involves a derivative of the response. Unfortunately, while it is possible to judge the quality of the estimation of the response (sum of squares, correlation coefficient, residual analysis etc.) it is extremely difficult to judge how well the derivative of the response is determined. Variations in the derivative of the response are directly related to the degree of smoothness of the curve used to estimate the response. Accordingly, the smoothness issue is *very* important in deconvolution.

2. Because of the irregularity and variable physiology of the gastrointestinal tract and other factors, the absorption input function is expected to be quite erratic. This in combination with an often irregular gastrointestinal release of the drug from the dosage form, will produce a complex input function of random irregularity. A specific modeling of the input function in such cases may not be possible. Instead, it is necessary to choose a nonparametric approach and use empirical flexible functions to represent the input function (e.g., by a nonparametric prescribed input function approach) or to represent the input–response nonparametrically in the direct deconvolution approach.

Smoothing in Deconvolution

The required flexibility of the empirical function used in deconvolution to estimate the input–response creates a difficult problem: if the flexibility is not controlled the result may be an exact fit to the data (zero residuals), which is undesirable since the data contain errors. Ideally, the fit should produce residuals consistent with the error structure in the data. It is desirable to be able to control the flexibility of the approximating empirical function through a smothering criterion to strike the right balance between degree of smoothness and proper fit to the data, i.e., providing proper residuals. In modern deconvolution methods, this is done through an objective function with two components for the curve fitting):[26,27]

$$\text{objective function } = \begin{array}{c} \text{metric for closeness of fit} \\ + \lambda \text{ metric degree of smoothing} \end{array} \qquad (16.135)$$

Typically, the metric for closeness of fit is the (weighted) sum of squared residuals in following Gaussian statistical principles, a likelihood function, or perhaps some robust statistical metric. The metric for the degree of smoothness is commonly some integral measure of the rate of change of the first derivative, e.g., typically[27]

$$\text{metric for degree of smoothing } = \int (g''(t))^2 dt \qquad (16.136)$$

where g'' is the second time derivative of the function used to estimate the input–response R. Alternatively, the smoothing metric may also be applied to the input function.[25,28,29] Many other metrics for smoothness may be applied such as other function forms of g or the use of entropy principles.[25,28] The statistical field of nonparametric estimation and regression is a rich field where smoothening, commonly referred to as regularization, is extensively discussed.[26]

The trade-off between closeness of fit and degree of smoothing is determined by the smoothing parameter λ. A too small λ value will favor the first metric and produce a too close fit, e.g., resulting in a too small sum of squared residuals. A too large λ value will favor the second metric, which will smooth and straighten out the function (g). The resulting reduced flexibility from a too large λ value produces too large residuals. The optimal λ value is the value that produces a fit that is optimal in a statistical sense, e.g., most consistent with the intrinsic error in the data.

The smoothing parameter λ may be readily determined in the ideal case when the error variance of the data is known. In the more common case of unknown data error structure, λ will have to be determined in a somewhat subjective manner by. visual inspection of the fits combined with some residual analysis. More appropriately, λ may be determined using cross-validation principles enabling a more objective and automatic procedure.[26,30]

Negative Input Problems

As often realized when applying "first-generation" deconvolution methods, deconvolution may produce an input function that in some time period(s), usually toward the end of the input process, can become negative. Evaluation of a negative input may occur even if both the response being deconvolved and the UIR are non-negative.

The problem may be caused by too little smoothing resulting from a locally poor estimation of the response. This is commonly because the derivative (slope) of the fitted function is estimated to be too negative. Increased smoothening may sometimes solve this problem. A significant intrinsic difference in the drug disposition (UIR) between the test and reference administration may also result in the calculation of a negative input. This problem may be solved by the disposition decomposition technique, if the disposition change is due to a change in the central elimination kinetics between the test and reference administration.[24] The use of a simultaneous IV administration of a tracer of the drug[31–35] or constraint of the input function may also solve this problem.[11,12]

A simple and reliable way of constraining the input function is to apply the prescribed input function deconvolution method because this method allows the input function to be directly constrained. For example, a simple linear spline may be used as an input function. The non-negativity constraint is introduced by a simple parameterization of the spline with parameters defined as the function values at the so-called knots where the linear line segments are joined. The input function will be non-negative by ensuring that all parameters, i.e., the function values at the knots are non-negative.

LSA AND DISPOSITION DECOMPOSITION ANALYSIS

The UIR or disposition function so much used in LSA considers all disposition processes by their net effect on the measured response, R, without any differentiation.

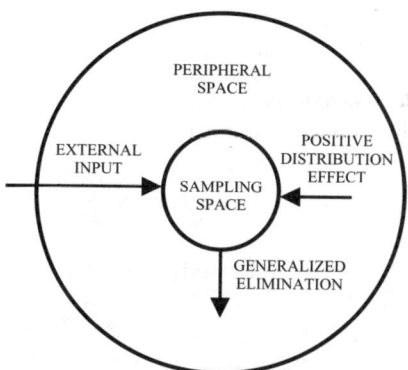

FIGURE 16.3 Fundamental processes of disposition decomposition.

This is sufficient in dealing with the important input–response relationship, but does not enable the effect of specific disposition changes to be considered. The disposition decomposition analysis (DDA) methodology overcomes this problem.[36] DDA is based on a most fundamental principle of dividing the disposition into two fundamental components; namely, an elimination component and a distribution component (Figure 16.3).

Following a bolus input, the drop in R is caused by "generalized elimination," i.e., drug leaving the sampling space due to irreversible elimination and due to distribution to the space external to the sampling space ("peripheral space"). The distribution effect is identified kinetically by the drug *returning* to the sampling space; i.e., the distribution effect counteracts the decrease in R caused by the generalized elimination and makes a positive contribution to R'. Thus, following an IV bolus input the rate of change in the response is

$$R' = - \text{(generalized elimination effect)} + \text{(positive distribution effect)} \quad (16.137)$$

The DDA assumes that the distribution effect is linear, which is consistent with stochastic molecular principles for the drug distribution (see the appendix).

LINEAR DISTRIBUTION

A linear distribution is a disposition where the distribution process makes a positive contribution to the rate of change in the response, R', that is linearly related to the response by a convolution:

$$\text{Positive distribution effect} = L(R) = h * R \quad (16.138)$$

where h is a characteristic, dose-independent distribution function.

Drug that is distributed to the peripheral space and eliminated there without returning will not make a contribution to the positive distribution effect. Such "distribution" is kinetically indistinguishable from a central irreversible elimination.

By central sampling it is only possible to determine the contribution of drug distribution to the generalized elimination by assuming all drug is eliminated centrally.

Theorem 16.7 Disposition Decomposition of Linear Dispositions For drugs with a linear disposition, the drug concentration response to an IV bolus input is given by

$$c' = -kc + h * c \qquad (16.139)$$

where k is the generalized elimination rate constant related to the unit impulse response, UIR, by

$$k = -\text{UIR}'(0)/\text{UIR}(0) \qquad (16.140)$$

and the distribution function $h(t)$ is given by

$$h(t) = L^{-1}\left[\frac{\left(s - \dfrac{\text{UIR}'(0)}{\text{UIR}(0)}\right)L(\text{UIR}) - \text{UIR}(0)}{L(\text{UIR})}\right] \qquad (16.141)$$

where L and L^{-1} denote Laplace transform and inverse Laplace transform, respectively.

Distribution Clearance, Cl_d

The distribution clearance is a distribution parameter that, when multiplied by the AUC of the drug concentration response, gives the total amount of drug returned from the peripheral distribution space. The distribution clearance is directly related to the distribution function, $h(t)$, by

$$Cl_d = V\int_0^\infty h(t)dt \qquad (16.142)$$

where V is the volume of distribution.

Generalized Elimination Clearance, Cl_e

The generalized elimination clearance is a disposition parameter that, when multiplied by the AUC of the drug concentration, gives the total amount of drug leaving the sampling space due to irreversible elimination and distribution. For a drug with a linear disposition, the generalized elimination clearance is given by

$$Cl_e = V \cdot k \qquad (16.143)$$

where V is the volume of distribution and k is the generalized elimination rate constant.

CLEARANCE, Cl

The (total) clearance may be defined in terms of the relationship between the total drug input and the resulting AUC of the drug concentration vs. time response:

$$Cl = \int_0^\infty I(t)dt/AUC \tag{16.144}$$

For drugs with a linear disposition, the clearance, the generalized elimination clearance, and the distribution clearance are simply related by

$$Cl = Cl_e - Cl_d \tag{16.145}$$

Consider, for example, a drug with a biexponential linear disposition function for which the concentration response from an IV bolus dose D is given by

$$c(t) = A_1 e^{-\alpha_1 t} + A_2 e^{-\alpha_2 t} \tag{16.146}$$

The distribution function is then given by

$$h(t) = A_1 A_2 \left(\frac{\alpha_1 - \alpha_2}{A_1 + A_2}\right)^2 \exp\left(-\frac{A_1 \alpha_2 + A_2 \alpha_1}{A_1 + A_2} t\right) \tag{16.147}$$

and

$$V = D/(A_1 + A_2) \tag{16.148}$$

Accordingly

$$Cl_d = \frac{DA_1 A_2}{A_1 \alpha_2 + A_2 \alpha_1} \left(\frac{\alpha_1 - \alpha_2}{A_1 + A_2}\right)^2 \tag{16.149}$$

$$Cl = D/(A_1/\alpha_1 + A_2/\alpha_2) \tag{16.150}$$

$$Cl_e = D(\alpha_1 A_1 + \alpha_2 A_2)/(A_1 + A_2)^2 \tag{16.151}$$

Theorem 16.8. Disposition Decomposition of Nonlinear Disposition with Linear Distribution Kinetics The drug concentration response, $c(t)$, resulting from a bolus input of a drug with a linear distribution kinetics can be decomposed as follows:

$$c' = -q(c) + h * c \tag{16.152}$$

where $h(t)$ is a unique distribution function and q is an elimination function which depends nonlinearly on c.

Disposition decomposition is useful to apply in three main areas:

1. Evaluation of drug absorption
2. Drug level predictions
3. Evaluation of nonlinear drug elimination

EVALUATION OF DRUG ABSORPTION

In the presence of a drug input, $I(t)$, of a drug with a linear distribution, the resulting drug concentration response is according to the disposition decomposition given by

$$c' = -q(c) + h * c + I/V \qquad (16.153)$$

Accordingly, the rate of input $I(t)$ is obtained from the following equation, which considers both linear and nonlinear elimination:

$$I(t) = V[c' + q(c) - h * c] \qquad (16.154)$$

In the specific case of a linear disposition, the above formula simply becomes equal to the direct deconvolution method previously described.

DRUG LEVEL PREDICTIONS

Most commonly, changes in the disposition kinetics arise from a change in the elimination kinetics rather than from changes in the distribution kinetics. This is to be expected because the distribution kinetics is largely determined by physical–chemical properties such as partition coefficients, solubilities, and diffusion coefficients. Such processes are less variable over time than biochemical and physiological processes influencing the elimination kinetics. A change in the elimination kinetics is readily considered by the disposition decomposition analysis. The perturbed unit impulse response, $\hat{U}IR$, corresponding to a perturbed clearance

$$\hat{Cl} = Cl + \Delta \qquad (16.155)$$

resulting from a positive or negative pertubation (Δ) is obtained from the unperturbed unit impulse response, UIR, according to

$$\hat{U}IR = L^{-1}\left[\frac{L(UIR)}{1 + \Delta L(UIR)}\right] \qquad (16.156)$$

where L and L^{-1} denote Laplace and inverse Laplace transforms. Once $\hat{U}IR$ is obtained, then drug level predictions in the presence of the perturbed clearance can readily be done by convolution or other methods, as previously discussed. Drug level predictions of drugs with nonlinear elimination and linear distribution can be done for arbitrary input, according to the numerical solution of the nonlinear disposition decomposition equation above. Steady state (SS) prediction of multiple IV bolus dosings can be done numerically using the following SS boundary condition:

$$C_{ss}(0) - C_{ss}(T) = D/V \qquad (16.157)$$

where T is the dosing time, D is the IV bolus dose, and V the volume of distribution. Thus, the SS drug level profile is obtained by solving the nonlinear disposition decomposition equation subject to the above SS boundary value condition. Prediction of the SS drug level c_{ss} corresponding to a constant-rate input (I_{ss} = constant) is obtained by numerically solving for C_{ss} in the following nonlinear equation:

$$-q(c_{ss}) + c_{ss}\int_0^\infty h(t)dt + I_{ss}/V = 0 \qquad (16.158)$$

Steady-state predictions resulting from multiple oral dosings can also be done by the disposition decomposition technique under the usual SS assumptions of reproducible drug absorption in the dosing intervals. Let C_1 denote the drug level response to a single dose; then the corresponding rate of input $I_1(t)$ $(0 < t < T)$ can be determined when q, h, and V have been determined from IV administrations.[9] The SS profile resulting from repeated dosing is then obtained by numerically solving:

$$C_{ss}'(t) = - q(C_{ss}(t)) + h(t) * C_{ss}(t) + I_1(t)/V(0 < t < T) \qquad (16.159)$$

subject to the SS boundary conditions $C_{ss}(0) = C_{ss}(T)$.

EVALUATION OF NONLINEAR DRUG ELIMINATION

In contrast to other modeling approaches, the disposition decomposition technique enables the elimination kinetics to be isolated and determined nonparametrically without the need for a specific structured modeling of the distribution kinetics.[9] Such objective determination of the nonlinear elimination kinetics is of significant value in the areas of nonlinear drug level predictions and absorption evaluations. Moreover, the nonparametric determination provides a useful visualization of the nonlinear elimination kinetics that may be quite helpful in determining the mechanism of the elimination process.

For example, if the nonparametrically determined q vs. C relationship shows an approximately straight-line relationship with a positive slope and intercept for large C values, it indicates that the drug is eliminated in a parallel fashion following a first-order process and a saturable process. Thus, fitting the following model,

$$q(C) = KC + \frac{V_m C}{K_m + C} \qquad (16.160)$$

to the q vs. c data may reveal directly if the elimination kinetics is well described by this elimination model consistent with parallel first-order and Michaelis–Menten kinetics.

LSA AND MEAN TIME PARAMETERS

Mean Time Parameters

A mean time (MT) parameter is the average time taken for one or more kinetic events to occur.

MT parameters are closely related to LSA because they share the same foundation (MSI) that make them quite independent of structural kinetic assumption.[37,38] Because of their simple "physical" definition, MT parameters are readily interpreted and in a meaningful way summarize important kinetic processes such as drug absorption, drug distribution, drug elimination, and drug release and delivery from a dosage form. The three most common MT parameters of general interest are mean residence time (MRT), mean transit time (MTT), and mean arrival time (MAT).[37-50] The mean absorption time often also denoted MAT is an example of a mean arrival time. The definition, scope, and calculations of MT parameters relate to the generalized concept of kinetic spaces.

Kinetic Space

A kinetic space defines the presence of a molecular moiety in one or more states. The states referred to are chemical (e.g., drug and metabolites) or physical states (e.g., distribution kinetic spaces) or both.

Transit Time

The transit time is the time from when a molecule enters a kinetic space to when it subsequently leaves the kinetic space.

Because of the general definition of kinetic spaces, the transit time may include biotransformation processes in addition to simple physical transport processes. For example, the kinetic space may be defined as a metabolite present in the body excluding the urine. In this context the transit time of a drug molecule through that kinetic space is the time from when the molecule is metabolized to a given metabolite to the time when the metabolite appears in the urine for subsequent elimination by micturation. The transit time will be randomly distributed according to a certain transit time distribution if the drug exhibits MSI. A valuable dose-independent mean transit time (MTT) parameter for the drug in the kinetic space may then be obtained.

Mean Transit Time

The MTT of drug molecules in a kinetic space is the average time taken by drug molecules from entering the kinetic space to subsequently leaving the kinetic space.

Mean Residence Time

The MRT of drug molecules in a kinetic space is the average total time the drug molecules spend in the kinetic space.

If all molecules exit irreversibly from a kinetic space, then MRT = MTT. If all the molecules enter the kinetic space at the same time ($t = 0$) and the drug leaves irreversibly from the kinetic space, then MTT and MRT are simply given by

$$\text{MRT} = \text{MTT} = \int_0^\infty F(t)\,dt \tag{16.161}$$

where $F(t)$ is the fraction of the molecules remaining in the kinetic space at time t. If the molecules can return to the kinetic space, then MRT is still obtained by the same formula but MTT will not be equal to MRT. If $f_e(t)$ denotes an arbitrary rate of elimination of the drug molecules from a kinetic space where the drug molecules are all introduced at $t = 0$, then

$$\text{MRT} = \int_0^\infty F(t)\,dt = \int_0^\infty \left[\int_0^\infty f_e(t)\,dt - \int_0^t f_e(t)\,dt \right] dt \Big/ \int_0^\infty f_e(t)\,dt \tag{16.162}$$
$$= \int_0^\infty t f_e(t)\,dt \Big/ \int_0^\infty f_e(t)\,dt$$

If the rate of elimination, $f_e(t)$, is proportional to the measured concentration, $c(t)$, then the above expression simply becomes

$$\text{MRT} = \text{AUMC/AUC} \tag{16.163}$$

where

$$\text{AUMC} \equiv \int_0^\infty t c(t)\,dt \tag{16.164}$$

Depending on the definition of the kinetic space, the above MRT has different interpretations. For example, the kinetic space may be defined as the unchanged drug present in the part of the body reached following an IV bolus injection excluding drug excreted irreversibly into the urine. This may be referred to as the disposition space, in which case MRT is the mean residence time in the disposition space. Alternatively, following an instantaneous oral dosing, the kinetic space may be defined as the unchanged drug in the whole body of bioavailable drug excluding drug excreted irreversibly into the urine. This space, which consists of the absorption space and the above disposition space, may be called the total body space. The gastrointestinal absorption space is the space the drug can occupy in the absorption processes prior to entering the disposition space via the first pass through the liver. The following theorem provides a simple way of relating the MRTs of kinetic spaces (Figure 16.4).

Theorem 16.9. MRT Partitioning Consider a kinetic space partitioned in an arbitrary way into any number of mutually exclusive kinetic spaces. The MRT of the kinetic space is equal to the sum of the MRTs of the mutually exclusive kinetic spaces. Applying the MRT partition theorem to bioavailable drug molecules gives:

$$MRT = MRT_1 + MRT_2 + MRT_3$$

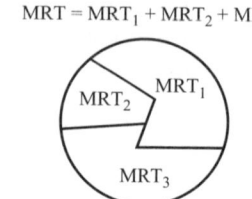

FIGURE 16.4 Arbitrary partitioning of kinetic space into mutually exclusive space.

MRT(Total Body) = MRT(Absorption Space) + MRT(Disposition Space)

Consequently, from the partition theorem, it then follows that

$$MRT(\text{Absorption Space}) = (AUMC/AUC)_{po} - (AUMC/AUC)_{\text{IV BOLUS}} \quad (16.165)$$

The partition theorem can also be applied to the disintegration and absorption kinetics of an oral tablet to determine the MRT of the drug in the various kinetic spaces in the drug delivery sequence (Figure 16.5A). The residence of the drug can be partitioned into four kinetic spaces (Figure 16.5):

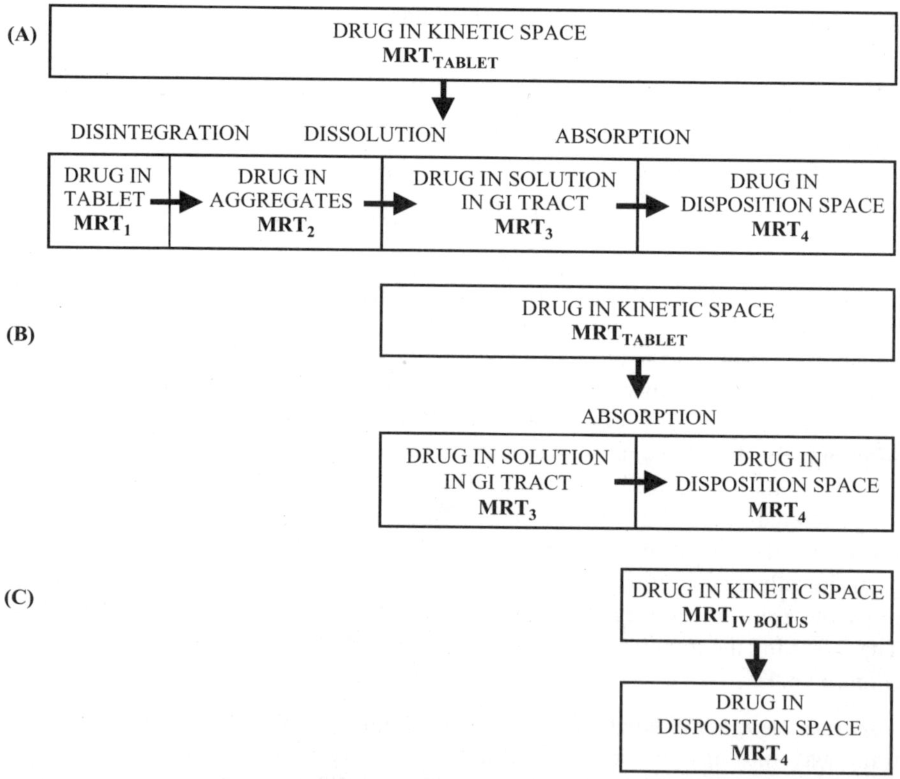

(A)

DRUG IN KINETIC SPACE
MRT_{TABLET}

DISINTEGRATION   DISSOLUTION   ABSORPTION

| DRUG IN TABLET MRT_1 | → | DRUG IN AGGREGATES MRT_2 | → | DRUG IN SOLUTION IN GI TRACT MRT_3 | → | DRUG IN DISPOSITION SPACE MRT_4 |

(B)

DRUG IN KINETIC SPACE
MRT_{TABLET}

ABSORPTION

| DRUG IN SOLUTION IN GI TRACT MRT_3 | → | DRUG IN DISPOSITION SPACE MRT_4 |

(C)

DRUG IN KINETIC SPACE
$MRT_{IV\ BOLUS}$

| DRUG IN DISPOSITION SPACE MRT_4 |

FIGURE 16.5 Partitioning of kinetic spaces in a subject receiving an oral tablet, a solution, and an IV bolus administration. GI = gastrointestinal.

1. Drug in tablet
2. Drug in aggregates formed by the *in vivo* disintegration of the tablet
3. Drug in solution in the gastrointestinal tract
4. Drug in the disposition space, i.e., in the general systemic circulation following the first-pass absorption step

From Theorem 16.9, it follows that

$$MRT_{TABLET} = (AUMC/AUC)_{TABLET} = MRT_1 + MRT_2 + MRT_3 + MRT_4 \quad (16.166)$$

where the MRT_{TABLET} is the MRT of the bioavailable drug molecules in the kinetic space consisting of the above four kinetic spaces. Similarly, an oral administration of a solution of the drug administered to the same subject results in

$$MRT_{SOLUTION} = (AUMC/AUC)_{SOLUTION} = MRT_3 + MRT_4 \quad (16.167)$$

where $MRT_{SOLUTION}$ is the MRT of bioavailable drug molecules in two of the previous four kinetic spaces, namely the drug in solution in the gastrointestinal tract, and the drug in the disposition space (Figure 16.5B). Finally, an IV bolus administration of the drug to the same subject gives

$$MRT_{IV} = (AUMC/AUC)_{IV} = MRT_4 = MRT(Disposition\ Space) \quad (16.168)$$

Because the transfer of the drug between the kinetic spaces (1 to 3 in Figure 16.5A) is largely irreversible, the MTTs through these spaces are very similar to the MRTs. From intrasubject administrations of a drug eliminated from the disposition space at a rate proportional to the measured drug concentration, it then follows that

$$
\begin{aligned}
(AUMC/AUC)_{TABLET} - (AUMC/AUC)_{SOLUTION} &= MRT_1 + MRT_2 \approx MTT_1 + MTT_2 \\
&= \text{mean } in\ vivo \text{ dissolution time}
\end{aligned}
\quad (16.169)
$$

$$
\begin{aligned}
(AUMC/AUC)_{TABLET} - (AUMC/AUC)_{IVBOLUS} &= MRT_1 + MRT_2 + MRT_3 \\
&\approx (MTT_1 + MTT_2) + MTT_3 \\
&= \text{mean } in\ vivo \text{ dissolution time} \\
&\quad + \text{mean absorption time from solution} \\
&= \text{total effective mean absorption time}
\end{aligned}
\quad (16.170)
$$

$$
\begin{aligned}
(AUMC/AUC)_{SOLUTION} - (AUMC/AUC)_{IV\ BOLUS} &= MRT_3 \approx MTT_3 \\
&= \text{mean absorption time from solution}
\end{aligned}
\quad (16.171)
$$

$$(AUMC/AUC)_{IV} = \text{mean residence time in the disposition space} \quad (16.172)$$

Mean Arrival Time

The MAT of drug molecules entering a kinetic space is the average time it takes the molecules to arrive in the kinetic space. The MAT for the absorption,

MAT(ABSORPTION), is the average time it takes the bioavailable drug molecules after entering the absorption space to enter the disposition space. This MAT is commonly referred to as the mean absorption time. The MAT(ABSORPTION) can generally be calculated from the absorption rate, $f(t)$, according to[37,38]

$$\text{MAT(ABSORPTION)} = \int_0^\infty tf(t)dt / \int_0^\infty f(t)dt \qquad (16.173)$$

If $f_{IN}(t)$ denotes the rate of input of drug into a kinetic space and $f_{OUT}(t)$ denotes the rate of irreversible elimination from that space, then the MRT of the drug in the kinetic space is

$$\text{MRT} = \int_0^\infty tf_{IN}(t)dt / \int_0^\infty f_{IN}(t)dt - \int_0^\infty tf_{OUT}(t)dt / \int_0^\infty f_{OUT}(t)dt \quad (16.174)$$

MRT relationships are also readily derived from linear input–response relationships from the fundamental convolution relationship involving the measured concentration $C(t)$, the rate of absorption $f(t)$, and the unit impulse response:

$$C(t) = f(t) * \text{UIR}(t) \qquad (16.175)$$

By using the Laplace transform technique, it can be shown that

$$\int_0^\infty tc(t)dt / \int_0^\infty c(t)dt = \int_0^\infty tf(t)dt / \int_0^\infty f(t)dt + \int_0^\infty t \cdot \text{UIR}(t)dt / \int_0^\infty \text{UIR}(t)dt \quad (16.176)$$

Thus, it follows from previous discussion that

$$(\text{AUMC/AUC})_{po} = \text{MAT(ABSORPTION)} + \text{AUMC}_{UIR}/\text{AUC}_{UIC} \qquad (16.177)$$

Thus, in comparing drug absorption from two products (denoted by subscripts A and B) in the same subject it follows from the above relationship that

$$(\text{AUMC/AUC})_A - (\text{AUMC/AUC})_B = \text{MAT(ABSORPTION)}_A - \text{MAT(ABSORPTION)}_B$$

The above simple expression for intrasubject comparison of MAT(ABSORPTION) does not require an assumption of an elimination rate proportional to the measured drug concentration $C(t)$. MRT parameters can also be derived on the basis of the corresponding residence time function, RT(t).

Residence Time Function

The residence time function RT(t) for a kinetic space describes the probabilitiy that a molecule that enters the kinetic space at the arbitrary time T is present in the kinetic space at time $T + t$. This definition allows the drug to reenter the kinetic space any number of times after the initial entry. The inclusion of the arbitrary entry time T is consistent with a time-invariant system.

Theorem 16.10 The MRT in a kinetic space is equal to the total integral of the residence time function of the kinetic space:

$$\text{MRT} = \int_0^\infty \text{RT}(t)\,dt \tag{16.178}$$

The residence time function of a kinetic space of a linear system is the UIR of the kinetic space expressed in terms of the dose-normalized amount vs. time function with respect to drug input into the kinetic space. For example, let the kinetic space be the unchanged drug molecules in the general systemic blood circulation, and let $c(t)$ denote the systemic drug concentration following an IV bolus injection; then,

$$\text{RT}(t) = [v \cdot c(t)/\text{DOSE}]_{\text{IV BOLUS}} = [c(t)/c(0)]_{\text{IV BOLUS}} \tag{16.179}$$

Thus,

$$\text{MRT(SYSTEMIC CIRCULATION)} = [\text{AUC}/C(0)]_{\text{IV BOLUS}} \tag{16.180}$$

One may define a tissue space as follows:

$$\begin{aligned} \text{TISSUE SPACE} = {}&\text{DISPOSITION SPACE} \\ &- \text{SYSTEMIC CIRCULATION SPACE} \end{aligned} \tag{16.181}$$

Then the MRT partition theorem gives

$$\begin{aligned} \text{MRT(TISSUE)} = {}&\text{MRT(DISPOSITION)} \\ &- \text{MRT(SYSTEMIC CIRCULATION)} \end{aligned} \tag{16.182}$$

From the above derivations, the MRT of drug molecules in the tissue is then readily obtained according to

$$\text{MRT(TISSUE)} = [\text{AUMC}/\text{AUC} - \text{AUC}/c(0)]_{\text{IV BOLUS}} \tag{16.183}$$

Theorem 16.11. The MTT of a kinetic space is related to the initial derivative of the residence time function.

$$\text{MTT} = -1/RT'(0) \tag{16.184}$$

Consider, for example, the systemic circulation space described above having $\text{RT}(t) = [c(t)/c(0)]_{\text{IV BOLUS}}$. The MTT of the drug molecules through the systemic circulation space is given by

$$\text{MTT(SYSTEMIC CIRCULATION)} = [-c(0)/c'(0)]_{\text{IV BOLUS}} \tag{16.185}$$

Mean Residence Number

The mean residence number (MRN) of a kinetic space is the average number of times the drug molecules enter a kinetic space.

The MTT of a kinetic space is related to the MRT and the MRN, as follows:

$$MTT = MRT/MRN \qquad (16.186)$$

Consider, for example, the systemic circulation space for which MTT and MRT are given above, leading to

$$MRN(SYSTEMIC\ CIRCULATION) = \frac{MRT}{MTT} = \left[\frac{AUC/c(0)}{-c(0)/c'(0)} \right]_{IV\ BOLUS} \qquad (16.187)$$

From the definition of MRT and the fact that the drug initially enters the systemic circulation (IV bolus input), it follows that

$$MRN(SYSTEMIC\ CIRCULATION) = MRN(TISSUE) + 1 \qquad (16.188)$$

Thus,

$$MRN(TISSUE) = \left[\frac{AUC \cdot (-c'(0))}{c(0)^2} - 1 \right]_{IV\ BOLUS} \qquad (16.189)$$

Combining this with the expression for MRT(TISSUE) from above gives

$$MTT(TISSUE) = \frac{MRT}{MRN} = \left[\frac{AUMC/AUC - AUC/C(0)}{\frac{AUC-(C'(0))}{C(0)^2} - 1} \right]_{IV\ BOLUS} \qquad (16.190)$$

Alternatively, this may also be written as

$$MTT(TISSUE) = \frac{MRT(TISSUE) - MRT(SYSTEMIC)}{\frac{MRT(SYSTEMIC)}{MTT(SYSTEMIC)} - 1} \qquad (16.191)$$

It can also be shown that the tissue transit time can be expressed solely in terms of the distribution function $h(t)$,

$$MTT(TISSUE) = AUMDC/AUDC \qquad (16.192)$$

where

$$AUMDC \equiv \int_0^\infty th(t)dt \qquad (16.193)$$

$$\text{AUDC} = \int_0^\infty h(t)\,dt \qquad (16.194)$$

Example 16.3: Biexponential Linear Drug Disposition. Consider a drug with a linear disposition and a biexponential concentration response resulting from an IV bolus injection,

$$c(t)_{\text{IV BOLUS}} = A_1 e^{-\alpha_1 t} + A_2 e^{-\alpha_2 t} \qquad (16.195)$$

Table 16.3 summarizes the calculations of the various relevant MT parameters using the above formula and gives an example for a drug with a linear biexponential disposition.

LSA AND DISTRIBUTION KINETICS

Traditionally, distribution kinetics has been described in terms of volume of distribution parameters (V and V_{ss}) and structure-specific distribution parameters, e.g., the k_{12}, k_{21} parameters of the two-compartment model. LSA offers a less-structured alternative that considers the net effect of the distribution kinetics based on the disposition decomposition analysis. For example, the partition/distribution properties of a drug may be expressed in terms of the affinity of the drug molecules to a kinetic space expressed as the MRT in the kinetic space. Accordingly, it is meaningful to consider ratios of MRTs for two kinetic spaces as a metric for the relative affinity. Thus, residence time coefficients (RTC) similar to a partition coefficient may be defined:

Residence Time Coefficient: The residence time coefficient RTC for the distribution of a drug in two kinetic spaces is the ratio of the MRT of the drug in the two kinetic spaces. It is readily shown that[37,38]

$$\text{RTC(TISSUE/SYSTEMIC CIRCULATION)} = \text{AUMDC} \qquad (16.196)$$

$$\text{RTC(SYSTEMIC CIRCULATION/DISPOSITION SPACE)}$$
$$= 1/(\text{AUMDC} + 1) \qquad (16.197)$$

$$\text{RTC(TISSUE/DISPOSITION SPACE)} = \text{AUMDC}/(\text{AUMDC} + 1) \qquad (16.198)$$

where

$$\text{AUMDC} \equiv \int_0^\infty t h(t)\,dt \qquad (16.199)$$

In addition to their simple and meaningful interpretation, the above RTCs have the distinct advantage that they are dose independent for linear disposition kinetics and nonlinear disposition kinetics as long as the nonlinearity is limited to a nonlinear central elimination. The RTC parameters above are simply related to the drug's volume of distribution, V, and volume of distribution at steady state V_{ss}.

TABLE 16.3
Mean Time Disposition Parameters Formulas and Example for a Linear, Biexponential IV Bolus Response: $c(t) = A_1\exp(-\alpha_1 t) + A_2\exp(-\alpha_2 t)$

Systemic Circulation		Tissue	
MRT[a]	MTT	MRT[a]	MTT[a]
AUC/$c(0)$	$-c(0)/c'(0)$	AUMC/AUC $-$ AUC/$c(0)$	$\dfrac{\text{AUMDC}}{\text{AUDC}} = \dfrac{\text{AUMC/AUC} - \text{AUC}/c(0)}{-c'(0)\text{AUC}/c(0)^2 - 1}$
$\dfrac{\dfrac{A_1}{\alpha_1}+\dfrac{A_2}{\alpha_2}}{A_1+A_2}$	$\dfrac{A_1+A_2}{\alpha_1 A_1+\alpha_2 A_2}$	$\dfrac{A_1 A_2(\alpha_1-\alpha_2)^2}{(A_1+A_2)\alpha_1\alpha_2(\alpha_2 A_1+\alpha_1 A_2)}$	$\dfrac{A_1+A_2}{\alpha_2 A_1+\alpha_1 A_2}$

[a] $\text{AUC} \equiv \int_0^\infty c(t)\,dt$; $\text{AUMC} \equiv \int_0^\infty tc(t)\,dt$; $\text{AUDC} \equiv \int_0^\infty h(t)\,dt$; $\text{AUMDC} \equiv \int_0^\infty th(t)\,dt$.

$$\text{RTC(TISSUE/SYSTEMIC CIRCULATION)} = V_{ss}/V - 1 \qquad (16.200)$$

$$\text{RTC(SYSTEMIC CIRCULATION/DISPOSITION SPACE)} = V/V_{ss} \qquad (16.201)$$

$$\text{RTC(TISSUE/DISPOSITION SPACE)} = 1 - V/V_{ss} \qquad (16.202)$$

The RTC parameters do not directly deal with the rate of distribution. However, this issue is addressed by the distribution clearance, Cl_d, defined as the ratio of the total amount of drug returned to the systemic circulation from the tissue and the total area under the drug concentration vs. time response in the systemic circulation. The distribution clearance is readily calculated from the distribution function, $h(t)$, and the volume of distribution.

$$Cl_d = V \cdot \text{AUDC} \qquad (16.203)$$

The distribution connection of the Cl_d parameter is also evident from the fact that if no elimination of drug takes place in the tissue, then for drugs with linear disposition kinetics,

$$\text{RATE OF DISTRIBUTION TO TISSUE} = Cl_d \cdot c(t) \qquad (16.204)$$

where $C(t)$ is the drug concentration in the systemic circulation.

Table 16.4 summarizes the calculation of the various LSA-based distribution parameters considered above and gives an example for a drug with a linear, biexponential disposition.

LSA AND PHARMACODYNAMICS

LSA principles have been applied in pharmacodynanics in situations where the drug concentration–effect transduction takes place at a site, commonly denoted the biophase, that is kinetically distinct from the sampling space. LSA-based PK/PD modeling is then based on the assumption that the nonmeasurable biophase concentration, $C_b(t)$, is linearly related to the drug concentration, $C(t)$, measured in the sampling space.

$$C_b(t) = L(C(t)) \qquad (16.205)$$

If the drug only enters the biophase via the systemic circulation and the $C(t)$ is equal to the drug concentration in the systemic circulation, then the relationship is properly described as a convolution relationship involving a so-called conduction function, $\phi(t)$.[51,52]

$$C_b(t) = \varphi(t) * C(t) \qquad (16.206)$$

TABLE 16.4

Formulae for LSA-Based Distribution Parameters and Example Considering Drugs with a Biexponential IV Bolus Response: $C(t) = A_1 \exp(-\alpha_1 t) + A_2 \exp(-\alpha_2 t)$

	Residence Time Coefficients (RTC)[a]			Distribution Clearance,[a] Cl_d
	Tissue/Systemic Circulation	Systemic Circulation/Disposition Space	Tissue/Disposition Space	
	AUMDC	1/(AUMDC + 1)	AUMD/(AUMD + 1)	$V \cdot AUDC$
	$A_1 A_2 \left(\dfrac{\alpha_1 - \alpha_2}{\alpha_2 A_1 + \alpha_1 A_2} \right)^2$	$\dfrac{\alpha_2 A_1 + \alpha_1 A_2}{(A_1 + A_2)(\alpha_2^2 A_1 + \alpha_1^2 A_2)}$	$\dfrac{A_1 A_2 (\alpha_1 - \alpha_2)^2}{(A_1 + A_2)(\alpha_2 A_1 + \alpha_1 A_2)(-\alpha_2^2 A_1 + \alpha_1^2 A_2)}$	$\dfrac{D A_1 A_2}{A_1 \alpha_2 + A_2 \alpha_1} \left(\dfrac{\alpha_1 - \alpha_2}{A_1 + A_2} \right)^2$

[a] $AUMDC = \int_0^\infty t h(t) dt$; $AUDC = \int_0^\infty h(t) dt$; $D =$ IV bolus dose.

If the measured effect, $E(t)$, is produced by a direct transduction from the biophase concentration (a so-called direct response model), then the PK/PD model simply becomes

$$E(t) = N(C_b) = N(\varphi\ (t) * C(t)) \tag{16.207}$$

where $N(C_b)$ is the transduction function. Thus, the LSA-based PK/PD model simply consists of two functions: the conduction function, $\varphi\ (t)$, dealing with the transport of the drug to the biophase, and the transduction function, $N(C_b)$, dealing with the pharmacodynamic biophase concentration-to-effect transduction. These two functions can be simultaneously determined by the powerful hysteresis minimization approaches.[51–53] These systems analysis approaches have the distinct advantage of providing a useful nonparametric estimation and visualization of the transduction relationship that greatly facilitates a subsequent analysis and possible specific (parametric) modeling of the PD transduction process. In its purely native form, the LSA-based hysteresis minimization approach provides a nonparametric predictive modeling of the PK/PD.[51]

LSA AND BIOPHASE DISTRIBUTION KINETICS — BIOPHASE EQUILIBRATION TIMES

The onset of drug effect for drugs with a "peripheral biophase" depends on the equilibration dynamics between the drug in the systemic circulation and the biophase. The equilibration dynamics can be summarized in the biophase equilibration times (BETs).[51] The steady-state (SS) biophase equilibration time, BET^p_{ss}, is defined as the time it takes to reach p percent of SS in the biophase after momentarily establishing a constant systemic drug level ($C(t)$ = constant) by a suitable IV bolus + infusion input. The SS-type BET parameter is obtained numerically from the conduction function according to the following relationship.

$$100\int_0^{\mathrm{BET}^p_{ss}}\varphi(t)dt \Big/ \int_0^\infty \varphi(t)dt\ =\ p \tag{16.208}$$

The R-type biophase equilibration time, BET^p_R, is defined as the time it takes to reach p percent of SS in the biophase when the drug is infused IV at a constant rate, R. For a drug with a linear disposition, this parameter depends on both the unit impulse response and the conduction function and can be obtained numerically from the following relationship:

$$100\varphi(t) * \int_0^t \mathrm{UIR}(t)dt \Big/ \Big[\int_0^\infty \varphi(t)dt \int_0^\infty \mathrm{UIR}(t)dt\Big]_{t\ =\ \mathrm{BET}^p_R}\ =\ p \tag{16.209}$$

LSA AND TOLERANCE

Although convolution has an exact kinetic foundation in PK (see the appendix), convolution has some basic properties that makes this operation valuable to consider in pharmacodynamics. A convolution involving the concentration profile of the drug, $C(t)$, depends on the whole prior "history" of the drug concentration response. This is similar to drug tolerance that depends on the prior exposure to the drug. Consider, for example, the following simple model for the amount (degree) of tolerance.

$$\text{Tolerance}(t) = e^{-K_T t} * c(t) \equiv \int_0^t e^{-K_T(t-u)} c(u) du \qquad (16.210)$$

This convolution-based PD tolerance model has the following properties:

- The tolerance depends on the intensity of drug exposure, $C(t)$, and its duration, t.
- The tolerance at time t depends on the whole prior exposure history.
- Discontinuation of drug exposure results in a reduction and eventual elimination of tolerance.
- Both tolerance formation and loss of tolerance are considered.

The tolerance can be incorporated in the PK/PD model in various ways consistent with the "rebound" effect commonly seen when the drug is quickly withdrawn following prolonged exposure.

LSA AND PHARMACODYNAMIC DECONVOLUTION

The PK/PD of centrally acting drugs (biophase = sampling space = general systemic circulation) with a direct effect and a linear disposition is described by the following input (I)–effect (E) relationship:

$$E = N(C(t)) = N(\text{UIR}) * I \qquad (16.211)$$

If the transduction function, $N(*)$, is monotonic, which is often the case, then an inverse function exists, $N^{-1}(*)$, which when applied to the model results in a transformed effect, E_T, that is linearly related to the drug input through a convolution:

$$E_T \equiv N^{-1}(E) = \text{UIR} * I \qquad (16.212)$$

Thus, if the transduction function (and thus its inverse) can be determined in addition to the UIR, then it is subsequently possible to do deconvolution based on effect vs. time data.[52]

LSA AND IN VIVO–IN VITRO CORRELATIONS

The use of LSA in in vivo–in vitro correlations (IVIVC) has mainly been in the analysis of level A IVIVC.

Level A IVIVC

USP: Level A *in vivo–in vitro* correlation "represents a point-to-point relationship between *in vitro* dissolution and the *in vivo* input rate of the drug from the dosage form." The IVIVC approach can either be implemented as a dual-step or single-step method.

Dual Step Method. Step 1, Phase 1

The *in vivo* release rate or input function is first determined. For drugs with a linear disposition kinetics this may be done using the LSA-based deconvolution methods previously discussed. Step 1 requires a suitable reference administration for the deconvolution. Dual-step methods are also denoted as reference-based methods. Three scenarios exist depending on the references given below.

1. **An IV reference is available.** The absorption rate may be determined by deconvolution using a direct deconvolution method or a prescribed input function method as described above.
2. **An oral solution reference is available.** The rate of release into the gastrointestinal tract may be determined by deconvolution as previously described.
3. **An IR reference is available.** If the absorption from an immediate-release (IR) formulation is a simple first-order process (i.e., described by a first-order absorption rate constant k_a), then it is possible by *a relative deconvolution method* or other means to obtain the shape of the absorption profile.[53]

The first scenario will produce the most accurate results in the first step of the dual-step procedure in determining the absorption rate. The second scenario will estimate the gastrointestinal release rate that more directly should link to the *in vitro* release/dissolution rate than the absorption rate, but the evaluation is confounded by the first-pass PK and possible violation of the deconvolution assumptions. The advantage of the oral solution reference may be largely offset to a negative extent by a violation of the assumptions specific to "GI deconvolution" (e.g., multiple input site, nonlinear absorption, etc., discussed above). Use of an IR reference is problematic in making the quite strict assumption of a simple, first-order absorption rate, i.e., assuming a single exponential absorption rate input.

Dual Step Method. Step 1, Phase 2

Once the release/input function has been determined, the connection (link modeling) to the *in vitro* function can be directly determined. The main advantage of the dual-step method is the fact that this linkage modeling can be done directly because of the *in vivo* function made available through deconvolution.

Dual Step Method. Step 2

Once the *in vivo* release/input function has been determined in step 1, the next step involves a PK-based prediction of the resulting measured drug concentration profile.

This may be done according to LSA principles by convolution assuming a linear input–response (concentration) relationship where the required UIR is determined in step 1 from the reference administration. Accordingly, the IVIVR model of the dual-step method may be summarized as follows:

$$\text{In vivo function = link model (in vitro function)} \qquad (16.213)$$

$$\text{In vivo drug concentration profile = UIR} * \text{in vivo function} \qquad (16.214)$$

Or, simply,

$$\text{In vivo drug concentration profile = UIR} * \text{link model (in vitro function)} \quad (16.215)$$

In summary, the dual-step method involves three separate procedures:

1. A deconvolution
2. A link modeling
3. A convolution

The Single-Step Method

The single-step method commonly does not make use of a reference administration. The method estimates the IVIVR in a single step; i.e., the IVIV link model is not determined separately in a preliminary first step as in the dual-step method. The single-step method does not require a deconvolution and, as such, does not need a reference administration. The method may be empirical and not require specific PK modeling.

However, more commonly, the single-step method is based on the assumption of a convolution linearly between the *in vivo* release function and the drug concentration response in following LSA-based convolution principles. Accordingly, a single step IVIVC model involving convolution may in the simplest form be described by

$$\text{In vivo drug concentration profile} = G(t) * H \text{ (in vitro function)} \quad (16.216)$$

where $H(*)$ is a transformation function that takes the place of a link model, and $G(t)$ is a function that takes the place of the unit impulse–response function when that is unavailable. However, the single-step method may also be applied to situations where the UIR is available. The IVIVC model in that case becomes the same as in the dual-step method, but the link model is not separately determined.

APPENDIX

MOLECULAR/PROBABILITY BASIS OF THE CONVOLUTION RELATIONSHIP IN LINEAR INPUT–RESPONSE SYSTEMS

Consider a drug with MSI (see the definition for MSI). Let $I = f(t)$ denote the rate of drug input (amount/time) at input site P (Figure 16.6). The number of drug

$$R = L(I)$$

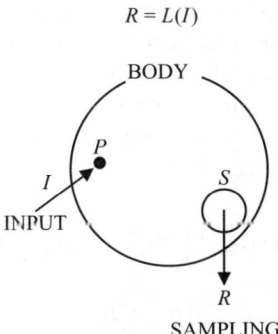

SAMPLING

FIGURE 16.6 Illustration of linear input–response system.

molecules entering P at the arbitrary time interval T to $T + \Delta T$ is then $n_T \cong k_1 f(t)\Delta t$, where k_1 is a conversion constant (units = number of molecules/unit amount) that depends on Avogadro's constant (K_A) and the molecular weight of the drug (MW): $k_1 = K_A/\text{MW}$. The drug molecules entering at P in the interval T to $T + \Delta T$ result in a corresponding number of molecules of the drug (or biotransformed drug) at the sampling site S at the arbitrary later time t, given by

$$n_{T,t} = \Pr(t - T)n_T \cong k_1 \Pr(t - T)f(T)\Delta T \tag{16.A1}$$

The total number, $N(t)$, of drug molecules (in original or biotransformed form) present at S at time t is obtained by summation of the individual numbers present at S resulting from all prior input at P in the time period from time $= T_1 = 0$ to time $= T_m = t$, i.e,

$$N(t) \cong n_{T_1,t} + n_{T_2,t} + \ldots + n_{T_m,t} \tag{16.A2}$$

$$N(t) \cong k_1 \Pr(t - T_1)f(T_1)\Delta T + k_1 \Pr(t - T_2)f(T_2)\Delta T + \ldots$$
$$+ k_1 \Pr(t - T_m)f(T_m)\Delta T \tag{16.A3}$$

When $\Delta T \to 0$ then $N(t)$ simply becomes equal to the following convolution integral relationship:

$$N(t) = k_1 \int_0^t \Pr(t - T)f(T)dT \tag{16.A4}$$

Converting from total numbers to total amount gives the following amount $x(t)$ of drug (in original or biotransformed form) at S at time t,

$$x(t) = k_b \int_0^t \Pr(t - T)f(T)dT \tag{16.A5}$$

where $k_b = 1$ if x is the untransformed drug and k_b is the ratio between the molecular weight of biotransformed drug and untransformed drug if x is the biotransformed drug. Let V_s denote the volume of the sampling site, defined as the volume withdrawn in each sampling. The concentration of drug (or biotransformed drug), $c(t) \equiv x(t)/V_s$, at the sampling site, S, at time t is then related to the rate of drug input, $f(t)$, at the input site, P, by the following equation:

$$c(t) = k\int_0^t \Pr(t - T)f(T)dT \qquad (16.A6)$$

$$k = k_b/V_s \qquad (16.A7)$$

If the input, I, is a bolus input of a unit amount of drug (Dose = 1), then the response $x(t)$ or $c(t)$ normalized with respect to the dose, denoted the unit impulse response, UIR(t), simply becomes

$$\text{UIR}(t)_x = k_b\Pr(t) \qquad (16.A8)$$

$$\text{UIR}(t)_c = k\Pr(t) \qquad (16.A9)$$

The UIR function is also commonly denoted the "characteristic function." This is an appropriate notation since it is a function that is characteristic for the given input–response system. In engineering, the Laplace transform of the unit impulse response is commonly called the "transfer function." The above equations for $x(t)$ and $c(t)$ can be written:

$$x(t) = \int_0^t \text{UIR}_x(t - T)f(T)dT \qquad (16.A10)$$

$$c(t) = \int_0^t \text{UIR}_c(t - T)f(T)dT \qquad (16.A11)$$

The integrals are so-called convolution integrals that in short-form notation often are written simply as

$$x(t) = \text{UIR}_x(t) * f(t) \qquad (16.A12)$$

$$c(t) = \text{UIR}_c(t) * f(t) \qquad (16.A13)$$

where the symbol $*$ is used to denote convolution. The latter two equations can be written in a more general form as

$$R(t) = \text{UIR}(t) * f(t) \qquad (16.A14)$$

where $R(t)$ denotes the PK response measured at the sampling site S. This response can be a concentration (or amount) of the drug or a biotransformed form of the drug, e.g., a metabolite. It is readily verified that convolution is a linear operator. Accordingly, the response, R, is linearly related to the input $I = f(t)$.

$$R = L(I) \qquad (16.A15)$$

More specifically, the linear relationship is give as a convolution relationship:

$$R = \text{UIR} * I \qquad (16.A16)$$

where

$$\text{UIR} \equiv L(I_{\text{bolus}})/D_{\text{bolus}} \qquad (16.A17)$$

$$I_{\text{bolus}} \equiv \text{bolus input of a dose } D_{\text{bolus}} \qquad (16.A18)$$

MOLECULAR/PROBABILITY BASIS OF LINEAR RESPONSE–RESPONSE RELATIONSHIPS

Let P denote the input site in the body and S_1, S_2 be sampling sites that may be the same or distinct, where pharmacokinetic responses R_1 and R_2 are measured. If the drug shows MSI, then the responses R_1 and R_2 are linearly related, which can be seen as follows.

Consider two sampling S_1 and S_2, where PK responses (concentration or amount of drug or metabolite) R_1 and R_2 are measured resulting from an input, I, at input site P. If the drug shows MSI, then R_1 and R_2 are linearly related to the input (Theorem 16.5).

$$R_1 = L_1(I) \qquad (16.A19)$$

$$R_2 = L_2(I) \qquad (16.A20)$$

Accordingly, two different, arbitrary inputs $I_{(1)}$ and $I_{(2)}$ result in the following responses:

$$R_{1(1)} = L_1(I_{(1)}) \qquad (16.A21)$$

$$R_{1(2)} = L_1(I_{(2)}) \qquad (16.A22)$$

$$R_{2(1)} = L_2(I_{(1)}) \qquad (16.A23)$$

$$R_{2(2)} = L_2(I_{(2)}) \qquad (16.A24)$$

Consider next a third input that is an arbitrary linear combination of first two inputs $I_{(1)}$ and $I_{(2)}$, i.e.,

$$I_{(3)} = k_1 I_{(1)} + k_2 I_{(2)} \qquad (16.\text{A}25)$$

where k_1 and k_2 are arbitrary constants. This third input results in the following responses:

$$R_{1(3)} = L_1(k_1 I_{(1)} + k_2 I_{(2)}) = k_1 L_1(I_{(1)}) + k_2 L_1(I_2) = k_1 R_{1(1)} + k_2 R_{1(2)} \quad (16.\text{A}26)$$

$$R_{2(3)} = L_2(k_1 I_{(1)} + k_2 I_{(2)}) = k_1 L_2(I_{(1)}) + k_2 L_2(I_{(2)}) = k_1 R_{2(1)} + k_2 R_{2(2)} \quad (16.\text{A}27)$$

From this, it is, according to Theorem 16.2, concluded that R_1, R_2, and I are linearly related.

REFERENCES

1. Cobelli, C. and Toffolo, G., Compartmental vs. noncompartmental modeling for two accessible pools, *Am. J. Physiol.*, 247:R488–496, 1984.
2. Cutler, D. J., Linear systems analysis in pharmacokinetics, *J. Pharmacokinet. Biopharm.*, 6:265–282, 1978.
3. Gillespie, W. R., Noncompartmental versus compartmental modeling in clinical pharmacokinetics, *Clin. Pharmacokinet.*, 20:253–262, 1991.
4. Landaw, E. M. and DiStefano, J. J., Multiexponential, multicompartmental, and noncompartmental modeling. II. Data analysis and statistical considerations, *Am. J. Physiol.*, 246:R665–677, 1984.
5. Piotrovskii, V. K., Model and model-independent methods of describing pharmacokinetics: the advantages, drawbacks and interrelationship, *Antibiot. Med. Biotekhnol.*, 32:492–497, 1987.
6. Strand, S. E., Zanzonico, P., and Johnson, T. K., Pharmacokinetic modeling, *Med. Phys.*, 20:515–527, 1993.
7. Veng-Pedersen, P., System approaches in pharmacokinetics: I. Basic concepts, *J. Clin. Pharmacol.*, 28:1–5, 1988.
8. Veng-Pedersen, P., System approaches in pharmacokinetics: II. Applications, *J. Clin. Pharmacol.*, 28:97–104, 1988.
9. Veng-Pedersen, P., Linear and nonlinear system approaches in pharmacokinetics: how much do they have to offer? I. General considerations, *J. Pharmacokinet. Biopharm.*, 16:413–472, 1988.
10. Veng-Pedersen, P., Linear and nonlinear system approaches in pharmacokinetics: how much do they have to offer? II. The response mapping operator (RMO) approach, *J. Pharmacokinet. Biopharm.*, 16:543–571, 1988.
11. Verotta, D., Two constrained deconvolution methods using spline functions, *J. Pharmacokinet. Biopharm.*, 21:609–636, 1993.
12. Verotta, D., Estimation and model selection in constrained deconvolution, *Ann. Biomed. Eng.*, 21:605–620, 1993.
13. Fattinger, K. E. and Verotta, D., A nonparametric subject-specific population method for deconvolution: II. External validation, *J. Pharmacokinet. Biopharm.*, 23:611–634, 1995.
14. Fattinger, K. E. and Verotta, D., A nonparametric subject-specific population method for deconvolution: I. Description, internal validation, and real data examples, *J. Pharmacokinet. Biopharm.*, 23:581–610, 1995.

15. Veng-Pedersen, P., An algorithm and computer program for deconvolution in linear pharmacokinetics [see comments], *J. Pharmacokinet. Biopharm.*, 8:463–481, 1980.

16. Gillespie, W. R., Veng-Pedersen, P., and Gibson, T. P., Deconvolution applied to the kinetics of extracorporal drug removal. Haemodialysis of cefsulodin, *Eur. J. Clin. Pharmacol.*, 29:503–509, 1985.

17. Gillespie, W. R. and Veng-Pedersen, P., Gastrointestinal bioavailability: determination of *in vivo* release profiles of solid oral dosage forms by deconvolution, *Biopharm. Drug Dispos.*, 6:351–355, 1985.

18. Gillespie, W. R. und Veng-Pedersen, P., A polyexponential deconvolution method. Evaluation of the "gastrointestinal bioavailability" and mean *in vivo* dissolution time of some ibuprofen dosage forms, *J. Pharmacokinet. Biopharm.*, 13:289–307, 1985.

19. Veng Pedersen, P., Pharmacokinetic analysis by linear system approach, I: Cimetidine bioavailability and second peak phenomenon, *J. Pharm. Sci.*, 70:32–38, 1981.

20. Cutler, D. J., Numerical deconvolution by least squares: use of prescribed input functions, *J. Pharmacokinet. Biopharm.*, 6:227–241, 1978.

21. Yu, Z. et al., Five modified numerical deconvolution methods for biopharmaceutics and pharmacokinetics studies, *Biopharm. Drug Dispos.*, 17:521–540, 1996.

22. Burm, A. G. et al., Pharmacokinetics of alfentanil after epidural administration. Investigation of systemic absorption kinetics with a stable isotope method, *Anesthesiology*, 81:308–315, 1994.

23. Karlsson, M. O. and Bredberg, U., Bioavailability estimation by semisimultaneous drug administration: a Monte Carlo simulation study, *J. Pharmacokinet. Biopharm.*, 18:103–120, 1990.

24. Veng-Pedersen, P., Drug absorption evaluation in the presence of changes in clearance: an algorithm and computer program for deconvolution with exact clearance correction, *J. Pharmacokinet. Biopharm.*, 8:185–203, 1987.

25. Madden, F. N. et al., A comparison of six deconvolution techniques, *J. Pharmacokinet. Biopharm.*, 24:283–299, 1996.

26. Wahba, G., A survey of some smoothing problems and the methods for solving them: University of Wisconsin-Madison Statistics Department, Report 347, 1980.

27. Craven, P. and Wahba, G., Smoothing noise data with spline functions, *Numer. Math.*, 31:377–403, 1979.

28. Paintaud, G. et al., Limitations of the maximum entropy principle in devising drug input rate, *Eur. J. Clin. Pharmacol.*, 49:139–143, 1995.

29. De Nicolao, G. and De Nicolao, A., WENDEC: a deconvolution program for processing hormone time-series, *Comput. Methods Programs Biomed.*, 47:237–252, 1995.

30. Hutchinson, M. F. and deHoog, F. R., Smoothing noise data with spline functions, *Numer. Math.*, 47:99–106, 1985.

31. Acuff, R. V. et al., Relative bioavailability of RRR- and all-rac-alpha-tocopheryl acetate in humans: studies using deuterated compounds, *Am. J. Clin. Nutr.*, 60:397–402, 1994.

32. Barrish, A. et al., The use of stable isotope labeling and liquid chromatography/tandem mass spectrometry techniques to study the pharmacokinetics and bioavailability of the antimigraine drug, MK-0462 (rizatriptan) in dogs, *Rapid Commun. Mass Spectrum.*, 10:1033–1037, 1996.

33. Bode, H. et al., Investigation of nifedipine absorption in different regions of the human gastrointestinal (GI) tract after simultaneous administration of 13C- and 12C-nifedipine, *Eur. J. Clin. Pharmacol.*, 50:195–201, 1996.

34. Hage, K. et al., Estimation of the absolute bioavailability of flecainide using stable isotope technique, *Eur. J. Clin. Pharmacol.*, 48:51–55, 1995.
35. Richard, J., Cardot, J. M., and Godbillon, J., Stable isotope methodology for studying the performance of metoprolol Oros tablets in comparison to conventional and slow release formulations, *Eur. J. Drug Metab. Pharmacokinet.*, 19:375–380, 1994.
36. Veng-Pedersen, P., Theorems and implications of a model independent elimination/distribution function decomposition of linear and some nonlinear drug dispositions. I. Derivations and theoretical analysis, *J. Pharmacokinet. Biopharm.*, 12:627–648, 1984.
37. Veng-Pedersen, P., Mean time parameters in pharmacokinetics. Definition, computation and clinical implications (Part II), *Clin. Pharmacokinet.*, 17:424–440, 1989.
38. Veng-Pedersen, P., Mean time parameters in pharmacokinetics. Definition, computation and clinical implications (Part I), *Clin. Pharmacokinet.*, 17:345–366, 1989.
39. Aarons, L., Mean residence time for drugs subject to reversible metabolism, *J. Pharm. Pharmacol.*, 39:565–567, 1987.
40. Bagli, M. et al., Mean input times of three oral chlorprothixene formulations assessed by an enhanced least-squares deconvolution method, *J. Pharm. Sci.*, 85:434–439, 1996.
41. Brockmeier, D. and von Hattingberg, H. M., Mean residence time, *Methods Find. Exp. Clin. Pharmacol.*, 8:309–312, 1986.
42. Cheng, H. and Jusko, W. J., Mean residence time of oral drugs undergoing first-pass and linear reversible metabolism, *Pharm. Res.*, 10:8–13, 1993.
43. Veng-Pedersen, P., A simple method for obtaining the mean residence time of metabolites in the body, *J. Pharm. Sci.*, 75:818–819, 1986.
44. Veng-Pedersen, P., Stochastic interpretation of linear pharmacokinetics: a linear system analysis approach, *J. Pharm. Sci.*, 80:621–631, 1991.
45. Veng-Pedersen, P. and Gillespie, W., Mean residence time in peripheral tissue: a linear disposition parameter useful for evaluating a drug's tissue distribution, *J. Pharmacokinet. Biopharm.*, 12:535–543, 1984.
46. Veng-Pedersen, P. and Gillespie, W., The mean residence time of drugs in the systemic circulation, *J. Pharm. Sci.*, 74:791–792, 1985.
47. Veng-Pedersen, P. and Gillespie, W. R., Single pass mean residence time in peripheral tissues: a distribution parameter intrinsic to the tissue affinity of a drug, *J. Pharm. Sci.*, 75:1119–1126, 1986.
48. Verotta, D., Sheiner, L. B., and Beal, S. L., Mean time parameters for generalized physiological flow models (semihomogeneous linear systems), *J. Pharmacokinet. Biopharm.*, 19:319–331, 1991.
49. Weiss, M., Washout time versus mean residence time, *Pharmazie*, 43:126–127, 1988.
50. Zierler, K. L., Occupancy principle: identity with that of mean transit time of tracers in biological systems, *Science*, 163:491–492, 1969.
51. Veng-Pedersen, P., Mandema, J. W., and Danhof, M., Biophase equilibration times, *J. Pharm. Sci.*, 80:881–886, 1991.
52. Smolen, V. F. and Schoenwald, R. D., Drug-absorption analysis from pharmacological data. I. Method and confirmation exemplified for the mydriatic drug tropicamide, *J. Pharm. Sci.*, 60:96–103, 1971.
53. Veng-Pedersen, P. and Miller, R., Relative deconvolution. An explicit method for bioavailability comparison not requiring intravenous administration, *Int. J. Clin. Pharmacol. Ther. Toxicol.*, 25:10–14, 1987.

Index

W

Z